Food Science Text Series

Third Edition

For other titles published in this series, go to
www.springer.com/series/5999

Series editor:

Dennis R. Heldman
Heldman Associates
Mason, Ohio, USA

The Food Science Text Series provides faculty with the leading teaching tools. The Editorial Board has outlined the most appropriate and complete content for each food science course in a typical food science program and has identified textbooks of the highest quality, written by the leading food science educators. Series Editor Dennis R. Heldman, Professor, Department of Food, Agricultural, and Biological Engineering, The Ohio State University. Editorial Board; John Coupland, Professor of Food Science, Department of Food Science, Penn State University, David A. Golden, Ph.D., Professor of Food Microbiology, Department of Food Science and Technology, University of Tennessee, Mario Ferruzzi, Professor, Food, Bioprocessing and Nutrition Sciences, North Carolina State University, Richard W. Hartel, Professor of Food Engineering, Department of Food Science, University of Wisconsin, Joseph H. Hotchkiss, Professor and Director of the School of Packaging and Center for Packaging Innovation and Sustainability, Michigan State University, S. Suzanne Nielsen, Professor, Department of Food Science, Purdue University, Juan L. Silva, Professor, Department of Food Science, Nutrition and Health Promotion, Mississippi State University, Martin Wiedmann, Professor, Department of Food Science, Cornell University, Kit Keith L. Yam, Professor of Food Science, Department of Food Science, Rutgers University

Food Analysis Laboratory Manual

Third Edition

edited by

S. Suzanne Nielsen
Purdue University
West Lafayette, IN, USA

S. Suzanne Nielsen
Department of Food Science
Purdue University
West Lafayette
Indiana
USA

ISSN 1572-0330   ISSN 2214-7799   (electronic)
Food Science Text Series
ISBN 978-3-319-44125-2   ISBN 978-3-319-44127-6   (eBook)
DOI 10.1007/978-3-319-44127-6

Library of Congress Control Number: 2017942968

This Springer imprint is published by Springer Nature
The registered company is Springer International Publishing AG
The registered company address is: Gewerbestrasse 11, 6330 Cham, Switzerland

Preface and Acknowledgments

This laboratory manual was written to accompany the textbook, *Food Analysis*, fifth edition. The laboratory exercises are tied closely to the text and cover 21 of the 35 chapters in the textbook. Compared to the second edition of this laboratory manual, this third edition contains four introductory chapters with basic information that compliments both the textbook chapters and the laboratory exercises (as described below). Three of the introductory chapters include example problems and their solutions, plus additional practice problems at the end of the chapter (with answers at the end of the laboratory manual). This third edition also contains three new laboratory exercises, and previous experiments have been updated and corrected as appropriate. Most of the laboratory exercises include the following: background, reading assignment, objective, principle of method, chemicals (with CAS number and hazards), reagents, precautions and waste disposal, supplies, equipment, procedure, data and calculations, questions, and resource materials.

Instructors using these laboratory exercises should note the following:

1. Use of Introductory Chapters:
   - Chap. 1, "Laboratory Standard Operating Procedures" – recommended for students prior to starting any food analysis laboratory exercises
   - Chap. 2, "Preparation of Reagents and Buffers" – includes definition of units of concentrations, to assist in making chemical solutions
   - Chap. 3, "Dilution and Concentration Calculations" – relevant for calculations in many laboratory exercises
   - Chap. 4, "Use of Statistics in Food Analysis" – relevant to data analysis
2. Order of Laboratory Exercises: The order of laboratory exercises has been changed to be fairly consistent with the reordering of chapters in the textbook, *Food Analysis*, fifth edition (i.e., chromatography and spectroscopy near the front of the book). However, each laboratory exercise stands alone, so they can be covered in any order.
3. Customizing Laboratory Procedures: It is recognized that the time and equipment available for teaching food analysis laboratory sessions vary considerably between schools, as do student numbers and their level in school. Therefore, instructors may need to modify the laboratory procedures (e.g., number of samples analyzed, replicates) to fit their needs and situation. Some experiments include numerous parts/methods, and it is not assumed that an instructor uses all parts of the experiment as written. It may be logical to have students work in pairs to make things go faster. Also, it may be logical to have some students do one part of the experiment/one type of sample and other students to another part of the experiment/type of sample.
4. Use of Chemicals: The information on hazards and precautions in the use of the chemicals for each experiment is not comprehensive but should make students and a laboratory assistant aware of major concerns in handling and disposing of the chemicals.
5. Reagent Preparation: It is recommended in the text of the experiments that a laboratory assistant prepare many of the reagents, because of the time limitations for students in a laboratory session. The lists of supplies and equipment for experiments do not necessarily include those needed by the laboratory assistant in preparing reagents for the laboratory session.
6. Data and Calculations: The laboratory exercises provide details on recording data and doing calculations. In requesting laboratory reports from students, instructors will need to specify if they require just sample calculations or all calculations.

Even though this is the third edition of this laboratory manual, there are sure to be inadvertent omissions and mistakes. I will very much appreciate receiving suggestions for revisions from instructors, including input from lab assistants and students.

I maintain a website with additional teaching materials related to both the *Food Analysis* textbook and laboratory manual. Instructors are welcome to contact me for access to this website. To compliment the laboratory manual, the website contains more detailed versions of select introductory chapters and Excel sheets related to numerous laboratory exercises.

I am grateful to the food analysis instructors identified in the text who provided complete laboratory experiments or the materials to develop the experiments. For this edition, I especially want to thank the authors of the new introductory chapters who used their experience from teaching food analysis to develop what I hope will be very valuable chapters for students and instructors alike. The input I received from other food analysis instructors, their students, and mine who reviewed these new introductory chapters was extremely valuable and very much appreciated. Special thanks go to Baraem (Pam) Ismail and Andrew Neilson for their input and major contributions toward this edition of the laboratory manual. My last acknowledgment goes to my former graduate students, with thanks for their help in working out and testing all experimental procedures written for the initial edition of the laboratory manual.

West Lafayette, IN, USA   S. Suzanne Nielsen

The original version of this book was revised.
The correction to this book can be found at DOI 10.1007/978-3-319-44127-6_32

Contents

Contributors

Charles E. Carpenter Department of Nutrition, Dietetics and Food Sciences, Utah State University, Logan, UT, USA

Young-Hee Cho Department of Food Science, Purdue University, West Lafayette, IN, USA

M. Monica Giusti Department of Food Science and Technology, The Ohio State University, Columbus, OH, USA

Y.H. Peggy Hsieh Department of Nutrition, Food and Exercise Sciences, Florida State University, Tallahassee, FL, USA

Baraem P. Ismail Department of Food Science and Nutrition, University of Minnesota, St. Paul, MN, USA

Helen S. Joyner School of Food Science, University of Idaho, Moscow, ID, USA

Dennis A. Lonergan The Vista Institute, Eden Prairie, MN, USA

Lloyd E. Metzger Department of Dairy Science, University of South Dakota, Brookings, SD, USA

Andrew P. Neilson Department of Food Science and Technology, Virginia Polytechnic Institute and State University, Blacksburg, VA, USA

S. Suzanne Nielsen Department of Food Science, Purdue University, West Lafayette, IN, USA

Sean F. O'Keefe Department of Food Science and Technology, Virginia Tech, Blacksburg, VA, USA

Oscar A. Pike Department of Nutrition, Dietetics, and Food Science, Brigham Young University, Provo, UT, USA

Michael C. Qian Department of Food Science and Technology, Oregon State University, Corvallis, OR, USA

Qinchun Rao Department of Nutrition, Food and Exercise Sciences, Florida State University, Tallahassee, FL, USA

Ann M. Roland Owl Software, Columbia, MO, USA

Daniel E. Smith Department of Food Science and Technology, Oregon State University, Corvallis, OR, USA

Denise M. Smith School of Food Science, Washington State University, Pullman, WA, USA

Stephen T. Talcott Department of Nutrition and Food Science, Texas A&M University, College Station, TX, USA

Catrin Tyl Department of Food Science and Nutrition, University of Minnesota, St. Paul, MN, USA

Robert E. Ward Department of Nutrition, Dietetics and Food Sciences, Utah State University, Logan, UT, USA

Ronald E. Wrolstad Department of Food Science and Technology, Oregon State University, Corvallis, OR, USA

Introductory Chapters

Laboratory Standard Operating Procedures

Andrew P. Neilson (✉)

Department of Food Science and Technology,
Virginia Polytechnic Institute and State University,
Blacksburg, VA, USA
e-mail: andrewn@vt.edu

Dennis A. Lonergan

The Vista Institute,
Eden Prairie, MN, USA
e-mail: dennis@thevistainstitute.com

S. Suzanne Nielsen

Department of Food Science, Purdue University,
West Lafayette, IN, USA
e-mail: nielsens@purdue.edu

S.S. Nielsen, *Food Analysis Laboratory Manual*, Food Science Text Series,
DOI 10.1007/978-3-319-44127-6_1, © Springer International Publishing 2017

1.1 INTRODUCTION

This chapter is designed to cover "standard operating procedures" (SOPs), or best practices, for a general food analysis laboratory. The topics covered in this chapter include balances, mechanical pipettes, glassware, reagents, precision and accuracy, data handling, data reporting, and safety. These procedures apply to all the laboratory experiments in this manual, and therefore a thorough review of general procedures will be invaluable for successful completion of these laboratory exercises.

This manual covers many of the basic skills and information that are necessary for one to be a good analytical food chemist. Much of this material is the type that one "picks up" from experience. Nothing can replace actual lab experience as a learning tool, but hopefully this manual will help students learn proper lab techniques early rather than having to correct improper habits later. When one reads this manual, your reaction may be "is all of this attention to detail necessary?" Admittedly, the answer is "not always." This brings to mind an old Irish proverb that "the best person for a job is the one that knows what to ignore." There is much truth to this proverb, but a necessary corollary is that one must know what they are ignoring. The decision to use something other than the "best" technique must be conscious decision and not one made from ignorance. This decision must be based not only upon knowledge of the analytical method being used but also on how the resulting data will be used. Much of the information in this manual has been obtained from an excellent publication by the US Environmental Protection Agency entitled *Handbook for Analytical Quality Control in Water and Wastewater Laboratories*.

1.2 PRECISION AND ACCURACY

To understand many of the concepts in this chapter, a rigorous definition of the terms "precision" and "accuracy" is required here. Precision refers to the **reproducibility** of replicate observations, typically measured as **standard deviation** (SD), **standard error** (SE), or **coefficient of variation** (CV). Refer to Chap. 4 in this laboratory manual and Smith, 2017, for a more complete discussion of precision and accuracy. The smaller these values are, the more reproducible or precise the measurement is. Precision is determined not on reference standards, but by the use of actual food samples, which cover a range of concentrations and a variety of interfering materials usually encountered by the analyst. Obviously, such data should not be collected until the analyst is familiar with the method and has obtained a reproducible standard curve (a mathematical relationship between the analyte concentration and the analytical response). There are a number of different methods available for the determination of precision. One method follows:

1. Three separate concentration levels should be studied, including a low concentration near the sensitivity level of the method, an intermediate concentration, and a concentration near the upper limit of application of the method.
2. Seven replicate determinations should be made at each of the concentrations tested.
3. To allow for changes in instrument conditions, the precision study should cover at least 2 h of normal laboratory operation.
4. To permit the maximum interferences in sequential operation, it is suggested that the samples be run in the following order: high, low, and intermediate. This series is then repeated seven times to obtain the desired replication.
5. The precision statement should include a range of standard deviations over the tested range of concentration. Thus, three standard deviations will be obtained over a range of three concentrations.

Accuracy refers to the degree (absolute or relative) of difference between observed and "actual" values. The "actual" value is often difficult to ascertain. It may be the value obtained by a standard reference method (the accepted manner of performing a measurement). Another means of evaluating accuracy is by the addition of a known amount of the material being analyzed for the food sample and then calculation of **% recovery**. This latter approach entails the following steps:

1. Known amounts of the particular constituent are added to actual samples at concentrations for which the precision of the method is satisfactory. It is suggested that amounts be added to the low-concentration sample, sufficient to double that concentration, and that an amount be added to the intermediate concentration, sufficient to bring the final concentration in the sample to approximately 75 % of the upper limit of application of the method.
2. Seven replicate determinations at each concentration are made.
3. Accuracy is reported as the percent recovery at the final concentration of the spiked sample. Percent recovery at each concentration is the mean of the seven replicate results.

A fast, less rigorous means to evaluate precision and accuracy is to analyze a food sample and replicate a spiked food sample, and then calculate the recovery of the amount spiked. An example is shown in Table 1.1.

| 1.1 table | Measured calcium content (g/L) of milk and spiked milk |

Replicate	Milk	Milk + 0.75 g Ca/L
1	1.29	2.15
2	1.40	2.12
3	1.33	2.20
4	1.24	2.27
5	1.23	2.07
6	1.40	2.10
7	1.24	2.20
8	1.27	2.07
9	1.24	1.74
10	1.28	2.01
11	1.33	2.12
Mean	1.2955	2.0955
SD	0.062	0.138
%CV	4.8	6.6

The accuracy can then be measured by calculating the % of the spike (0.75 g/L) detected by comparing the measured values from the unspiked and spiked samples:

$$\text{accuracy} \approx \% \text{ recovery} = \quad (1.1)$$

$$\frac{\text{measured spiked sample}}{\text{measured sample} + \text{amount of spike}} \times 100\%$$

$$\text{accuracy} \approx \% \text{ recovery} = \frac{2.0955 \text{ g/L}}{1.2955 \text{ g/L} + 0.75 \text{ g/L}}$$
$$\times 100\% = 102.44\%$$

The method measured the spike to within 2.44%. By adding 0.75 g/L Ca to a sample that was measured to have 1.2955 g/L Ca, a perfectly accurate method would result in a spiked sample concentration of 1.295 5 g/L + 0.75 g/L = 2.0455 g/L. The method actually measured the spiked sample at 2.0955 g/L, which is 2.44% greater than it should be. Therefore, the accuracy is estimated at ~2.44% relative error.

1.3 BALANCES

1.3.1 Types of Balances

Two general types of balances are used in most laboratories. These are **top loading balances** and **analytical balances**. Top loading balances usually are sensitive to 0.1–0.001 g, depending on the specific model in use (this means that they can measure differences in the mass of a sample to within 0.1–0.001 g). In, general, as the **capacity** (largest mass that can be measured) increases, the sensitivity decreases. In other words, balances that can measure larger masses generally measure differences in those masses to fewer decimal places. Analytical

balances are usually sensitive to 0.001–0.00001 g, depending on the specific model. It should be remembered, however, that **sensitivity** (ability to detect small differences in mass) is not necessarily equal to **accuracy** (the degree to which the balances correctly report the actual mass). The fact that a balance can be read to 0.01 mg does not necessarily mean it is accurate to 0.01 mg. What this means is that the balance can distinguish between masses that differ by 0.01 mg, but may not accurately measure those masses to within 0.01 mg of the actual masses (because the last digit is often rounded). The accuracy of a balance is independent of its sensitivity.

1.3.2 Choice of Balance

Which type of balance to use depends on "how much accuracy" is needed in a given measurement. One way to determine this is by calculating how much **relative (%) error** would be introduced by a given type of balance. For instance, if 0.1 g of a reagent was needed, weighing it on a top loading balance accurate to within only ±0.02 g of the actual mass would introduce approximately 20% error:

$$\% \text{ error in measured mass} =$$
$$\frac{\text{absolute error in measured mass}}{\text{measured mass}} \times 100\% \quad (1.2)$$

$$\% \text{ error in measured mass} = \frac{0.02 \text{ g}}{0.1 \text{ g}} \times 100\% = 20\%$$

This would clearly be unacceptable in most situations. Therefore, a more accurate balance would be needed. However, the same balance (with accuracy to within ±0.02 g) would probably be acceptable for weighing out 100 g of reagent, as the error would be approximately 0.02%:

$$\% \text{ error in measured mass} = \frac{0.02 \text{ g}}{100 \text{ g}} \times 100\% = 0.02\%$$

The decision on "how much accuracy" is needed can only be answered when one knows the function of the reagent in the analytical method. This is one reason why it is necessary to understand the chemistry involved in an analytical method, and not to simply approach an analytical method in a cookbook fashion. Therefore, a general guideline regarding which balance to use is hard to define.

Another situation in which care must be exercised in determining what type of balance to use is when a difference in masses is to be calculated. For instance, a dried crucible to be used in a total ash determination may weigh 20.05 g on a top loading balance, crucible plus sample = 25.05 g, and the ashed crucible 20.25 g. It may appear that the use of the top loading balance with its accuracy of ±0.02 g would introduce approximately 0.1% error, which would often be acceptable. Actually, since a difference in weight (0.20 g) is being determined, the error would be approximately 10% and thus unacceptable. In this case, an analytical balance is definitely required because sensitivity is required in addition to accuracy.

1.3.3 Use of Top Loading Balances

These instructions are generalized but apply to the use of most models of top loading balances:

1. Level the balance using the bubble level and the adjustable feet (leveling is required so that the balance performs correctly).
2. Either *zero* the balance (so the balance reads 0 with nothing on the pan) or *tare* the balance so that the balance reads 0 with a container that will hold the sample (empty beaker, weighing boat, etc.) on the weighing pan. The tare function is conveniently used for "subtracting" the weight of the beaker or weighing boat into which the sample is added.
3. Weigh the sample.

1.3.4 Use of Analytical Balances

It is always wise to consult the specific instruction manual for an analytical balance before using it. Speed and accuracy are both dependent on one being familiar with the operation of an analytical balance. If it has been a while since you have used a specific type of analytical balance, it may be helpful to "practice" before actually weighing a sample by weighing a spatula or other convenient article. The following general rules apply to most analytical balances and should be followed to ensure that accurate results are obtained and that the balance is not damaged by improper use:

1. Analytical balances are expensive precision instruments; treat them as such.
2. Make sure that the balance is level and is on a sturdy table or bench free of vibrations.
3. Once these conditions are met, the same procedure specified above for top loading balances is used to weigh the sample on an analytical balance.
4. Always leave the balance clean.

1.3.5 Additional Information

Other points to be aware of regarding the use of balances are the following:

1. Many analyses (moisture, ash, etc.) require weighing of the final dried or ashed sample

with the vessel. The mass of the vessel must be known so that it can be subtracted from the final mass to get the mass of the dried sample or ash. Therefore, make sure to obtain the mass of the vessel before the analysis. This can be done by either weighing the vessel before taring the balance and then adding the sample or obtaining the mass of the vessel and then the mass of the vessel plus the sample.

2. The accumulation of moisture from the air or fingerprints on the surface of a vessel will add a small mass to the sample. This can introduce errors in mass that affect analytical results, particularly when using analytical balances. Therefore, beakers, weigh boats, and other weighing vessels should be handled with tongs or with gloved hands. For precise measurements (moisture, ash, and other measurements), weighing vessels should be pre-dried and stored in a desiccator before use, and then stored in a desiccator after drying, ashing, etc. prior to weighing the cooled sample.
3. Air currents or leaning on the bench can cause appreciable error in analytical balances. It is best to take the reading after closing the side doors of an analytical balance.
4. Most balances in modern laboratories are electric balances. Older lever-type balances are no longer in wide use, but they are extremely reliable.

1.4 MECHANICAL PIPETTES

Mechanical pipettes (i.e., **automatic pipettors**) are standard equipment in many analytical laboratories. This is due to their convenience, precision, and acceptable accuracy *when used properly and when calibrated*. Although these pipettes may be viewed by many as being easier to use than conventional glass volumetric pipettes, this does not mean that the necessary accuracy and precision can be obtained without attention to proper pipetting technique. Just the opposite is the case; if mechanical pipettes are used incorrectly, this will usually cause greater error than the misuse of glass volumetric pipettes. The proper use of glass volumetric pipettes is discussed in the section on glassware. The PIPETMAN mechanical pipette (Rainin Instrument Co., Inc.) is an example of a continuously adjustable design. The proper use of this type of pipette, as recommended by the manufacturer, will be described here. Other brands of mechanical pipettes are available, and although their specific instructions should be followed, their proper operation is usually very similar to that described here.

1.4.1 Operation

1. Set the desired volume on the digital micrometer/volumeter. For improved precision, always approach the desired volume by dialing downward from a larger volume setting. Make sure not to wind it up beyond its maximum capacity; this will break it beyond repair.
2. Attach a disposable tip to the shaft of the pipette and press on firmly with a slight twisting motion to ensure a positive, airtight seal.
3. Depress the **plunger** to the **first positive stop**. This part of the stroke is the calibrated volume displayed. Going past the first positive stop will cause inaccurate measurement.
4. Holding the mechanical pipette vertically, immerse the disposable tip into sample liquid to a depth indicated (Table 1.2), specific to the maximum volume of the pipette (P-20, 100, 200, 500, 1000, 5000, correspond to maximum volumes of 20, 100, 200, 500, 1000, and 5000 µL, respectively).
5. Allow plunger to *slowly* return to the "up" position. *Never permit it to snap up* (this will suck liquid up into the pipette mechanism, causing inaccurate measurement and damaging the pipette).
6. Wait 1–2 s to ensure that full volume of sample is drawn into the tip. If the solution is viscous such as glycerol, you need to allow more time.
7. Withdraw tip from sample liquid. Should any liquid remain on outside of the tip, wipe carefully with a lint-free cloth, taking care not to touch the tip opening.
8. To dispense sample, place tip end against side wall of vessel and depress plunger *slowly* past the first stop until the **second stop** (fully depressed position) is reached.
9. Wait (Table 1.3).

10. With plunger fully depressed, withdraw mechanical pipette from vessel carefully with tip sliding along wall of vessel.
11. Allow plunger to return to top position.
12. Discard tip by depressing **tip-ejector button** smartly.
13. A fresh tip should be used for the next measurement if:

    (a) A different solution or volume is to be pipetted.
    (b) A significant residue exists in the tip (not to be confused with the visible "film" left by some viscous or organic solutions).

1.4.2 Pre-rinsing

Pipetting very viscous solutions or organic solvents will result in a significant film being retained on the inside wall of the tip. This will result in an error that will be larger than the tolerance specified if the tip is only filled once. Since this film remains relatively constant in successive pipettings with the same tip, accuracy may be improved by filling the tip, dispensing the volume into a waste container, refilling the tip a second time, and using this quantity as the sample. This procedure is recommended in all pipetting operations when critical reproducibility is required, whether or not tips are reused (same solution) or changed (different solutions/different volumes). Note that the "non-wettability" of the polypropylene tip is not absolute and that pre-rinsing will improve the precision and accuracy when pipetting any solution.

1.4.3 Pipetting Solutions of Varying Density or Viscosity

Compensation for solutions of varying viscosity or density is possible with any adjustable pipette by setting the digital micrometer slightly higher or lower than the required volume. The amount of compensation is determined empirically. Also, when dispensing viscous liquids, it will help to *wait* 1 s longer at the first stop before depressing to the second stop.

1.4.4 Performance Specifications

The manufacturer of PIPETMAN mechanical pipettes provides the information in Table 1.4, on the precision and accuracy of their mechanical pipettes.

1.4.5 Selecting the Correct Pipette

Although automatic pipettes can dispense a wide range of volumes, you may often have to choose the "best" pipette with the most accuracy/precision from among several choices. For example, a P5000

1.2 table — Appropriate pipette depth for automatic pipettors

Pipette	Depth (mm)
P-20D, P-100D, P-200D	1–2
P-500D, P-1000D	2–4
P-5000D	3–6

1.3 table — Appropriate dispense wait time for automatic pipettors

Pipette	Time (s)
P-20D, P-100D, P-200D	1
P-500D, P-1000D	1–2
P-5000D	2–3

1.4 table Accuracy and precision of PIPETMAN mechanical pipettes

Model	Accuracy[a]	Reproducibility[a] (standard deviation)
P-2OD	<0.l µL @ 1–10 µL <1% @ 10–20 µL	<0.04 µL @ 2 µL <0.05 µL @ 10 µL
P-200D	<0.5 µL @ 20–60 µL <0.8% @ 60–200 µL	<0.15 µL @ 25 µL <0.25 µL @ 100 µL <0.3 µL @ 200 µL
P-1000D	<3 µL @ 100–375 µL <0.8% @ 375–l000 µL	<0.6 µL @ 250 µL <1.0 µL @ 500 µL <1.3 µL @ 1000 µL
P-5000D	<12 µL @ 0.5–2 mL <0.6% @ 2.0–5.0 mL	<3 µL @ 1.0 mL <5 µL @ 2.5 mL <8 µL @ 5.0 mL

[a]Aqueous solutions, tips prerinsed once

1.5 table Recommended volume ranges for mechanical pipettors

Maximum volume	Lowest recommended volume
5 mL (5000 µL)	1 mL (1000 µL)
1 mL (1000 µL)	0.1–0.2 mL (100–200 µL)
0.2 mL (200 µL)	0.02–0.04 mL (20–40 µL)
0.1 mL (100 µL)	0.01–0.02 mL (10–20 µL)
0.05 mL (50 µL)	0.005–0.01 mL (5–10 µL)
0.02 mL (20 µL)	0.002–0.004 mL (2–4 µL)
0.01 mL (10 µL)	0.001–0.002 mL (1–2 µL)

(i.e., 5 mL) automatic pipettor could theoretically pipette anywhere between 0 and 5 mL. However, there are several limitations that dictate which pipettes to use. The first is a practical limitation: mechanical pipettes are limited by the graduations (the increments) of the pipette. The P5000 and P1000 are typically adjustable in increments of 0.01 mL (10 µL). Therefore, these pipettes cannot dispense volumes of <10 µL, nor can they dispense volumes with more precision that of 10 µL. However, just because these pipettes can technically be adjusted to 10 µL does not mean that they should be used to measure volumes anywhere near this small. Most pipettes are labeled with a working range that lists the minimum and maximum volume, but this is not the range for ideal performance. Mechanical pipettes should be operated from 100% down to 10–20% of their maximum capacity (Table 1.5). Below 10–20% of their maximum capacity, performance (accuracy and precision) suffers. A good way of thinking of this is to use the largest pipette capable of dispensing the volume in a single aliquot.

Mechanical pipettes are invaluable pieces of laboratory equipment. If properly treated and maintained, they can last for decades. However, improper use can destroy them in seconds. Mechanical pipettes should be calibrated, lubricated, and maintained at least yearly by a knowledgeable pipette technician. Weighing dispensed water is often a good check to see if the pipette needs calibration.

1.5 GLASSWARE

1.5.1 Types of Glassware/Plasticware

Glass is the most widely used material for construction of laboratory vessels. There are many grades and types of glassware to choose from, ranging from student grade to others possessing specific properties such as resistance to thermal shock or alkali, low boron content, and super strength. The most common type is a highly resistant borosilicate glass, such as that manufactured by Corning Glass Works under the name "Pyrex" or by Kimble Glass Co. as "Kimax." Brown/amber actinic glassware is available, which blocks UV and IR light to protect light-sensitive solutions and samples. The use of vessels, containers, and other apparatus made of Teflon, polyethylene, polystyrene, and polypropylene is common. Teflon stopcock plugs have practically replaced glass plugs in burets, separatory funnels, etc., because lubrication to avoid sticking (called "freezing") is not required. Polypropylene, a methylpentene polymer, is available as laboratory bottles, graduated cylinders, beakers, and even volumetric flasks. It is crystal clear, shatterproof, autoclavable, chemically resistant, but relatively expensive as compared to glass. Teflon (polytetrafluoroethylene, PTFE) vessels are available, although they are very expensive. Finally, most glassware has a polar surface. Glassware can be treated to derivatize the surface (typically, tetramethylsilane, or TMS) to make it nonpolar, which is required for some assays. However, acid washing will remove this nonpolar layer.

1.5.2 Choosing Glassware/Plasticware

Some points to consider in choosing glassware and/or plasticware are the following:

1. Generally, special types of glass are not required to perform most analyses.
2. Reagents and standard solutions should be stored in borosilicate or polyethylene bottles.
3. Certain dilute metal solutions may plate out on glass container walls over long periods of storage. Thus, dilute metal standard solutions should be prepared fresh at the time of analysis.
4. Strong mineral acids (such as sulfuric acid) and organic solvents will readily attack polyethylene; these are best stored in glass or a resistant plastic.

5. Borosilicate glassware is not completely inert, particularly to alkalis; therefore, standard solutions of silica, boron, and the alkali metals (such as NaOH) are usually stored in polyethylene bottles.

6. Certain solvents dissolve some plastics, including plastics used for pipette tips, serological pipettes, etc. This is especially true for acetone and chloroform. When using solvents, check the compatibility with the plastics you are using. Plastics dissolved in solvents can cause various problems, including binding/precipitating the analyte of interest, interfering with the assay, clogging instruments, etc.

7. Ground-glass stoppers require care. Avoid using bases with any ground glass because the base can cause them to "freeze" (i.e., get stuck). Glassware with ground-glass connections (burets, volumetric flasks, separatory funnels, etc.) are very expensive and should be handled with extreme care.

For additional information, the reader is referred to the catalogs of the various glass and plastic manufacturers. These catalogs contain a wealth of information as to specific properties, uses, sizes, etc.

1.5.3 Volumetric Glassware

Accurately calibrated glassware for accurate and precise measurements of volume has become known as **volumetric glassware**. This group includes **volumetric flasks, volumetric pipettes**, and accurately **calibrated burets**. Less accurate types of glassware, including **graduated cylinders, serological pipettes**, and **measuring pipettes**, also have specific uses in the analytical laboratory when exact volumes are unnecessary. Volumetric flasks are to be used in preparing standard solutions, but not for storing reagents. The precision of an analytical method depends in part upon the accuracy with which volumes of solutions can be measured, due to the inherent parameters of the measurement instrument. For example, a 10 mL volumetric flask will typically be more precise (i.e., have smaller variations between repeated measurements) than a 1000 mL volumetric flask, because the neck on which the "fill to" line is located is narrower, and therefore smaller errors in liquid height above or below the neck result in smaller volume differences compared to the same errors in liquid height for the larger flask. However, accuracy and precision are often independent of each other for measurements on similar orders of magnitude. In other words, it is possible to have precise results that are relatively inaccurate and vice versa. There are certain sources of error, which must be carefully considered. The volumetric apparatus must be read correctly; the bottom of the **meniscus** should be tangent to the calibration mark. There are other sources of error, however, such as changes in temperature, which result in changes in the actual capacity of glass apparatus and in the volume of the solutions. The volume capacity of an ordinary 100 mL glass flask increases by 0.025 mL for each 1° rise in temperature, but if made of borosilicate glass, the increase is much less. One thousand mL of water (and of most solutions that are ≤ 0.1 N) increases in volume by approximately 0.20 mL per 1 °C increase at room temperature. Thus, solutions must be measured at the temperature at which the apparatus was calibrated. This temperature (usually 20 °C) will be indicated on all volumetric ware. There may also be errors of calibration of the adjustable measurement apparatus (e.g., measuring pipettes), that is, the volume marked on the apparatus may not be the true volume. Such errors can be eliminated only by recalibrating the apparatus (if possible) or by replacing it.

A volumetric apparatus is calibrated "**to contain**" or "**to deliver**" a definite volume of liquid. This will be indicated on the apparatus with the letters "TC" (to contain) or "**TD**" (to deliver). Volumetric flasks are calibrated to contain a given volume, which means that the flask contains the specified volume ± a defined tolerance (error). The certified TC volume only applies to the volume contained by the flask and it does not take into account the volume of solution that will stick to the walls of the flask if the liquid is poured out. Therefore, for example, a TC 250 mL volumetric flask will hold 250 mL ± a defined tolerance; if the liquid is poured out, slightly less than 250 mL will be dispensed due to solution retained on the walls of the flask (this is the opposite of "to deliver" or TD, glassware discussed below). They are available in various shapes and sizes ranging from 1 to 2000 mL capacity. Graduated cylinders, on the other hand, can be either TC or TD. For accurate work the difference may be important.

Volumetric pipettes are typically calibrated to deliver a fixed volume. The usual capacities are 1–100 mL, although micro-volumetric pipettes are also available. The proper technique for using volumetric pipettes is as follows (this technique is for TD pipettes, which are much more common than TC pipettes):

1. Draw the liquid to be delivered into the pipette above the line on the pipette. Always use a pipette bulb or pipette aid to draw the liquid into the pipette. Never pipette by mouth.

2. Remove the bulb (when using the pipette aid, or bulbs with pressure release valves, you can deliver without having to remove it) and replace it with your index finger.

3. Withdraw the pipette from the liquid and wipe off the tip with tissue paper. Touch the tip of the pipette against the wall of the container from which the liquid was withdrawn (or a spare

beaker). Slowly release the pressure of your finger (or turn the scroll wheel to dispense) on the top of the pipette and allow the liquid level in the pipette to drop so that the bottom of the meniscus is even with the line on the pipette.

4. Move the pipette to the beaker or flask into which you wish to deliver the liquid. Do not wipe off the tip of the pipette at this time. Allow the pipette tip to touch the side of the beaker or flask. Holding the pipette in a vertical position, allow the liquid to drain from the pipette.

5. Allow the tip of the pipette to remain in contact with the side of the beaker or flask for several seconds. Remove the pipette. There will be a small amount of liquid remaining in the tip of the pipette. Do not blow out this liquid with the bulb, as TD pipettes are calibrated to account for this liquid that remains.

Note that some volumetric pipettes have calibration markings for both TC and TD measurements. Make sure to be aware which marking refers to which measurement (for transfers, use the TD marking). The TC marking will be closer to the dispensing end of the pipette (TC does not need to account for the volume retained on the glass surface, whereas TD does account for this).

Measuring and serological pipettes should also be held in a vertical position for dispensing liquids; however, the tip of the pipette is only touched to the wet surface of the receiving vessel *after* the outflow has ceased. Some pipettes are designed to have the small amount of liquid remaining in the tip blown out and added to the receiving container; such pipettes have a frosted band near the top. If there is no frosted band near the top of the pipette, do not blow out any remaining liquid.

1.5.4 Using Volumetric Glassware to Perform Dilutions and Concentrations

Typically, dilutions are performed by adding a liquid (water or a solvent) to a sample or solution. Concentrations may be performed by a variety of methods, including rotary evaporation, shaking vacuum evaporation, vacuum centrifugation, boiling, oven drying, drying under N_2 gas, or freeze drying.

For bringing samples or solutions up to a known volume, the "gold standard" providing maximal accuracy and precision is a **Class A** glass volumetric flask (Fig. 1.1a). During manufacture, glassware to be certified as Class A is calibrated and tested to comply with tolerance specifications established by the American Society for Testing and Materials (ASTM, West Conshohocken, PA). These specifications are the

standard for laboratory glassware. Class A glassware has the tightest tolerances and therefore the best accuracy and precision. These flasks are rated TC. Therefore, volumetric flasks are used to bring samples and solutions up to a defined volume. They are not used to quantitatively deliver or transfer samples because the delivery volume is not known. Other types of glassware (non-Class A flasks, graduated cylinders, Erlenmeyer flasks, round-bottomed flasks, beakers, bottles, etc., Fig. 1.1b) are less accurate and less precise. They should not be used for quantitative volume dilutions or concentrations if Class A volumetric flasks are available.

For transferring a known volume of a liquid sample for a dilution or concentration, the "gold standard" providing maximal accuracy and precision is a Class A glass volumetric pipette (Fig. 1.2a). These pipettes are rated "to deliver" (TD), which means that the pipette will deliver the specified volume ± a defined tolerance (error). The certified TD volume takes into account the volume of solution that will stick to the walls of the pipette as well as the volume of the drop of solution that typically remains in the tip of the pipette after delivery (again, you should not attempt to get this drop out, as it is already accounted for). Therefore, for example, a TD 5 mL pipette will hold slightly more than 5 mL but will deliver (dispense) 5 mL ± a defined tolerance (the opposite of TC glassware). It is important to note that volumetric pipettes are used only to deliver a known amount of solution. Typically they should not be used to determine the final volume of the solution unless the liquids dispensed are the only components of the final solution. For example, if a sample is dried down and then liquid from a volumetric pipette is used to resolubilize the solutes, it is unknown if the solutes significantly affect the volume of the resulting solution, unless the final volume is measured, which may be difficult to do. Although the effect is usually negligible, it is best to use volumetric glassware to assure that the final volume of the resulting solution is known (the dried solutes could be dissolved in a few mL of solvent and then transferred to a volumetric flask for final dilution). However, it is acceptable to add several solutions together using volumetric pipettes and then add the individual volumes together to calculate the final volume. However, using a single volumetric flask to dilute to a final volume is still the favored approach, as using one measurement for the final volume reduces the uncertainty. (The errors, or tolerances, of the amounts added are also added together; therefore, using fewer pieces of glassware lowers the uncertainty of the measurement even if the tolerances of the glassware are the same.) For example, suppose you need to measure out 50 mL of solution. You have access to a 50 mL volumetric flask and a 25 mL volumetric pipette, both of which have tolerances of ± 0.06 mL. If you obtain 50 mL by filling

Class A volumetric flask (**a**) and other types of non-Class A volume measuring glassware: graduated cylinder
(**b**), Erlenmeyer flask (**c**), beaker (**d**), and bottle (**e**)

the volumetric flask, the measured volume is
50 mL±0.06 mL (or somewhere between 49.94 and
50.06 mL). If you pipette 25 mL twice into a beaker, the
tolerance of each measurement is 25 mL±0.06 mL, and
the tolerance of the combined volume is the sum of the
means and the errors:

$$(25\,mL \pm 0.06\,mL) + (25\,mL \pm 0.06\,mL) =$$
$$50\,mL \pm 0.12\,mL = 49.88 - 50.12\,mL$$

This additive property of tolerances, or errors, compounds further as more measurements are combined;
conversely, when the solution is brought to volume
using a volumetric flask, only a single tolerance factors
into the error of the measurement.

Other types of pipettes (non-Class A volumetric
glass pipettes, adjustable pipettors, automatic pipettors, reed pipettors, serological pipettes, etc., Fig. 1.2b)
and other glassware (graduated cylinders, etc.) are less
accurate and less precise. They should not be used for
quantitative volume transfers. Pipettes are available
(but rare) that are marked with lines for both TC and
TD. For these pipettes, the TD line would represent the
volume delivered when the drop at the tip is dispensed
and TC when the drop remains in the pipette.

Information typically printed on the side of the
pipette or flask includes the class of the pipette or
flask, whether the glassware is TD or TC, the TC or TD
volume, and the defined tolerance (error) (Fig. 1.3).
Note that the specifications are typically valid at a
specified temperature, typically 20 °C. Although it is
rare that scientists equilibrate solutions to exactly
20 °C before volume measurement, this temperature is
assumed to be approximate room temperature. Be
aware that the greater the deviation from room temperature, the greater the error in volume measurement. The specific gravity (density) of water at 4, 20,
60, and 80 °C relative to 4 °C is 1.000, 0.998, 0.983, and
0.972. This means that a given mass of water has lower
density (greater volume for given mass) at temperatures above 20 °C. This is sometimes seen when a volumetric flask is brought exactly to volume at room
temperature and then is placed in an ultrasonic bath to
help dissolve the chemicals, warming the solution. A
solution that was exactly at the volume marker at
room temperature will be above the volume when the
solution is warmer. To minimize this error, volumes
should be measured at room temperature.

Volumetric glassware (flasks and pipettes) should
be used for quantitative volume measurements during

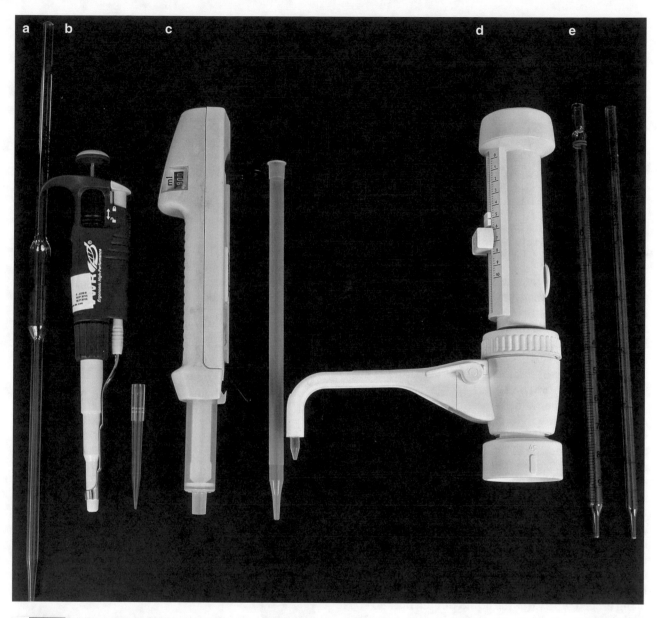

.2 Class A volumetric pipette (**a**) and non-volumetric pipettes: adjustable pipettors (**b**), reed pipettor (**c**), serological
figure pipettes (**d**)

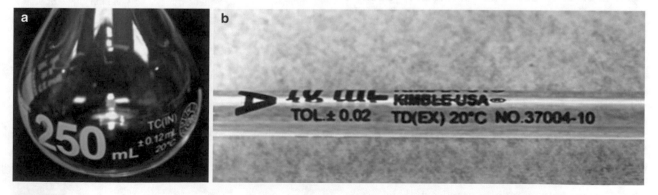

.3 Image of the label on a Class A volumetric flask pipette (**a**) and Class A volumetric pipette (**b**)
figure

1.4 figure — Image of a liquid meniscus at the line for a Class A volumetric flask

dilutions and concentrations whenever possible to maximize the accuracy and precision of the procedure. For both volumetric flasks and pipettes, the level of the liquid providing the defined volume is indicated by a line (usually white or red) etched or printed on the neck of the glassware. To achieve the TD or TC volume, the bottom of the meniscus of the liquid should be at the line as shown in Fig. 1.4.

For a volumetric flask, the proper technique for achieving the correct volume is to pour the liquid into the flask until the meniscus is close to the marking line, and then add additional liquid dropwise (with a manual pipette or Pasteur pipette) until the bottom (NOT the top or middle) of the meniscus is at the line with your eye level to the line. (If you do not look straight at the line, occur, making it appear that so that your eye and the line are at the same level, a phenomenon known as "parallax" can occur, making it appear that the bottom of the meniscus is at the line when in fact it is not, resulting in errors in volume measurement.) If the level of the liquid is too high, liquid can be removed using a clean pipette (or the liquid poured out and start again). However, be aware that this cannot be done when preparing a reagent for which the solutes were accurately measured into the flask and you are adding liquid to make up to volume. In this case, you must start over. For this reason, the best practice is to add liquid slowly, and then use a pipette to add liquid dropwise when approaching the desired volume.

For a volumetric pipette, the proper technique for achieving the correct volume is to draw liquid into the pipette until the meniscus is above the line, and then withdraw the pipette from the liquid and dispense the excess liquid from the pipette until the bottom of the meniscus is at the line. It is critical that the pipette be withdrawn from the solution for this step. If the level of the liquid goes below the line, additional liquid is drawn up, and the process is repeated. Proper volumetric measurements require practice and should be repeated until they are performed correctly. Improper

volumetric measurements can result in significant error being introduced into the measurement.

Typical tolerances for lab glassware are presented in Tables 1.6 and 1.7. References for ASTM specifications are found at http://www.astm.org/.

A comparison of Tables 1.6 and 1.7 reveals some important points. First, even for Class A glassware, the tolerances for volumetric transfer pipettes (pipettes with a single TD measurement) are much tighter than for graduated measuring pipettes (pipettes with graduations that can be used to measure a wide range of volumes) of the same volume. Second, even for Class A glassware, the tolerances for volumetric transfer pipettes and volumetric flasks are much tighter than

1.6 table — Volume tolerances of Class A glassware required by ASTM specifications

Volume (mL)	Tolerance (± mL)				
	Buret	Volumetric (transfer) pipette	Measuring (graduated) pipettes	Volumetric flask	Graduated cylinder
0.5		0.006			
1		0.006		0.010	
2		0.006	0.01	0.015	
3		0.01	0.02	0.015	
4		0.01	0.03	0.020	
5		0.01	0.05	0.020	0.05
10	0.02	0.02	0.08	0.020	0.10
25	0.03	0.03	0.10	0.030	0.17
50	0.05	0.05		0.050	0.25
100	0.10	0.08		0.080	0.50
250				0.012	1.00
500				0.013	2.00
1000				0.015	3.00

1.7 table — Volume tolerances of non-Class A glassware required by ASTM specifications

Volume (mL)	Tolerance (± mL)			
	Buret	Volumetric (transfer) pipette	Volumetric flask	Graduated cylinder
0.5		0.012		
1		0.012		
2		0.012		
3		0.02		
4		0.02		
5		0.02		0.10
10	0.04	0.04	0.04	0.20
25	0.06	0.06	0.06	0.34
50	0.10	0.10	0.24	0.50
100	0.20	0.16	0.40	1.00
250			0.60	2.00
500				4.00
1000				6.00

for graduated cylinders of the same volume. Therefore, volumetric transfer pipettes and volumetric flasks are preferred for dilutions and concentrations. For example, a 1000 mL Class A volumetric flask has a tolerance of ±0.015 mL (the actual TC volume is somewhere between 999.985 and 1000.015 mL), while a 1000 mL graduated cylinder has a tolerance of ±3.00 mL (the actual TC volume is somewhere between 997 and 1003 mL). This is a 200-fold larger potential error in the measurement of 1000 mL! Finally, tolerances for non-Class A glassware are much broader than for Class A, and thus Class A should be used if available.

1.5.5   Conventions and Terminology

To follow the analytical procedures described in this manual and perform calculations correctly, common terminology and conventions (a convention is a standard or generally accepted way of doing or naming something) must be understood. A common phrase in dilutions and concentrations is **"diluted to"** or **"diluted to a final volume of."** This means that the sample or solution is placed in a volumetric flask, and the final volume is adjusted to the specified value. In contrast, the phrase **"diluted with"** means that the specified amount is added to the sample or solution. In this latter case, the final mass/volume must be calculated by adding the sample mass/volume and the amount of liquid added. For example, suppose you take a 1.7 mL volume and either (1) dilute to 5 mL with methanol or (2) dilute with 5 mL methanol. In the first case, this means that the sample (1.7 mL) is placed in a volumetric flask and methanol (~3.3 mL) is added so that the final volume is 5 mL total. In the second case, the sample (1.7 mL) is combined with 5 mL methanol, and the final volume is 6.7 mL. As you can see, these are very different values. This will always be the case except when one of the volumes is much larger than the other. For example, if you were working with a 10 μL sample, diluting it "to 1 L" or "with 1 L" would result in final volumes of 1 L and 1.00001 L, respectively. It is important to understand the differences between these two conventions to perform procedures correctly and interpret data accurately.

Another common term in dilutions/concentrations is the term **"fold"** or **"X."** This refers to the ratio of the final and initial concentrations (or volumes and masses) of the sample or solution during each step. An "X-fold dilution" means that the concentration of a sample decreases (and typically the volume increases) by a given factor. For example, if 5 mL of an 18.9% NaCl solution is diluted tenfold (or 10X) with water, 45 mL water is added so that the final volume is 50 mL (tenfold or 10X greater than 5 mL) and the final concentration is 1.89% NaCl (tenfold or 10X less than 18.9%). Conversely, an "X-fold concentration" means that the concentration of a sample increases (and typically the volume decreases) by the stated factor. For example, if 90 mL of a 0.31 ppm

salt solution is concentrated tenfold (10X), the volume is decreased to 9 mL (either by reducing to 9 mL or drying completely and reconstituting to 9 mL, tenfold or 10X lower than 90 mL), and the final concentration is 3.1 ppm salt (tenfold or 10X more than 0.31 ppm). Although tenfold or 10X was used for these examples, any value can be used. In microbiology, values of 10X, 100X, 1000X, etc. are commonly used due to the log scale used in that field. However, less standard dilutions of any value are routinely used in analytical chemistry.

The last terminology system for dilutions and concentrations involves **ratios**. This system is somewhat ambiguous and is not used in the *Food Analysis* text or lab manual. This system refers to dilutions as "X:Y," where X and Y are the masses or volumes of the initial and final solutions/samples. For example, it may be stated that "the solution was diluted 1:8." This system is ambiguous for the following reasons:

1. The first and last numbers typically refer to the initial and final samples, respectively (therefore, a 1:8 dilution would mean 1 part initial sample and 8 parts final sample). However, there is no standard convention. Therefore, an "X:Y" dilution could be interpreted either way.
2. There is no standard convention as to whether this system describes the "diluted to" or "diluted with" (as described above) approach. Therefore, diluting a sample 1:5 could be interpreted as either (1) diluting 1 mL sample with 4 mL for a final volume of 5 mL ("diluted to") or (2) diluting 1 mL sample with 5 mL for a final volume of 6 mL ("diluted with").

Because of these ambiguities, the ratio system is discouraged in favor of the "X-fold" terminology. However, ratio dilutions still appear in some literature. If possible, it is recommended that you investigate to clarify what is meant by this terminology.

Another factor to consider is that liquid volumes are often not strictly additive. For example, exactly 500 ml 95% v/v ethanol aq. added to 500 ml distilled water will not equal 1000 ml; in fact, the new volume will be closer to 970 ml. Where did the missing 30 ml go? Polar molecules such as water undergo different three-dimensional intermolecular bonding in a pure solution versus in a mixture with other solute or chemicals such as ethanol. The difference in bonding causes an apparent contraction in this case. As well, addition of solute to an exact volume of water will change the volume after dissolved. To account for this effect, volumetric glassware is used to bring mixed solutions up to a final volume after initial mixing. When two liquids are mixed, the first liquid is volumetrically transferred into a volumetric flask, and then the second liquid is added to volume, with intermittent swirling or vortexing to mix the liquids as they are being combined. For mixing solids into solvents, the chemicals are first placed in a volumetric flask, dissolved in a partial volume, and then brought to exact volume with additional solvent.

1.5.6 Burets

Burets are used to deliver definite volumes. The more common types are usually of 25 or 50 ml capacity, graduated to tenths of a milliliter, and are provided with stopcocks. For precise analytical methods in microchemistry, microburets are also used. Microburets generally are of 5 or 10 ml capacity, graduated in hundredths of a milliliter division. General rules in regard to the manipulation of a buret are as follows:

1. Do not attempt to dry a buret that has been cleaned for use, but rather rinse it two or three times with a small volume of the solution with which it is to be filled.
2. Do not allow alkaline solutions to stand in a buret, because the glass will be attacked, and the stopcock, unless made of Teflon, will tend to freeze.
3. A 50 ml buret should not be emptied faster than 0.7 ml per second; otherwise, too much liquid will adhere to the walls; as the solution drains down, the meniscus will gradually rise, giving a high false reading.

It should be emphasized that improper use of and/or reading of burets can result in serious calculation errors.

1.5.7 Cleaning of Glass and Porcelain

In the case of all apparatus for delivering liquids, the glass must be absolutely clean so that the film of liquid never breaks at any point. Careful attention must be paid to this fact or the required amount of solution will not be delivered. The method of cleaning should be adapted to both the substances that are to be removed and the determination to be performed. Water-soluble substances are simply washed out with hot or cold water, and the vessel is finally rinsed with successive small amounts of distilled water. Other substances more difficult to remove, such as lipid residues or burned material, may require the use of a detergent, organic solvent, nitric acid, or aqua regia (25 % v/v conc. HNO_3 in conc. HCl). In all cases it is good practice to rinse a vessel with tap water as soon as possible after use. Material allowed to dry on glassware is much more difficult to remove.

1.6 REAGENTS

Chemical reagents, solvents, and gases are available in a variety of **grades of purity**, including technical grade, analytical reagent grade, and various

"ultrapure" grades. The purity of these materials required in analytical chemistry varies with the type of analysis. The parameter being measured and the sensitivity and specificity of the detection system are important factors in determining the purity of the reagents required. **Technical grade** is useful for making cleaning solutions, such as the nitric acid and alcoholic potassium hydroxide solutions mentioned previously. For many analyses, **analytical reagent grade** is satisfactory. Other analyses, e.g., trace organic and HPLC, frequently require special "ultrapure" reagents and solvents. In methods for which the purity of reagents is not specified, it is intended that analytical reagent grade be used. Reagents of lesser purity than that specified by the method should not be used.

There is some confusion as to the definition of the terms **analytical reagent grade**, **reagent grade**, and **ACS analytical reagent grade**. A review of the literature and chemical supply catalogs indicates that the three terms are synonymous. National Formulary (NF), US Pharmaceutical (USP), and Food Chemicals Codex (FCC) are grades of chemicals certified for use as food ingredients. It is important that only NF, USP, or FCC grades be used as food additives if the product is intended for consumption by humans, rather than for chemical analysis.

1.6.1 Acids

The concentration of common commercially available acids is given in Table 1.8.

1.6.2 Distilled Water

Distilled or **demineralized water** is used in the laboratory for dilution, preparation of reagent solutions, and final rinsing of washed glassware.

| 1.8 table | Concentration of common commercial strength acids |

Acid	Molecular weight (g/mol)	Concentration (M)	Specific gravity
Acetic acid, glacial	60.05	17.4	1.05
Formic acid	46.02	23.4	1.20
Hydriodic acid	127.9	7.57	1.70
Hydrochloric acid	36.5	11.6	1.18
Hydrofluoric acid	20.01	32.1	1.167
Hypophosphorous acid	66.0	9.47	1.25
Lactic acid	90.1	11.3	1.2
Nitric acid	63.02	15.99	1.42
Perchloric acid	100.5	11.65	1.67
Phosphoric acid	98.0	14.7	1.70
Sulfuric acid	98.0	18.0	1.84
Sulfurous acid	82.1	0.74	1.02

Ordinary distilled water is usually not pure. It may be contaminated by dissolved gases and by materials leached from the container in which it has been stored. Volatile organics distilled over from the original source feed water may be present, and nonvolatile impurities may occasionally be carried over by the steam, in the form of a spray. The concentration of these contaminants is usually quite small, and distilled water is used for many analyses without further purification. There are a variety of methods for purifying water, such as distillation, filtration, and ion exchange. Distillation employs boiling of water and condensation of the resulting steam, to eliminate nonvolatile impurities (such as minerals). Ion exchange employs cartridges packed with ionic residues (typically negatively charged) to remove charged contaminants (typically positively charged minerals) when water is passed through the cartridge. Finally, filtration and reverse osmosis remove insoluble particulate matter above a specific size.

1.6.3   Water Purity

Water purity has been defined in many different ways, but one generally accepted definition states that high purity water is water that has been distilled and/or deionized so that it will have a specific resistance of $500,000\ \Omega$ ($2.0\ \mu\Omega$/cm conductivity) or greater. This definition is satisfactory as a base to work from, but for more critical requirements, the breakdown shown in Table 1.9 has been suggested to express degrees of purity.

Distilled water is usually produced in a steam-heated metal still. The feed water is (or should be) softened to remove calcium and magnesium to prevent scale (Ca or Mg carbonate) formation. Several companies produce ion-exchange systems that use resin-packed cartridges for producing "distilled water." The lifespan of an ion-exchange cartridge is very much a function of the mineral content of the feed water. Thus, the lifespan of the cartridge is greatly extended by using distilled or reverse osmosis-treated water as the incoming stream. This procedure can also be used for preparing ultrapure water, especially if a low flow rate is used and the ion-exchange cartridge is of "research" grade.

1.9

table   Classification of water purity

Degree of purity	Maximum conductivity ($\mu\Omega$/cm)	Approximate concentration of electrolytes (mg/L)
Pure	10	2–5
Very pure	1	0.2–0.5
Ultrapure	0.1	0.01–0.02
Theoretically pure	0.055	0.00

1.6.4   Carbon Dioxide-Free Water

Carbon dioxide (CO_2) dissolved in water can interfere with many chemical measurements. Thus, CO_2-free water may need to be produced. CO_2-free water may be prepared by boiling distilled water for 15 min and cooling to room temperature. As an alternative, distilled water may be vigorously aerated with a stream of inert gas (e.g., N_2 or He_2) for a period sufficient to achieve CO_2 removal. The final pH of the water should lie between 6.2 and 7.2. It is not advisable to store CO_2-free water for extended periods. To ensure that CO_2-free water remains that way, an **ascarite trap** should be fitted to the container such that air entering the container (as boiled water cools) is CO_2-free. Ascarite is silica coated with NaOH, and it removes CO_2 by the following reaction:

$$2NaOH + CO_2 \rightarrow Na_2CO_3 + H_2O$$

Ascarite should be sealed from air except when water is being removed from the container.

1.6.5   Preparing Solutions and Reagents

The accurate and reproducible preparation of laboratory reagents is essential to good laboratory practice. Liquid reagents are prepared using volumetric glassware (pipettes and flasks) as appropriate.

To prepare solutions from solid reagents (such as sodium hydroxide):

1. Determine the amount of solid reagent needed.
2. Fill the TC volumetric flask ~¼–½ full with the solvent.
3. Add the solid reagent (it is best to pre-dissolve solids in a beaker with a small amount of liquid, and then add this to the flask; rinse the smaller beaker thoroughly and also put the rinses into flask).
4. Swirl to mix until essentially dissolved.
5. Fill the flask to volume with the solvent.
6. Cap and invert the flask ~10–20 times to completely mix the solution.

Note that it is not appropriate to simply combine the solid reagent with the final volume and assume that the final volume does not change. This is particularly true for high % concentrations. For example, 1 L of a 10 % aqueous NaOH solution is correctly made by filling a 1 L flask with ~25–500 mL water, adding 100 g NaOH, mixing until dissolved, and diluting to 1 L. It would be incorrect to simply combine 100 g NaOH with 1 L water, as the dissolved solid will take up some volume in solution. (Note that solid NaOH is difficult to dissolve, requires a stir bar, and is exothermic, releasing heat upon dissolution; therefore, do not handle the glass with bare hands.) Additionally, if a stir bar is used, make sure to remove this after the solution is dissolved but BEFORE diluting to volume. Note that sonication is preferred to using a stir bar in a volumetric flask.

The following similar procedures are used to prepare reagents from two or more liquids:

1. Determine the total volume of the final reagent.
2. Obtain a TC volumetric flask (if possible) equal to the final volume.
3. Use TD volumetric glassware to add the correct amount of the liquids with the smallest volumes.
4. Dilute to volume with the liquid with the largest volume, gently swirling during addition.
5. Cap and invert the flask ~10–20 times to completely mix the solution.

Note that a TC volumetric flask should be used whenever possible to bring the solution to final volume. For example, the correct way to prepare 1 L of a 5% ethanol in water solution is to use a 50 mL TD pipette to dispense 50 mL ethanol into a 1L TC flask and then fill the flask to volume with water. It would be incorrect to simply combine 50 mL ethanol and 950 mL water, since complex physical properties govern the volume of a mixture of liquids, and it cannot be assumed that two liquids of different densities and polarities will combine to form a volume equal to the sum of their individual volumes. If the final volume is not a commonly available TC flask size, then use TD glassware to deliver all reagents.

The use of graduated cylinders and beakers should be avoided for measuring volumes for reagent preparation.

1.7 DATA HANDLING AND REPORTING

1.7.1 Significant Figures

The term **significant figure** is used rather loosely to describe some judgment of the number of reportable digits in a result. Often the judgment is not soundly based and meaningful digits are lost or meaningless digits are accepted. Proper use of significant figures gives an indication of the reliability of the analytical method used. Thus, reported values should contain only significant figures. A value is made up of significant figures when it contains all digits known to be true and one last digit in doubt. For example, if a value is reported at 18.8 mg/l, the "18" must be a firm value, while the "0.8" is somewhat uncertain and may be between "0.7" or "0.9." The number zero may or may not be a significant figure:

1. Final zeros after a decimal point are always significant figures. For example, 9.8 g to the nearest mg is reported as 9.800 g.

2. Zeros before a decimal point with other preceding digits are significant. With no preceding digit, a zero before the decimal point is not significant.
3. If there are no digits preceding a decimal point, the zeros after the decimal point but preceding other digits are not significant. These zeros only indicate the position of the decimal point.
4. Final zeros in a whole number may or may not be significant. In a conductivity measurement of 1000 $\mu\Omega/cm$, there is no implication that the conductivity is 1000 ± 1 $\mu\Omega/cm$. Rather, the zeros only indicate the magnitude of the number.

A good measure of the significance of one or more zeros before or after another digit is to determine whether the zeros can be dropped by expressing the number in exponential form. If they can, the zeros are not significant. For example, no zeros can be dropped when expressing a weight of 100.08 g is exponential form; therefore the zeros are significant. However, a weight of 0.0008 g can be expressed in exponential form as 8×10^{-4} g, and the zeros are not significant. Significant figures reflect the limits of the particular method of analysis. If more significant figures are needed, selection of another method will be required to produce an increase in significant figures.

Once the number of significant figures is established for a type of analysis, data resulting from such analyses are reduced according to the set rules for rounding off.

1.7.2 Rounding Off Numbers

Rounding off numbers is a necessary operation in all analytical areas. However, it is often applied in chemical calculations incorrectly by blind rule or prematurely and, in these instances, can seriously affect the final results. Rounding off should normally be applied only as follows:

1. If the figure following those to be retained is less than 5, the figure is dropped, and the retained figures are kept unchanged. As an example, 11.443 is rounded off to 11.44.
2. If the figure following those to be retained is greater than 5, the figure is dropped, and the last retained figure is raised by 1. As an example, 11.446 is rounded off to 11.45.
3. When the figure following those to be retained is 5 and there are no figures other than zeros beyond the 5, the figure is dropped, and the last place figure retained is increased by 1 if it is an odd number, or it is kept unchanged if an even number. As an example, 11.435 is rounded off to 11.44, while 11.425 is rounded off to 11.42.

1.7.3 Rounding Off Single Arithmetic Operations

Addition: When adding a series of numbers, the sum should be rounded off to the same numbers of decimal places as the addend with the smallest number of places. However, the operation is completed with all decimal places intact and rounding off is done afterward. As an example:

$$11.1 + 11.12 + 11.13 = 33.35$$

The sum is rounded off to 33.4

Multiplication: When two numbers of unequal digits are to be multiplied, all digits are carried through the operation, and then the product is rounded off to the number of significant digits of the less accurate number.

Division: When two numbers of unequal digits are to be divided, the division is carried out on the two numbers using all digits. Then the quotient is rounded off to the lower number of significant digits between the two values.

Powers and roots: When a number contains n significant digits, its root can be relied on for n digits, but its power can rarely be relied on for n digits.

1.7.4 Rounding Off the Results of a Series of Arithmetic Operations

The rules for rounding off are reasonable for simple calculations. However, when dealing with two nearly equal numbers, there is a danger of loss of all significance when applied to a series of computations that rely on a relatively small difference in two values. Examples are calculation of variance and standard deviation. The recommended procedure is to carry several extra figures through the calculation and then to round off the final answer to the proper number of significant figures. This operation is simplified by using the memory function on calculators, which for most calculators is a large number, often 10 or more, digits.

1.8 BASIC LABORATORY SAFETY

1.8.1 Safety Data Sheets

Safety Data Sheets (SDSs), formerly called Material Safety Data Sheets (MSDSs), are informational packets that are "intended to provide workers and emergency personnel with procedures for handling or working with that substance in a safe manner and include information such as physical data (melting point, boiling point, flash point, etc.), toxicity, health effects, first aid,

reactivity, storage, disposal, protective equipment, and spill-handling procedures" (http://en.wikipedia.org/wiki/Material_safety_data_sheet#United_States).

SDSs are available for all reagents, chemicals, solvents, gases, etc. used in your laboratory. You can consult these documents if you have questions regarding how to safely handle a material, the potential risks of the material, how to properly clean up a spill, etc. They should be available to you in a centralized location (typically, a binder) in the lab. If not available, you may request these from your instructor or find them online. Generally, the following information is available on a MSDS or SDS in a 16-section format:

1. Identification of the substance/mixture
2. Hazard identification
3. Composition/information on ingredients
4. First aid measures
5. Firefighting measures
6. Accidental release measures
7. Handling and storage
8. Exposure controls/personal protection
9. Physical and chemical properties
10. Stability and reactivity
11. Toxicological information
12. Ecological information
13. Disposal considerations
14. Transport information
15. Regulatory information
16. Other information

1.8.2 Hazardous Chemicals

Food analysis laboratories, like any chemical laboratory, often contain hazardous compounds, including:

1. Acids (hydrochloric acid, sulfuric acid, etc.)
2. Bases (e.g., sodium hydroxide)
3. Corrosives and oxidizers (sulfuric acid, nitric acid, perchloric acid, etc.)
4. Flammables (organic solvents such as hexane, ether, alcohols)

1.8.3 Personal Protective Equipment and Safety Equipment

It is important to understand the location and use of lab safety equipment. The purpose of this is threefold:

1. To prevent accidents and/or injuries in the lab
2. To quickly and effectively respond to any accident and/or injury in the lab
3. Be able to perform laboratory procedures without excessive worrying about lab hazards

Your laboratory instructor should provide instruction regarding basic laboratory safety equipment. You should be aware of these general rules and the existence of this equipment.

Proper clothing is required to work in any chemical laboratory. The following standards and rules regarding dress are generally applicable, although standards may vary between laboratories:

1. Close-toed shoes (no flip-flops, sandals, or other "open" footwear).
2. Long pants (dresses, skirts, and shorts may be allowed in some laboratories).
3. No excessively loose clothing or accessories.
4. Long hair should be pulled back from the face into a ponytail or otherwise restrained.

You should be able to obtain and wear the following personal protective equipment (PPE) and understand their proper use:

1. Safety glasses, goggles, and face shields
2. Lab coat or apron
3. Shoe covers
4. Latex or acetonitrile gloves
5. Puncture-resistant gloves
6. Heat-resistant gloves

You should be aware of the locations of the following safety equipment items and their proper use:

1. First aid kit
2. Bodily fluids cleanup kit
3. Acid, base, and solvent spill kits
4. Fire extinguisher and fire blanket
5. Safety shower and eyewash station
6. Solid, liquid, chlorinated, and biohazard waste disposal containers, if applicable
7. Sharps and broken glass disposal containers, if applicable

1.8.4 Eating, Drinking, Etc.

Your hands may become contaminated with substances used in the lab simply by touching lab benches, glassware, etc. This may happen even without your knowledge. Even if you are not handling hazardous substances, previous lab occupants may not have cleaned benches and glassware, leaving behind hazardous substances that you are unaware of. To avoid spreading potentially harmful substances from your hands to your face, eyes, nose, and mouth (where they may irritate sensitive or be introduced to circulation by mucus membranes, ingestion, or inhalation), the following activities are prohibited in chemical laboratories: eating, drinking, smoking, chewing tobacco or snuff, and applying cosmetics (e.g., lip balm). The following should not even be brought into a chemical laboratory: food, water, beverages, tobacco, and cosmetics. Some unconscious activities (e.g., touching your face and eyes) are difficult to avoid. However, wearing gloves in the laboratory may minimize these actions.

1.8.5 Miscellaneous Information

The following general rules and guidelines apply to working in the laboratory:

1. When combining acid and water, always add the acid to water (instead of adding water to the acid). When acid dissolves in water, heat is released. This can cause splattering of the solution. By adding the acid to water, the heat is dissipated, and splattering is reduced or eliminated.
2. Be aware that dissolving sodium hydroxide in water generates heat. Making high concentrations of aqueous sodium hydroxide can lead to very hot solutions that can burn bare hands. Allow these solutions to cool, or handle with heat-resistant gloves.
3. Broken glass and other sharps (razor blades, scalpel blades, needles, etc.) should be disposed of in puncture-resistant sharps containers.
4. Do not pour waste or chemicals down the drain. This practice can damage the building's plumbing and harm the environment. Dispose of liquid, solid, chlorinated, radioactive, and biohazard wastes into the appropriate containers provided by the lab instructor. If you are unsure how to properly dispose of waste, ask your instructor or teaching assistant.
5. Handle volatile, noxious, or corrosive compounds in the fume hood with appropriate PPE.

RESOURCE MATERIALS

Analytical Quality Control Laboratory. 2010. Handbook for analytical quality control in water and wastewater laboratories. U.S. Environmental Protection Agency, Technology Transfer.

Anonymous. 2010. Instructions for Gilson Pipetman. Rainin Instrument Co., Inc., Washburn, MA.

Applebaum, S.B. and Crits, G.J. 1964. "Producing High Purity Water". Industrial Water Engineering.

Smith JS. 2017. Evaluation of analytical data, Ch. 4, In: Nielsen SS (ed.) Food analysis, 5th edn. Springer, New York.

Willare, H.H. and Furman, W.H. 1947. Elementary Quantitative Analysis – Theory and Practice. Van Norstrand Co., Inc., New York.

Preparation of Reagents and Buffers

Catrin Tyl (✉) • Baraem P. Ismail

Department of Food Science and Nutrition, University of Minnesota,
St. Paul, MN, USA
e-mail: tylxx001@umn.edu; bismailm@umn.edu

2.1   PREPARATION OF REAGENTS OF SPECIFIED CONCENTRATIONS

Virtually every analytical method involving wet chemistry starts with preparing reagent solutions. This usually involves dissolving solids in a liquid or diluting from stock solutions. The concentration of analytes in solution can be expressed in weight (kg, g, or lower submultiples) or in the amount of substance (mol), per a unit volume (interchangeably L or dm^3, mL or cm^3, and lower submultiples). Preparing reagents of correct concentrations is crucial for the validity and reproducibility of any analytical method. Below is a sample calculation to prepare a calcium chloride reagent of a particular concentration.

Example A1   How much calcium chloride do you need to weigh out to get 2 L of a 4 mM solution?

Solution

The molarity (M) equals the number of moles (n) in the volume (v) of 1 L:

$$M\left(\frac{\text{mol}}{\text{L}}\right) = \frac{n(\text{mol})}{v[\text{L}]} \qquad (2.1)$$

The desired molarity is 4 mM = 0.004 M; the desired volume is 2 L. Rearrange Eq. 2.1 so that:

$$n(\text{mol}) = M\left(\frac{\text{mol}}{\text{L}}\right) \times v(\text{L}) \qquad (2.2)$$

The mass (m) of 1 mole $CaCl_2$ (110.98 g/mol) is specified by the molecular weight (MW) as defined through Eq. 2.3. The mass to be weighed is calculated by rearranging into Eq. 2.4:

$$\text{MW}\left(\frac{\text{g}}{\text{mol}}\right) = \frac{m(\text{g})}{n(\text{mol})} \qquad (2.3)$$

$$m(\text{g}) = n(\text{mol}) \times \text{MW}\left(\frac{\text{g}}{\text{mol}}\right) \qquad (2.4)$$

Now substitute Eq. 2.2 into Eq. 2.4. Some of the units then cancel out, indicated as strike-throughs:

$$m(\text{g}) = M\left(\frac{\text{mol}}{\text{L}}\right) \times v(\text{L}) \times \text{MW}\left(\frac{\text{g}}{\text{mol}}\right) \qquad (2.5)$$

$$m(\text{g}) = 0.004\left(\frac{\text{mol}}{\text{L}}\right) \times 2(\text{L}) \times 110.98\left(\frac{\text{g}}{\text{mol}}\right)$$

$$= 0.888[\text{g}] \qquad (2.6)$$

Example A2   How much would you need to weigh out if you were using the dihydrate (i.e., $CaCl_2.2H_2O$) to prepare 2 L of a 4 mM solution of calcium chloride?

Solution

During crystallization of salts, water may be incorporated into the crystal lattice. Examples include phosphate salts, calcium chloride, and certain sugars. The names of these compounds are amended by the number of bound water molecules. For example, $Na_2HPO_4.7H_2O$ is called sodium monophosphate heptahydrate, and $CaCl_2.2H_2O$ is called calcium chloride dihydrate. The water is tightly bound and is not visible (the dry reagents do not look clumped). Many commercially available salts are sold as hydrates that can be used analogously to their dry counterparts after adjusting for their increased molecular weight, which includes the bound water molecules. The molecular weight of $CaCl_2$ changes from 110.98 g/mol to 147.01 g/mol to account for two molecules of water at about 18 g/mol. Hence, Eq. 2.6 needs to be modified to:

$$m(\text{g}) = 0.004\left(\frac{\text{mol}}{\text{L}}\right) \times 2(\text{L}) \times 147.01\left(\frac{\text{g}}{\text{mol}}\right)$$

$$= 1.176[\text{g}] \qquad (2.7)$$

Thus, 1.176 g of $CaCl_2$ would be weighed out and the volume made up to 2 L in a volumetric flask.

Other commonly used ways to express concentrations are listed in Table 2.1. For instance, for very low concentrations as encountered in residue analysis, parts per million (e.g., µg/mL or mg/L) and parts per billion (e.g., µg/L) are preferred units. Concentrated acids and bases are often labeled in percent mass by mass or percent mass per volume. For instance, a 28 % wt/wt solution of ammonia in water contains 280 g ammonia per 1000 g of solution. On the other hand, 32 % wt/vol NaOH solution contains 320 g of NaOH per L. For dilute solutions in water, the density is approximately 1 kg/L at room temperature (the density of water is exactly 1 kg/L only at 5 °C), and thus wt/wt and wt/vol are almost equal. However, concentrated solutions or solutions in organic solvents can deviate substantially in their density. Therefore, for concentrated reagents, the correct amount of reagent needed for dilute solutions is found by accounting for the density, as illustrated in Example A3 below:

2.1 table Concentration expression terms

Unit	Symbol	Definition	Relationship
Molarity	M	Number of moles of solute per liter of solution	$M = \dfrac{\text{mol}}{\text{liter}}$
Normality	N	Number of equivalents of solute per liter of solution	$N = \dfrac{\text{equivalents}}{\text{liter}}$
Percent by weight (parts per hundred)	wt %	Ratio of weight of solute to weight of solute plus weight of solvent × 100	$\text{wt \%} = \dfrac{\text{wt solute} \times 100}{\text{total wt}}$
	wt/vol %	Ratio of weight of solute to total volume × 100	$\text{wt / vol \%} = \dfrac{\text{wt solute} \times 100}{\text{total volume}}$
Percent by volume	vol %	Ratio of volume of solute to total volume × 100	$\text{vol \%} = \dfrac{\text{vol of solute} \times 100}{\text{total volume}}$
Parts per million	ppm	Ratio of solute (wt or vol) to total weight or volume × 1,000,000	$\text{ppm} = \dfrac{\text{mg solute}}{\text{kg solution}}$ $= \dfrac{\mu\text{g solute}}{\text{g solution}}$ $= \dfrac{\text{mg solute}}{\text{L solution}}$ $= \dfrac{\mu\text{g solute}}{\text{mL solution}}$
Parts per billion	ppb	Ratio of solute (wt or vol) to total weight or volume × 1,000,000,000	$\text{ppb} = \dfrac{\mu\text{g solute}}{\text{kg solution}}$ $= \dfrac{\text{ng solute}}{\text{g solution}}$ $= \dfrac{\mu\text{g solute}}{\text{L solution}}$ $= \dfrac{\text{ng solute}}{\text{mL solution}}$

Example A3 Prepare 500 mL of a $6M$ sulfuric acid solution from concentrated sulfuric acid. The molecular weight is 98.08 g/mol, and the manufacturer states that it is 98 % wt/wt and has a density (d) of 1.84 g/mL.

Solution

The density specifies the mass per volume of the concentrated H_2SO_4:

$$d = \frac{m}{v}\left(\frac{\text{g}}{\text{mL}}\right) \qquad (2.8)$$

The molarity of the concentrated sulfuric acid is determined by rearranging Eqs. 2.3 and 2.8 to express the unknown number of moles (n) in terms of known quantities (MW and d). However, because molarity is specified in moles per L, the density needs to be multiplied by 1000 to obtain the g per L (the unit of density is g per mL or kg per L).

Rearrange Eq. 3: $n(\text{mol}) = \dfrac{m(\text{g})}{MW\left(\dfrac{\text{g}}{\text{mol}}\right)}$ \qquad (2.9)

Rearrange Eq. 8: $m(\text{g}) = d\left(\dfrac{\text{g}}{\text{L}}\right) \times 1000 \times v(\text{L})$ (2.10)

Substitute Eq. 10 into Eq. 9:

$$n(\text{mol}) = \frac{d \times 1000\left(\dfrac{\text{g}}{\text{L}}\right) \times v(\text{L})}{\text{MW}\left(\dfrac{\text{mol}}{\text{g}}\right)} \quad (2.11)$$

Substitute Eq. 2.11 into Eq. 2.1. The 98 % wt/wt (see Table 2.1) can be treated like a proportionality factor: every g of solution contains 0.98 g H_2SO_4. To account for this, multiply Eq. 2.1 with the wt %:

$$M\left(\frac{\text{mol}}{\text{L}}\right) = \frac{d \times \cancel{v} \times 1000}{\text{MW}} \times \frac{1}{\cancel{v}} \times \%w\left(\frac{\text{g}}{\text{g}}\right) = \quad (2.12)$$

$$M\left(\frac{\text{mol}}{\text{L}}\right) = \frac{d \times 1000\left(\dfrac{\text{g}}{\text{L}}\right)}{\text{MW}\left(\dfrac{\text{g}}{\text{mol}}\right)} \times \%w\left(\frac{\text{g}}{\text{g}}\right) \quad (2.13)$$

$$M\left(\frac{\text{mol}}{\text{L}}\right) = \frac{1.84\left(\dfrac{\cancel{\text{g}}}{\text{L}}\right) \times 1000 \times 0.98\left(\dfrac{\cancel{\text{g}}}{\cancel{\text{g}}}\right)}{98.08\left(\dfrac{\cancel{\text{g}}}{\text{mol}}\right)}$$

$$= 18.39\left(\frac{\text{mol}}{\text{L}}\right) \quad (2.14)$$

The necessary volume to supply the desired amount of moles for 500 mL of a 6 M solution is found by using Eq. 2.15:

$$v_1(\text{L}) \times M_1\left(\frac{\text{mol}}{\text{L}}\right) = v_2(\text{L}) \times M_2\left(\frac{\text{mol}}{\text{L}}\right) \quad (2.15)$$

v stock solution (L) =

$$\frac{\begin{array}{c} v \text{ of diluted} \\ \text{solution (L)} \end{array} \times \begin{array}{c} M \text{ of diluted} \\ \text{solution}\left(\cancel{\text{mol}} / \text{L}\right) \end{array}}{M \text{ of stock solution}\left(\dfrac{\cancel{\text{mol}}}{\text{L}}\right)} \quad (2.16)$$

$$v(\text{L}) = \frac{0.5(\text{L}) \times 6}{18.39}$$

$$= 0.163 \text{ L or } 163 \text{ mL} \quad (2.17)$$

Hence, to obtain 500 mL, 163 mL concentrated sulfuric acid would be combined with 337 mL of water (500–163 mL). The dissolution of concentrated sulfuric acid in water is an exothermic process, which may cause splattering, and glassware can get very hot (most plastic containers are not suited for this purpose!). The recommended procedure would be to add some water to a 500 mL volumetric glass flask, e.g., ca. 250 mL, then add the concentrated acid, allow the mixture to cool down, mix, and bring up to volume with water.

2.2 USE OF TITRATION TO DETERMINE CONCENTRATION OF ANALYTES

A wide range of standard methods in food analysis, such as the iodine value, peroxide value and titratable acidity, involve the following concept:

- A **reagent** of **known** concentration (i.e., the titrant) is titrated into a solution of **analyte** with **unknown** concentration. The used up volume of the reagent solution is measured.
- The ensuing reaction converts the reagent and analyte into products.
- When all of the analyte is used up, there is a measurable change in the system, e.g., in color or pH.
- The concentration of the titrant is known; thus the amount of converted reactant can be calculated. The stoichiometry of the reaction allows for the calculation of the concentration of the analyte, for example, in the case of iodine values, the absorbed grams of iodine per 100 g of sample; in the case of peroxide value, milliequivalents of peroxide per kg of sample; or in the case of titratable acidity, % wt/vol acidity.

There are two principle types of reactions for which this concept is in widespread use: acid-base and redox reactions. For both reaction types, the concept of normality (N) plays a role, which signifies the number of equivalents of solute per L of solution. The number of equivalents corresponds to the number of transferred H^+ for acid-base reactions and transferred electrons for redox reactions. Normality equals the product of the molarity with the number of equivalents (typically 1–3 for common acids and bases with low molecular weight), i.e., it is equal to or higher than the molarity. For instance, a 0.1 M sulfuric acid solution would be 0.2 N, because two H^+ are donated per molecule H_2SO_4. On the other hand, a 0.1 M NaOH solution would still be 0.1 N, as indicated by Eq. 2.18:

$$\text{Normality} = \text{Molarity} \times \text{number of equivalents} \quad (2.18)$$

Substituting N for M, reaction equivalence can be expressed through a modified version of Eq. 2.15 (which is the same as Refr. [4], Sect. 22.2.2, Eq. 2.1):

$$v \text{ of titrant} \times N \text{ of titrant} = v \text{ of analyte solution} \times N \text{ of analyte solution} \quad (2.19)$$

Some analyses (such as in titratable acidity) require the use of **equivalent weights** instead of molecular weights. These can be obtained by dividing the molecular weight by the number of equivalents transferred over the course of the reaction.

Equivalent weight (g) =

$$\frac{\text{molecular weight}\left(\dfrac{\text{g}}{\cancel{\text{mol}}}\right)}{\text{number of equivalents}\left(\cancel{\text{mol}}\right)} \quad (2.20)$$

Using equivalent weights can facilitate calculations, because it accounts for the number of reactive groups of an analyte. For H_2SO_4, the equivalent weight would be $\frac{98.08}{2} = 49.04$ g/mol, whereas for NaOH it would be equal to the molecular weight since there is only one OH group.

To illustrate the concept, the reaction of acetic acid with sodium hydroxide in aqueous solution is stated below:

$$CH_3COOH + NaOH \rightarrow CH_3COO^- + H_2O + Na^+$$

This reaction can form the basis for quantifying acetic acid contents in vinegar (for which it is the major acid):

Example B1 100 mL of vinegar is titrated with a solution of NaOH that is exactly $1\,M$. (Chapter 21, Sect. 21.2, in this laboratory manual describes how to standardize titrants.) If 18 mL of NaOH are used up, what is the corresponding acetic acid concentration in vinegar?

Solution The reaction equation shows that both NaOH and CH_3COOH have an equivalence number of 1, because they each have only one reactive group. Thus, their normality and molarity are equal. Equation 2.19 can be used to solve Example B1, and the resulting N will, in this case, equal the M:

$$18(mL) \times 1\left(\frac{mol}{L}\right) = 100(mL) \times \text{Normality}$$

$$\text{of acetic acid solution}\left(\frac{mol}{L}\right) \quad (2.21)$$

$$M \text{ and } N \text{ of acetic acid in vinegar}\left(\frac{mol}{L}\right) =$$

$$\frac{18(\cancel{mL})}{100(\cancel{mL})} \times 1\left(\frac{mol}{L}\right) = 0.18\left(\frac{mol}{L}\right) \quad (2.22)$$

Sometimes the stoichiometry of a reaction is different, such as when NaOH reacts with malic acid, the main acid found in apples and other fruits.

Example B2 Assume that 100 mL of apple juice are titrated with $1\,M$ NaOH, and the volume used is 36 mL. What is the molarity of malic acid in the apple juice?

Solution Malic acid contains two carboxylic groups, and thus for every mole of malic acid, two moles of NaOH are needed to fully ionize it. Therefore, the normality of malic acid is two times the molarity. Again, use Eq. 2.19 to solve Example B2:

$$36(mL) \times 1\left(\frac{mol}{L}\right) = 100(mL) \times 2$$

$$\times M \text{ of malic acid solution}\left(\frac{mol}{L}\right) \quad (2.23)$$

$$M \text{ of malic acid in apple juice}\left(\frac{mol}{L}\right) =$$

$$\frac{36(\cancel{mL})}{100(\cancel{mL})} \times 1\left(\frac{mol}{L}\right) \times \frac{1}{2} = 0.18\left(\frac{mol}{L}\right) \quad (2.24)$$

This is the same value as obtained above for acetic acid. However, as a direct consequence of malic acid having two carboxyl groups instead of one, twice the amount of NaOH was needed.

Tables, calculators, and other tools for calculating molarities, normalities, and % acidity are available in print and online literature. However, it is important for a scientist to know the stoichiometry of the reactions involved, and the reactivity of reaction partners to correctly interpret these tables.

The concept of normality also applies to redox reactions; only electrons instead of protons are transferred. For instance, potassium dichromate, $K_2Cr_2O_7$, can supply six electrons, and therefore the normality of a solution would be six times its molarity. While the use of the term normality is not encouraged by the IUPAC, the concept is ubiquitous in food analysis because it can simplify and speed up calculations.

For a practical application of how normality is used to calculate results of titration experiments, see Chap. 21 in this laboratory manual.

2.3 PREPARATION OF BUFFERS

A **buffer** is an aqueous solution containing **comparable molar amounts** of either **a weak acid and its corresponding base** or a **weak base and its corresponding acid**. A buffer is used to keep a pH constant. In food analysis, buffers are commonly used in methods that utilize enzymes, but they also arise whenever weak acids or bases are titrated. To explain how a weak acid or base and its charged counterpart manage to maintain a certain pH, the "comparable molar amounts" part of the definition is key: For a buffer to be effective, its components must be present in a certain molar ratio. This section is intended to provide guidance on how to solve calculation and preparation problems relating to buffers. While it is important for food scientists to master calculations of buffers, the initial focus will be on developing an understanding of the chemistry. Figures 2.1 and 2.2 show an exemplary buffer system, and how the introduction of strong acid would affect it.

Only **weak acids and bases** can form buffers. The distinction of strong versus weak acids/bases is made based on how much of either H_3O^+ or OH^- is generated,

respectively. Most acids found in foods are weak acids; hence, once the equilibrium has been reached, only trace amounts have dissociated, and the vast majority of the acid is in its initial, undissociated state. For the purpose of this discussion, we will refer to the buffer components as **acid AH, undissociated state**, and corresponding **base A$^-$**, concentrations can be measured and are published in the form of **dissociation constants, K_a,** or their **negative logarithms to base 10, pK_a** values. These values can be found on websites of reagent manufacturers as well as numerous other websites and textbooks. Table 2.2 lists the pK_a values of some common acids either present in foods or often used to prepare buffers, together with the acid's molecular and equivalent weights. However, reported literature values for K_a/pK_a can be different for the same compound. For instance, they range between 6.71 and 7.21 for $H_2PO_4^-$. Dissociation constants depend on the ionic strength of the system, which is influenced by the concentrations of all ions in the system, even if they do not buffer. In addition, the pH in a buffer system is, strictly speaking, determined by activities, not by concentrations (see Refr. [4], Sect. 22.3.2.1). However, for concentrations $<0.1M$, activities are approximately equal to concentrations, especially for monovalent ions. For $H_2PO_4^-$, the value 7.21 is better suited for very dilute systems. For buffers intended for media or cell culture, the system typically contains several salt components, and 6.8 would be a commonly used value. If literature specifies several pK_a values, try finding information on the ionic strength where these values were obtained and calculate/estimate the ionic strength of the solution where the buffer is to be used. However, even for the same ionic strength, there may be different published values, depending on the analytical method. When preparing a buffer (see notes below), *always test* and, if necessary, *adjust the pH*, even for commercially available dry buffer mixes that only need to be dissolved. The values listed in Table 2.2 are for temperatures of 25 °C and ionic strengths of $0M$, as well as $0.1M$, if available.

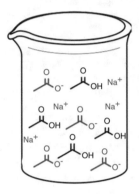

A buffer solution composed of the weak acetic acid, CH_3COOH, and a salt of its corresponding base, sodium acetate, $CH_3COO^-Na^+$. In aqueous solution, the sodium acetate dissociates into CH_3COO^- (acetate ions) and Na^+ (sodium ions), and for this reason, these ions are drawn spatially separated. The Na^+ ions do not participate in buffering actions and can be ignored for future considerations. Note that the ratio of acetic acid and acetate is equal in our example. Typical buffer systems have concentrations between 1 and 100 mM

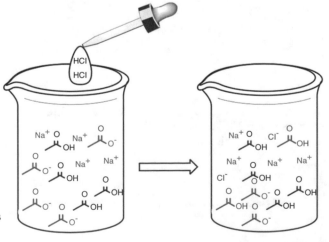

Changes induced by addition of the strong acid HCl, to the buffer from Fig. 2.1. HCl can be considered as completely dissociated into Cl^- and H^+. The H^+ combines with CH_3COO^- instead of H_2O, because CH_3COO^- is the stronger base. Thus, instead of H_3O^+, additional CH_3COOH is formed. This alters the ratio between CH_3COOH and CH_3COO^-, resulting in a different pH as illustrated in problem C2. The Cl^- ions are merely counterions to balance charges but do not participate in buffer reactions and can thus be ignored

2.2 table Properties of common food acids

	Formula	pK_{a1}	pK_{a2}	pK_{a3}	Molecular weight	Equivalent weight
Acetic acid	CH_3COOH	4.75^b	–	–	60.06	60.06
Carbonic acid	H_2CO_3	6.4^b 6.1^c	10.3^b 9.9^c	–	62.03	31.02
Citric acid	$HOOCCH_2C(COOH)OHCH_2COOH$	3.13^b 2.90^c	4.76^b 4.35^c	6.40^b 5.70^c	192.12	64.04
Formic acid	$HCOOH$	3.75^b	–	–	46.03	46.03
Lactic acid	$CH_3CH(OH)COOH$	3.86	–	–	90.08	90.08
Malic acid	$HOOCCH_2CHOHCOOH$	3.5^b 3.24^c	5.05^b 4.68^c	–	134.09	67.05
Oxalic acid	$HOOCCOOH$	1.25^b 1.2^c	4.27^b 3.80^c	–	90.94	45.02
Phosphoric acid	H_3PO_4	2.15^b 1.92^c	7.20^b 6.71^c	12.38^b 11.52^c	98.00	32.67
Potassium acid phthalate	$HOOCC_6H_4COO^-K^+$	5.41^b	–	–	204.22	204.22
Tartaric acid[a]	$HOOCCH(OH)CH(OH)COOH$	3.03	4.45	–	150.09	75.05

Data from Refs. [1, 2]

[a]Dissociation constants depend on the stereoisomer (D/L vs. meso-form). Values given for the naturally occurring R,R enantiomer (L-form)

[b]At ionic strength of $0M$

[c]At ionic strength of $0.1M$

Maximum buffering capacity always occurs around the pK_a value of the acid component. At the pK_a, the ratio of acid/corresponding base is 1:1 (not, as often erroneously assumed, 100:0). Therefore, a certain acid/corresponding base pair is suitable for buffering a pH range of $pK_a \pm 1$. The **molarity** of a buffer refers to the sum of concentration for acid and corresponding base. The resulting pH of a buffer is governed by their concentration ratio, as described through the **Henderson-Hasselbalch equation**:

$$pH = pK_a + \log \frac{[A^-]}{[AH]} \qquad (2.25)$$

Weak bases such as ammonia, NH_3, can also form buffers with their corresponding base, in this case NH_4^+. The Henderson-Hasselbalch equation would actually not change, since NH_4^+ would serve as the acid, denoted BH, and the base NH_3 is denoted as B. The acid component is always the form with more H^+ to donate. However, you may find the alternative equation:

$$pOH = pK_b + \log \frac{[BH^+]}{[B]} \qquad (2.26)$$

The pH would then be calculated as $14 - pOH$. The term pOH refers to the concentration of OH^-, which increases when bases are present in the system (see Ref. [4], Sect. 22.3, for details). However, using Eq. 2.25 and BH^+ instead of AH as well as B in place

of A^- gives the same result, as the pK_b is related to the pK_a through $14 - pK_b = pK_a$.

Below are several examples of buffer preparation using the Henderson-Hasselbalch equation:

Example C1 What is the pH of a buffer obtained by mixing 36 mL of a $0.2M$ Na_2HPO_4 solution and 14 mL of a $0.2M$ NaH_2PO_4 solution, after adding water to bringing the volume to 100 mL to obtain a $0.1M$ buffer?

Solution The Henderson-Hasselbalch equation requires knowledge of the acidic component's pK_a value and the concentrations of both buffer components. Rearrange Eq. 2.15 so that:

$$M \text{ in buffer} \left(\frac{mol}{L} \right) =$$

$$\frac{M \text{ of stock solutions} \left(\frac{mol}{L} \right) \times v \text{ of stock solutions} \left(\frac{mol}{L} \right)}{v \text{ of buffer} (L)} \qquad (2.27)$$

$$M \text{ of } Na_2HPO_4 \left(\frac{mol}{L}\right) = \frac{0.2 \left(\frac{mol}{L}\right) \times 0.036 (\cancel{L})}{0.1 (\cancel{L})}$$

$$= 0.072 \left(\frac{mol}{L}\right) \quad (2.28)$$

$$M \text{ of } NaH_2PO_4 \left(\frac{mol}{L}\right) = \frac{0.2 \left(\frac{mol}{L}\right) \times 0.014 (\cancel{L})}{0.1 (\cancel{L})}$$

$$= 0.028 \left(\frac{mol}{L}\right) \quad (2.29)$$

The other necessary step to solve Example C1 is finding the correct pK_a.

$$H_3PO_4 \underset{k_{-1}}{\overset{k_1}{\rightleftharpoons}} H_2PO_4^- + H^+$$

$$\underset{k_{-2}}{\overset{k_2}{\rightleftharpoons}} HPO_4^{2-} + H^+ \underset{k_{-3}}{\overset{k_3}{\rightleftharpoons}} PO_4^{3-} + H^+$$

In this buffer, $H_2PO_4^-$ acts as the acid, as it donates H^+ more strongly than HPO_4^{2-}. The acid component in a buffer is always the one with more acidic H^+ attached. Hence the relevant pK_a value is pk_2. As listed in Table 2.2, this value is 6.71.

The solution to Example C1 is now only a matter of inserting values into Eq. 2.25:

$$pH = 6.71 + \log \frac{[0.072]}{[0.028]} \quad (2.30)$$

$$pH = 6.71 + 0.41 = 7.12 \quad (2.31)$$

Example C2 How would the pH of the buffer in Example C1 change upon addition of 1 mL of $2M$ HCl? Note: You may ignore the slight change in volume caused by the HCl addition.

Solution Calculate the moles of HCl supplied using Eq. 2.1. HCl converts HPO_4^{2-} into $H_2PO_4^-$, because HPO_4^{2-} is a stronger base than $H_2PO_4^-$, as apparent by its higher pK_a value. This changes the ratio of $[AH]:[A^-]$. The new ratio needs to be inserted into Eq. 2.24. To calculate the new ratio, account for the volume of the buffer, 0.1 L, and calculate the amount of HPO_4^{2-} and $H_2PO_4^-$ present:

$$n \text{ of buffer components} (mol) = M \times v \quad (2.2)$$

$$n \text{ of } Na_2HPO_4 = 0.072 \times 0.1 = 0.0072 (mol) \quad (2.32)$$

$$n \text{ of } NaH_2PO_4 = 0.028 \times 0.1 = 0.0028 (mol) \quad (2.33)$$

$$n \text{ of } HCl (mol) = M \times v = 2 \times 0.001$$

$$= 0.002 (mol) \quad (2.34)$$

$$n \text{ of } Na_2HPO_4 \text{ after HCl addition} (mol):$$
$$0.0072 - 0.002 = 0.0052 \quad (2.35)$$

$$n \text{ of } NaH_2PO_4 \text{ after HCl addition} (mol):$$
$$0.0028 + 0.002 = 0.0048 \quad (2.36)$$

pH of buffer after HCl addition and conversion of n into M through Eq. 2.1:

$$pH = 6.71 + \log \frac{[0.052]}{[0.048]} = 6.74 \quad (2.37)$$

Note: The ratio of A^- to AH stays the same whether concentrations or amounts are inserted, since units are canceled, so technically n does not need to be converted into M to obtain the correct result.

Example C3 Prepare 250 mL of $0.1M$ acetate buffer with pH 5. The pK_a of acetic acid is 4.76 (see Table 2.2). The molecular weights of acetic acid and sodium acetate are 60.06 and 82.03 g/mol, respectively.

Solution This example matches the tasks at hand in a lab better than Example C1. One needs a buffer to work at a certain pH, looks up the pK_a value, and decides on the molarity and volume needed. The molarity of a buffer equals $[A^-] + [AH]$. To solve Example C3, one of those concentrations needs to be expressed in terms of the other, so that the equation only contains one unknown quantity. For this example, it will be $[A^-]$, but the results would be the same if $[AH]$ had been chosen. Together with the target pH (5) and the pK_a (4.76), the values are inserted into Eq. 2.25:

$$\text{Molarity of buffer} \left(\frac{mol}{L}\right) = [AH] + [A^-] \quad (2.38)$$

$$0.1 = [A^-] + [AH] \quad (2.39)$$

$$[A^-] = 0.1 - [AH] \quad (2.40)$$

$$5 = 4.76 + \log \frac{0.1 - [AH]}{[AH]} \tag{2.41}$$

$$0.24 = \log \frac{0.1 - [AH]}{[AH]} \tag{2.42}$$

$$1.7378 = \frac{0.1 - [AH]}{[AH]} \tag{2.43}$$

$$1.7378 \times [AH] = 0.1 - [AH] \tag{2.44}$$

$$1.7378 \times [AH] + [AH] = 0.1 \tag{2.45}$$

$$[AH] \times (1.7378 + 1) = 0.1 \tag{2.46}$$

$$[AH] = \frac{0.1}{1.7378 + 1} \tag{2.47}$$

$$[AH] = 0.0365 \left(\frac{mol}{L} \right) \tag{2.48}$$

$$[A^-] = 0.1 - 0.0365 = 0.0635 \left(\frac{mol}{L} \right) \tag{2.49}$$

There are three ways to prepare such a buffer:

1. Prepare $0.1\,M$ acetic acid and $0.1\,M$ sodium acetate solutions. For our example, 1 L will be prepared. Sodium acetate is a solid and can be weighed out directly using Eq. 2.5. For acetic acid, it is easier to pipet the necessary amount. Rearrange Eq. 2.8 to calculate the volume of concentrated acetic acid (density = 1.05) needed to prepare a volume of 1 L. Then express the mass through Eq. 2.5.

$$m \text{ of sodium acetate}(g) = M \times v \times MW \tag{2.5}$$

$$m \text{ of sodium acetate}(g) =$$
$$0.1 \left(\frac{mol}{L} \right) \times 1(L) \times 82 \left(\frac{g}{mol} \right) = 8.2(g) \tag{2.50}$$

$$v \text{ of acetic acid}(mL) = \frac{m(g)}{d \left(\frac{g}{mL} \right)} \tag{2.51}$$

$$v \text{ of acetic acid}(mL) =$$
$$\frac{M \left(\frac{mol}{L} \right) \times v(L) \times MW \left(\frac{g}{mol} \right)}{d \left(\frac{g}{mL} \right)} \tag{2.52}$$

$$v \text{ of acetic acid}(mL) =$$
$$\frac{60.06 \left(\frac{g}{mol} \right) \times 0.1 \left(\frac{mol}{L} \right) \times 1(L)}{1.05 \left(\frac{g}{mL} \right)} = 5.72(mL) \tag{2.53}$$

Dissolving each of these amounts of sodium acetate and acetic acid in 1 L of water gives two 1 L stock solutions with a concentration of $0.1\,M$. To calculate how to mix the stock solutions, use their concentrations in the buffer, i.e., 0.0365 as obtained from Eqs. 2.48 and 2.49 – $0.0365 \left(\frac{mol}{L} \right)$ for acetic acid and $0.0635 \left(\frac{mol}{L} \right)$ for sodium acetate and Eq. 2.15:

$$v \text{ of acetic acid stock solution}(L)$$
$$= \frac{M \text{ of acetic acid in buffer} \left(\frac{mol}{L} \right) \times v \text{ of buffer}(L)}{M \text{ of stock solution} \left(\frac{mol}{L} \right)} \tag{2.54}$$

$$v \text{ of acetic acid stock solution}(L) =$$
$$\frac{0.0365 \times 0.25}{0.1} = 0.091(L) \tag{2.55}$$

$$v \text{ of sodium acetate stock solution}(L) =$$
$$\frac{0.0635 \times 0.25}{0.1} = 0.159(L) \tag{2.56}$$

Combining the 0.091 L of acetic acid stock and 0.159 L of sodium acetate stock solution gives 0.25 L of buffer with the correct pH and molarity.

2. Directly dissolve appropriate amounts of both components in the same container. Equations 2.48 and 2.49 yield the molarities of acetic acid and sodium acetate in a buffer, i.e., the moles per 1 liter. Just like for approach 1 above, use Eqs. 2.5 and 2.52 to calculate the m for sodium acetate and v for acetic acid, but this time use the buffer volume of 0.25 [L] to insert for v:

For sodium acetate :

$$m(g) = M \left(\frac{mol}{L} \right) \times v(L) \times MW \left(\frac{g}{mol} \right)$$

$$= 0.0635 \times 0.25 \times 82.03 = 1.3(g) \tag{2.57}$$

For acetic acid :

$$v(mL) = \frac{M \left(\frac{mol}{L} \right) \times v(L) \times MW \left(\frac{g}{mol} \right)}{d \left(\frac{g}{mL} \right)}$$

$$= \frac{0.0365 \times 0.25 \times 60.02}{1.05} = 0.52(mL) \tag{2.58}$$

Dissolve both reagents in the same glassware in 200 mL water, adjust the pH if necessary, and bring up to 250 mL after transferring into a volumetric flask.

Note: It does not matter in how much water you initially dissolve these compounds, but it should be >50% of the total volume. Up to a degree, buffers are independent of dilution; however, you want to ensure complete solubilization and leave some room to potentially adjust the pH.

3. Pipet the amount of acetic acid necessary for obtaining 250 mL of a 0.1 M acetic acid solution, but dissolve in <250 mL, e.g., 200 mL. Then add concentrated NaOH solution drop-wise until pH 5 is reached, and make up the volume to 250 mL. The amount of acetic acid is found analogously to Eq. 2.52:

$$v \text{ acetic acid} (\text{mL}) = \frac{M \times v \times \text{MW}}{d} =$$

$$\frac{0.1 \times 0.25 \times 60.02}{1.05} = 1.43 (\text{mL}) \quad (2.59)$$

You will find all three approaches described above if you search for buffer recipes online and in published methods. For instance, AOAC Method 991.43 for total dietary fiber involves approach 3. It requires dissolving 19.52 g of 2-(N-morpholino)ethanesulfonic acid (MES) and 12.2 g of 2-amino-2-hydroxymethyl-propane-1,3-diol (Tris) in 1.7 L water, adjusting the pH to 8.2 with 6 M NaOH, and then making up the volume to 2 L.

However, the most common approach for preparing buffers is approach 1. It has the advantage that once stock solutions are prepared, they can be mixed in different ratios to obtain a range of pH values, depending on the experiment. One disadvantage is that if the pH needs to be adjusted, then either some acid or some base needs to be added, which slightly alters the volume and thus the concentrations. This problem can be solved by preparing stock solutions of higher concentrations and adding water to the correct volume, like in Example C1, for which 36 mL and 14 mL of 0.2 M stock solutions were combined and brought to a volume of 100 mL to give a 0.1 M buffer. This way, one also corrects for potential volume contraction effects that may occur when mixing solutions. Another potential disadvantage of stock solutions is that they are often not stable for a long time (see Notes below).

2.4   NOTES ON BUFFERS

When choosing an appropriate buffer system, the most important selection criterion is the pK_a of the acid component. However, depending on the system, additional factors may need to be considered, as detailed below (listed in no particular order of importance):

1. The buffer components need to be well soluble in water. Some compounds require addition of acids or bases to fully dissolve.

2. Buffer recipes may include salts that do not participate in the buffering process, such as sodium chloride for phosphate-buffered saline. However, the addition of such salts changes the ionic strength and affects the acid's pK_a. Therefore, combine all buffer components before adjusting the pH.

3. If a buffer is to be used at a temperature other than room temperature, heat or cool it to this intended temperature before adjusting the pH. Some buffer systems are more affected than others, but it is always advisable to check. For instance, 2-[4-(2-hydroxyethyl)piperazin-1-yl] ethanesulfonic acid [HEPES] is a widely used buffer component for cell culture experiments. At 20 °C, its pK_a is 7.55, but its change in pH from 20 to 37 °C is $-0.014 \, \Delta pH/°C$ [3]. Thus, the pK_a at 37 °C would be:

$$pK_a \text{ at } 37 °C = pH \text{ at } 20 °C - \Delta pH \times (T1 - T2) \quad (2.60)$$

$$7.55 - 0.014 \times (37 - 20) = 7.31 \quad (2.61)$$

4. Do not bring the buffer up to volume before adjusting the pH. Use relatively concentrated acids and bases for this purpose, so that the volumes needed for pH adjustment are small.

5. Ensure that buffer components do not interact with the test system. This is especially important when performing experiments on living systems, such as cell cultures, but even in vitro systems are affected, particularly when enzymes are used. For instance, phosphate buffers tend to precipitate with calcium salts or affect enzyme functionality. For this reason, a range of zwitterionic buffers with sulfonic acid and amine groups has been developed for use at physiologically relevant pH values.

6. Appropriate ranges for pH and molarities of buffer systems that can be described through the Henderson-Hasselbalch equation are roughly 3–11 and 0.001–0.1 M, respectively.

7. The calculations and theoretical background described in this chapter apply to aqueous systems. Consult appropriate literature if you wish to prepare a buffer in an organic solvent or water/organic solvent mixtures.

8. If you store non-autoclaved buffer or salt solutions, be aware that over time microbial growth or precipitation may occur. Visually inspect the

solutions before use, and discard if cloudy or discolored. If autoclaving is an option, check if the buffer components are suitable for this process.

9. When developing a standard operating procedure for a method that includes a buffer, include all relevant details (e.g., reagent purity) and calculations. This allows tracing mishaps back to an inappropriate buffer.

10. There are numerous online tools that can help with preparing buffer recipes. They can save time and potentially allow you to verify calculations. However, they are no substitute for knowing the theory and details about the studied system.

2.5 PRACTICE PROBLEMS

(Note: Answers to problems are in the last section of the laboratory manual.)

1. (a) How would you prepare 500 mL of $0.1\,M$ NaH_2PO_4 starting with the solid salt?

   (b) When you look for NaH_2PO_4 in your lab, you find a jar with $NaH_2PO_4 \cdot 2H_2O$. Can you use this chemical instead, and if so, how much do you need to weigh out?

2. How many g of dry NaOH pellets (molecular weight: 40 g/mol) would you weigh out for 150 mL of 10% wt/vol sodium hydroxide?

3. What is the normality of a 40% wt/vol sodium hydroxide solution?

4. How many mL of 10 N NaOH would be required to neutralize 200 mL of 2 M H_2SO_4?

5. How would you prepare 250 mL of 2N HCl starting with concentrated HCl? The supplier states that its concentration is 37% wt/wt, and the density is 1.2.

6. How would you prepare 1 L of $0.04\,M$ acetic acid starting with concentrated acetic acid (density = 1.05)? The manufacturer states that the concentrated acetic acid is >99.8%, so you may assume that it is 100% pure.

7. Is a 1% wt/vol acetic acid solution the same as a $0.1\,M$ solution? Show calculations.

8. Is a 10% wt/vol NaOH solution the same as a 1N solution? Show calculations.

9. What would the molarity and normality of a solution of 0.2 g potassium dichromate (molecular weight = 294.185) in 100 mL water be? It acts as an oxidizer that can transfer six electrons per reaction.

10. How would you make 100 mL of a 0.1N KHP solution? Note: For this practice problem, you only need to calculate the amount to weigh out, not explain how to standardize it. Chapter 21 of this laboratory manual provides further information about standardization of acids and bases.

11. You want to prepare a standard for atomic absorption spectroscopy measurements containing 1000 ppm Ca. How much $CaCl_2$ do you need to weigh out for 1000 mL stock solution? The atomic mass unit of Ca is 40.078, and the molecular weight of $CaCl_2$ is 110.98 g/mol.

12. Outline how you would prepare 250 mL of $0.1\,M$ acetate buffer at pH 5.5 for enzymatic glucose analysis.

13. Complexometric determination of calcium requires an ammonium buffer with 16.9 g ammonium chloride (molecular weight, 53.49) in 143 mL concentrated ammonium hydroxide solution (28% wt/wt, density = 0.88; molecular weight = 17; pK_b = 4.74). It is also to contain 1.179 g of $Na_2EDTA \cdot 2H_2O$ (molecular weight = 372.24 g/mol) and 780 mg $MgSO_4 \cdot 7H_2O$ (molecular weight = 246.47 g/mol). After combining all reagents, the volume is brought up to 250 mL. What are the molarities for EDTA and $MgSO_4$, and what pH does this buffer have?

14. Your lab uses $0.2\,M$ stock solutions of NaH_2PO_4 and Na_2HPO_4 for buffer preparation.

   (a) Calculate the amounts weighed out for preparing 0.5 L of fresh $0.2\,M$ stock solutions of $NaH_2PO_4 \cdot H_2O$ (molecular weight, 138) and $Na_2HPO_4 \cdot 7H_2O$ (molecular weight, 268).

   (b) You want to make 200 mL of a $0.1\,M$ buffer with pH 6.2 from these stock solutions. How many mL of each stock solution do you have to take?

   (c) How would the pH change if 1 mL of a $6\,M$ NaOH solution was added (note: you may ignore the volume change)?

15. Tris (2-amino-2-hydroxymethyl-propane-1,3-diol) is an amino compound suitable for preparation of buffers in physiological pH ranges such as for the dietary fiber assay; however, its pK_a is highly affected by temperature. The reported pKa at 25 °C is 8.06 [3]. Assuming a decline in pK_a of approximately 0.023 $\Delta pH/°C$, what pH would a Tris buffer with a molar acid/base ratio of 4:1 have at 60 °C versus at 25 °C?

16. Ammonium formate buffers are useful for LC-MS experiments. How would you prepare 1 L of a 0.01 M buffer of pH 3.5 with formic acid (pK_a = 3.75, 98% wt/wt, density = 1.2) and ammonium formate (molecular weight 63.06 g/mol)? Note: You may ignore the contribution of ammonium ions to the pH and focus on the ratio of formic acid/formate anion. You may treat the formic acid as pure for the calculation, i.e., ignore the % wt/wt.

REFERENCES

1. Albert A, Serjeant EP (1984) The determination of ioniza-tion constants. A laboratory manual, 3rd edn. Chapman and Hall, New York
2. Harris, D (2015) Quantitative Chemical Analysis, 9th edn W. H. Freeman and Company, New York
3. Harakany AA, Abdel Halim FM, Barakat AO (1984) Dissociation constants and related thermodynamic quan-tities of the protonated acid form of tris-(hydroxymethyl)-aminomethane in mixtures of 2-methoxyethanol and water at different temperatures. J. Electroanalytical Chemistry and Interfacial Electrochemistry 162:285–305
4. Tyl C, Sadler GD (2017) pH and titratable acidity. Ch. 22. In: Nielsen SS(ed) Food analysis 5th edn. Springer, New York

Dilutions and Concentrations

Andrew P. Neilson (✉) • Sean F. O'Keefe

Department of Food Science and Technology,
Virginia Polytechnic Institute and State University,
Blacksburg, VA, USA
e-mail: andrewn@vt.edu; okeefes@vt.edu

S.S. Nielsen, *Food Analysis Laboratory Manual*, Food Science Text Series,
DOI 10.1007/978-3-319-44127-6_3, © Springer International Publishing 2017

3.1   INTRODUCTION

This chapter builds upon the information presented in the previous Chap. 1 (Laboratory Standard Operating Procedures). This chapter covers the following topics:

1. Reasons for performing dilutions and concentrations in food analysis
2. Basic calculations and strategies for calculating the final concentration, given the initial concentration and a known dilution/concentration scheme (and vice versa)
3. Strategies for designing and performing dilutions to obtain standard curves
4. Example and practice problems

This information will be used repeatedly in a food analysis course (e.g., homework, laboratories, and exams). More importantly, the principles described in this chapter are essential for virtually all laboratory or bench work in food science, including quality assurance/quality control, analysis for food labeling, and product formulation.

3.2   REASONS FOR DILUTIONS AND CONCENTRATIONS

There are various reasons why dilutions and concentrations are performed, including:

1. Dilution:
   (a) To reduce high analyte concentrations in the sample down to levels within the operating range or optimal range of a method/instrument
   (b) To reduce high analyte concentrations in the sample down to levels within the linear region of a method/instrument or to levels within a defined standard curve (see *Practice Problems 5 and 7*)
   (c) To dilute the background matrix to levels that do not interfere with the analysis
   (d) Reagent addition, which dilutes the sample by increasing the volume
   (e) Solvent extraction, which dilutes the sample by using large volumes of solvent to favor transfer of the analyte from sample to solvent (see *Practice Problem 1*)
2. Concentration:
   (a) To increase low analyte concentrations in the sample up to levels within the operating range or optimal range of a method/instrument
   (b) To increase low analyte concentrations in the sample up to levels in the linear region of a method or to within a standard curve (see *Practice Problem 1*)

(c) Post-extraction evaporation of solvent (see *Practice Problem 1*)

3.3   USING VOLUMETRIC GLASSWARE TO PERFORM DILUTIONS AND CONCENTRATIONS

The use of laboratory glassware was covered in Chap. 1. However, a few key points are sufficiently important for calculations regarding dilutions and concentrations to mention again:

1. Class A glass volumetric flasks are used for bringing samples or solutions up to a known "**to contain**" (*TC*) volume.
2. Class A glass volumetric pipettes are used for transferring (delivering) a known "**to deliver**" (*TD*) volume of sample.
3. Other types of pipettes (non-class A volumetric glass pipettes, adjustable pipettors, automatic pipettors, reed pipettors, graduated measuring pipettes) and other glassware (graduated cylinders, beakers, etc.) are less accurate and less precise and should not be used for quantitative volume measurements.
4. Dilutions can be performed as "**diluted to**" or "**diluted to a final volume of**" or as "**diluted with.**" These are very different terms (see Chap. 1 for definitions). It is critical to understand the differences between these types of dilutions to perform lab procedures and calculations correctly.

3.4   CALCULATIONS FOR DILUTIONS AND CONCENTRATIONS

3.4.1   Introduction

Calculations involving dilutions or concentrations are critical for quantitative measurements. Many analytical methods (such as spectrophotometric assays, chromatography, titrations, protein determinations, etc.) are performed on diluted or concentrated samples, and the analytical result must be converted to the concentration in the undiluted or unconcentrated sample for labeling or research purposes. Instances for which these calculations are used include, but are not limited to:

1. Preparation of standard curves (see *Practice Problems 3 and 4*).
2. Determining the necessary dilution required to obtain a sample concentration within a specified range (see *Practice Problems 5 and 7*)
3. Determining the necessary range of dilutions for preparing a range of concentrations for a standard curve from a stock standard solution (see *Practice Problems 5 and 7*)

4. For converting an analytical result obtained from a diluted or concentrated sample to the undiluted or concentrated food (see *Practice Problems 1 and 6*)

3.4.2 Expressing Concentration

Recall that the concentration of an analyte in a sample or solution is defined as follows:

$$C = \frac{X}{m} \quad \text{or} \quad C = \frac{X}{V} \qquad (3.1)$$

C = concentration

X = amount of analyte (g, mol, etc.)

V = sample volume

m = mass

Note that concentrations also can be expressed in terms of **percentages** (%), parts per million (ppm), etc. (see Chap. 2, Table 2.1). Percentage is a particular problem as it can refer to ratios expressed as weight per weight, weight per volume, or volume per volume. Simply saying 5 % ethanol is unclear as it could represent 5 % w/w, w/v, or v/v. Therefore, % are typically expressed with the accompanying notation (w/w, w/v, etc.) to clarify the meaning of the % value. Rearranging Eq. 3.1, the amount of analyte can be expressed as:

$$X = Cm \quad \text{or} \quad X = CV \qquad (3.2)$$

For each step in a dilution or concentration, a portion (mass or volume) of a sample or solution is either diluted with additional liquid or reduced in volume. The mass, volume, and concentration of the sample change, but the amount (mol, g, etc.) of the analyte present in the amount of the sample that is diluted or concentrated does not change. Therefore, the following is true:

$$X_1 = X_2 \qquad (3.3)$$

X_1 = the amount of analyte in the sample before the dilution / concentration step

X_2 = the amount of analyte in the sample after the dilution / concentration step

Example D1 Suppose that a stock solution of thiamine is made by dissolving 168 mg thiamine to a volume of 150 mL, and then 0.25 mL of that stock solution is added to a volumetric flask and water is added so that the final volume is 200 mL. The concentration of thiamine in the stock solution is:

$$C = \frac{X}{V} = \frac{168\,\text{mg thiamine}}{150\,\text{mL solution}} = \frac{1.12\,\text{mg thiamine}}{\text{mL solution}}$$

The amount of thiamine in the 0.25 mL that is diluted is:

$$X = CV = \left(0.25\,\text{mL solution}\right)\left(\frac{1.12\,\text{mg thiamine}}{\text{mL solution}}\right)$$
$$= 0.28\,\text{mg thiamine}$$

Note that we are only concerned with the amount of analyte in the portion of the sample that is diluted/concentrated (0.25 mL in this example) and *not* the amount of analyte in the whole initial sample (150 mL in this case). When the 0.25 mL of the stock solution (containing a total of 0.28 mg thiamine) is diluted to a final volume of 200 mL, the total amount of thiamine (0.28 mg) does not change because the added water does not contain any thiamine. Therefore:

$$X_1 = X_2 = 0.28\,\text{mg thiamine}$$

However, the concentration of thiamine changes, because now the 0.28 mg thiamine is present in 200 mL instead of 0.25 mL:

$$C = \frac{X}{V} = \frac{0.28\,\text{mg thiamine}}{200\,\text{mL solution}} = \frac{0.0014\,\text{mg thiamine}}{\text{mL solution}}$$

Therefore, the thiamine solution has been diluted, as the final concentration (0.0014 mg/mL) is less than the initial concentration (1.68 mg/mL).

"Concentration" and "amount" are very distinct concepts. Concentration is a ratio of the amount of analyte to the amount of sample. The concentration (mg/mL, M, N, %, ppm, etc.) of an analyte present in a sample is not dependent on the amount (g, mL, etc.) of sample.

Example D2 If a pesticide is present at 2.8 parts per billion (ppb) (note that 1 ppb = 1 μg/kg) in a sample of applesauce, the pesticide is uniformly distributed throughout the applesauce, and therefore the concentration is 2.8 ppb regardless of whether 1 μg, 1 mg, 1 g, 1 kg, 1 mL, or 1 L of applesauce is analyzed. However, the total amount of the analyte present does depend on the size of the sample. We can easily see that

0.063 and 2.2 kg of applesauce contain the same concentration, but vastly different total amounts, of the pesticide:

$$X = Cm \qquad (3.4)$$

$$0.063\,\text{kg applesauce}\left(\frac{2.8\,\mu\text{g pesticide}}{\text{kg applesauce}}\right) =$$

$$0.176\,\mu\text{g pesticide}$$

$$2.2\,\text{kg applesauce}\left(\frac{2.8\,\mu\text{g pesticide}}{\text{kg applesauce}}\right)$$

$$= 6.16\,\mu\text{g pesticide}$$

Once we know the concentration of a sample or solution, we can take any part, or all, of the sample. This changes the amount of the analyte but not the concentration. Then, diluting or concentration the aliquot chosen changes the concentration, but not the amount, of the analyte. These are two critical concepts that must be implicitly understood to master the concepts of dilution and concentrations. Since the total amount of analyte does not change during each step, the following equations can be derived:

$$X_i = X_f \qquad (3.5)$$

and $X = Cm$ or $X = CV$, therefore:

$$C_iV_i = C_fV_f \quad \text{or} \quad C_im_i = C_fm_f \qquad (3.6)$$

$$i = \text{initial}\left(\text{before dilution or concentration}\right)$$

$$f = \text{final}\left(\text{after dilution or concentration}\right)$$

3.4.3 Forward Calculations

The relationship between the initial and final concentration can be used to determine the starting and/or final concentration if one concentration and the masses or volumes used in the dilution are known. If the starting concentration is known, the final concentration can be calculated (a **"forward" calculation**):

$$C_f = \frac{C_iV_i}{V_f} \quad \text{or} \quad C_f = \frac{C_im_i}{m_f} \qquad (3.7)$$

Similarly, if the final concentration is known, the starting concentration can be calculated (a **"back" calculation**). From this relationship, it becomes evident that the final concentration can be expressed as the initial concentration multiplied by the ratio of the initial and final concentrations (or vice versa for the initial concentration):

$$C_f = \frac{C_iV_i}{V_f} = C_i\left(\frac{V_i}{V_f}\right) \quad \text{or} \quad C_f = \frac{C_im_i}{m_f} = C_i\left(\frac{m_i}{m_f}\right) \qquad (3.8)$$

This is intuitive, as the ratio of the final to the initial volumes, masses, or concentrations is referred to as the "fold" of the dilution. The ratio of the starting and final masses or volumes for each step is referred to as the **dilution factor** *(DF)* for that step (see *Practice Problem 2*):

$$\text{DF} = \frac{V_i}{V_f} \qquad (3.9)$$

The final concentration is the product of the initial concentration and the DF (the DF is a "multiplier" that can be used to convert the initial concentration to the final concentration):

$$C_f = \frac{C_iV_i}{V_f} = C_i\left(\frac{V_i}{V_f}\right) = C_i(DF) \qquad (3.10)$$

Another way to think of the DF is the ratio of the final and initial concentrations for each step:

$$\text{DF} = \frac{C_f}{C_i} \qquad (3.11)$$

Some conventions that apply to the term "dilution factor" are as follows:

1. DF always refers to the forward direction (initial mass or volume divided by final mass or volume for each step).
2. The term DF is usually used even if the step is a concentration; less commonly, the term "concentration factor" may be used.
3. Since DF is a multiplier of the initial concentration:

    (a) If the step is a dilution, DF < 1.

    (b) If the step is a concentration, DF > 1.

The "fold" or "X" of a dilution is defined as follows:

$$\text{dilution "fold" or "X"} = \frac{1}{\text{DF}} \qquad (3.12)$$

Example D3 For the thiamine solution described earlier, if 1 mL of the solution is added to a 10 mL volumetric flask and diluted to volume, the ratio of the final to initial volumes is 10:1 or a "tenfold" or "10X" dilution, and therefore the final concentration is tenfold (10X) lower than, or 1/10th of, the initial concentration:

$$C_f = \frac{C_i V_i}{V_f} = C_i\left(\frac{V_i}{V_f}\right) = \frac{1.12\,\text{mg thiamine}}{\text{mL solution}}\left(\frac{1\,\text{mL}}{10\,\text{mL}}\right)$$

$$= \frac{0.112\,\text{mg thiamine}}{\text{mL solution}}$$

For a tenfold dilution:

$$\text{dilution"fold"or"X"} = \frac{1}{\text{DF}}$$

$$\text{Therefore, DF} = \frac{1}{\text{dilution"fold"or"X"}}$$

$$\text{DF} = \frac{1}{10} = 0.1$$

$$C_f = C_i(\text{DF}) = \frac{1.12\,\text{mg thiamine}}{\text{mL solution}}(0.1)$$

$$= \frac{0.112\,\text{mg thiamine}}{\text{mL solution}}$$

As discussed above, note that this is different than adding 10 mL water to 1 mL of the thiamine solution, which makes the final volume 11 mL for an 11-fold or 11X dilution:

$$C_f = C_i\left(\frac{V_i}{V_f}\right) = \frac{1.12\,\text{mg thiamine}}{\text{mL solution}}\left(\frac{1\,\text{mL}}{1\,\text{mL} + 10\,\text{mL}}\right)$$

$$= \frac{1.12\,\text{mg thiamine}}{\text{mL solution}}\left(\frac{1\,\text{mL}}{11\,\text{mL}}\right)$$

$$= \frac{0.102\,\text{mg thiamine}}{\text{mL solution}}$$

For concentration processes, the calculation is the same except the analyte concentration increases from initial to final. For example, suppose that 12.8 mL of the thiamine solution is reduced (by boiling, freeze drying, rotary evaporation, etc.) to 3.9 mL. The final concentration is:

$$C_f = C_i\left(\frac{V_i}{V_f}\right) = \frac{1.12\,\text{mg thiamine}}{\text{mL solution}}\left(\frac{12.8\,\text{mL}}{3.9\,\text{mL}}\right)$$

$$= \frac{3.68\,\text{mg thiamine}}{\text{mL solution}}$$

Note that masses can be used instead of volumes:

$$C_i m_i = C_f m_f \tag{3.13}$$

Furthermore, both masses and volumes can be used:

$$C_i m_i = C_f V_f \quad \text{or} \quad C_i V_i = C_f m_f \tag{3.14 and 3.15}$$

Example D4 Suppose that soybean oil contains 175 mg oleic acid per g, and 2.5 g of the oil is diluted with hexane to a final volume of 75 mL. The concentration of oleic acid in the final solution can be determined as follows:

$$C_i m_i = C_f V_f$$

$$C_f = C_i\left(\frac{m_i}{V_f}\right) = \frac{175\,\text{mg oleic acid}}{\text{g oil}}\left(\frac{2.5\,\text{g oil}}{75\,\text{mL}}\right)$$

$$= \frac{5.83\,\text{mg oleic acid}}{\text{mL}}$$

3.4.4 Back Calculations

The examples worked up to this point involve "forward" calculations (calculating the final concentration from a known initial concentration, see *Practice Problems 3 and 4*). However, many food analysis calculations will involve "back" calculations (calculating the initial sample concentration from a final concentration obtained by analysis of the diluted or concentrated sample, see *Practice Problems 1 and 6*). This calculation is simply the reverse of the "forward" calculation, for which the initial concentration is solved for as opposed to the final concentration:

$$C_i V_i = C_f V_f \quad \text{or} \quad C_i m_i = C_f m_f, \text{ therefore :}$$

$$C_i = \frac{C_f V_f}{V_i} = C_f\left(\frac{V_f}{V_i}\right) \quad \text{or}$$

$$C_i = \frac{C_f m_f}{m_i} = C_f\left(\frac{m_f}{m_i}\right) \tag{3.16}$$

Note that for back calculations, we flip the DF:

$$C_i = \frac{C_f V_f}{V_i} = C_f\left(\frac{V_f}{V_i}\right) = C_f\left(\frac{1}{\text{DF}}\right) \quad \text{or}$$

$$C_i = \frac{C_f m_f}{m_i} = C_f\left(\frac{1}{\text{DF}}\right) \tag{3.17}$$

Example D5 Suppose that 50 mL of water is added to a 35 mL sample of apple juice (therefore, initial volume = 35 mL and final volume = 85 mL), and titration indicates that the concentration of malic acid in the diluted sample is 0.24 % w/v. What is the concentration of malic acid in the undiluted juice? This can be determined using Eq. 3.17 above:

$$C_i = C_f\left(\frac{V_f}{V_i}\right) = 0.24\%\,\text{w/v}\left(\frac{85\,\text{mL}}{35\,\text{mL}}\right) = 0.583\%\,\text{w/v}$$

A special case of concentration is when a solution is completely evaporated or dried (so that there is no liquid left, only nonvolatile solutes) and then reconstituted with a volume of liquid smaller than the initial volume. This is technically composed of two distinct steps: a concentration step (drying the sample or solution until only solutes, usually a very small mass, are left) followed by a dilution step (diluting the residual dry mass to a define volume).

Example D6 Suppose that 10 mL apple juice, containing 0.025 g malic acid/mL, is freeze dried, leaving a residue of 0.09 g. This residue is then reconstituted to a final volume of 3 mL with water. What is the concentration of malic acid in the final solution? For the first step, the problem is set up as:

$$C_f = C_i \left(\frac{V_i}{V_f} \right) = \frac{0.025 \, \text{g malic acid}}{\text{mL juice}} \left(\frac{10 \, \text{mL juice}}{0.09 \, \text{g residue}} \right)$$

$$= \frac{2.78 \, \text{g malic acid}}{\text{g residue}}$$

For the second step, the problem is set up as:

$$C_f = C_i \left(\frac{V_i}{V_f} \right) = \frac{2.78 \, \text{g malic acid}}{\text{g residue}} \left(\frac{0.09 \, \text{g residue}}{3 \, \text{mL}} \right)$$

$$= \frac{0.0833 \, \text{g malic acid}}{\text{mL}}$$

However, in actual laboratory practice, this is typically treated as a single step (starting with the initial volume and ending with the final diluted volume) for two reasons. First, the whole dried sample is typically reconstituted, so the final mass for the first step and the initial mass of the second step are the same and thus cancel each other out in the calculation. Second, for most concentrations involving solutions, the mass of the solutes remaining after complete drying is much too small to be measured accurately with balances commonly found in most food analysis laboratories, and it is difficult to get all of the residue out accurately for weighing. Therefore, these problems are typically solved in a single step as follows:

$$C_f = C_i \left(\frac{V_i}{V_f} \right) = \frac{0.025 \, \text{g malic acid}}{\text{mL juice}} \left(\frac{10 \, \text{mL juice}}{3 \, \text{mL}} \right)$$

$$= \frac{0.0833 \, \text{g malic acid}}{\text{mL}}$$

As long as the entire residue is reconstituted, the calculation can be treated as a single step.

3.4.5 Multiple-Step Dilutions and Concentrations

Up to this point, we have dealt exclusively with single-step dilutions and concentrations. However, real-world analytical practice often requires that several dilutions and/or concentrations be performed (see *Practice Problems 1, 2, and 6*). There are two ways to approach this problem. Suppose that three dilutions are used as follows:

$$\text{Step 1}: C_1 \rightarrow C_2 \text{ by performing } V_1 \rightarrow V_2$$

$$\text{Step 2}: C_2 \rightarrow C_3 \text{ by performing } V_3 \rightarrow V_4$$

$$\text{Step 3}: C_3 \rightarrow C_4 \text{ by performing } V_5 \rightarrow V_6$$

The obvious approach is to solve for the final (or initial) concentration of the first step, use the final (or initial) concentration of the first step as the initial (or final) concentration of the second step, and so forth:

$$C_2 = C_1 \left(\frac{V_1}{V_2} \right) \quad \text{then } C_3 = C_2 \left(\frac{V_3}{V_4} \right) \quad \text{then } C_4 = C_3 \left(\frac{V_5}{V_6} \right)$$

This can be time-consuming. Furthermore, performing multiple calculations can introduce rounding errors and increases the probability of making other errors. Notice that we can substitute the formula used to calculate a concentration in one step for that value in the next step:

$$C_2 = C_1 \left(\frac{V_1}{V_2} \right)$$

$$C_3 = C_2 \left(\frac{V_3}{V_4} \right) = \left[C_1 \left(\frac{V_1}{V_2} \right) \right] \left(\frac{V_3}{V_4} \right)$$

$$C_4 = C_3 \left(\frac{V_5}{V_6} \right) = \left[C_1 \left(\frac{V_1}{V_2} \right) \left(\frac{V_3}{V_4} \right) \right] \left(\frac{V_5}{V_6} \right)$$

Therefore, as long as we have the volumes (or masses) from each step, we can reduce this multistep calculation to a single-step calculation that goes directly from the initial to the final concentration (or vice versa), keeping in mind that when going forward (from initial sample to final solution), the initial concentration is multiplied by a series of dilution factors for each step, with the starting mass/volume divided by the final mass/volume for each step. If we have a dilution or concentration scheme with n steps, we can set up the calculations so that:

$$\text{Step 1}: C_i \rightarrow C_2 \text{ by performing } V_1 \rightarrow V_2$$

$$\text{Step 2}: C_2 \rightarrow C_3 \text{ by performing } V_3 \rightarrow V_4$$

and so forth until the last step, where:

$$\text{Step } n: C_3 \rightarrow C_f \text{ by performing } V_{k-1} \rightarrow V_k$$

Then, a general formula for multistep forward calculations is:

$$C_f = C_i \left(\frac{m \, \text{or} \, V_1}{m \, \text{or} \, V_2} \right) \left(\frac{m \, \text{or} \, V_3}{m \, \text{or} \, V_4} \right) \cdots \left(\frac{m \, \text{or} \, V_{k-1}}{m \, \text{or} \, V_k} \right)$$
$$= C_i \left(\text{DF}_1 \right) \left(\text{DF}_2 \right) \cdots \left(\text{DF}_n \right) \tag{3.18}$$

We can simplify this by calculating the "overall DF" (DF_Σ), the product of DFs from each step:

$$\text{DF}_\Sigma = \left(\text{DF}_1 \right) \left(\text{DF}_2 \right) \cdots \left(\text{DF}_n \right) \tag{3.19}$$

$$C_f = C_i \left(\text{DF}_\Sigma \right) \tag{3.19}$$

When going from final solution to initial sample, the calculation is reversed: the initial concentration is multiplied by dilution factors for each step, with the final mass/volume divided by the initial mass/volume for each step. A general formula for multistep back calculations is:

$$C_i = C_f \left(\frac{m \, \text{or} \, V_2}{m \, \text{or} \, V_1} \right) \left(\frac{m \, \text{or} \, V_4}{m \, \text{or} \, V_3} \right) \cdots \left(\frac{m \, \text{or} \, V_k}{m \, \text{or} \, V_{k-1}} \right)$$
$$= C_f \left(\frac{1}{\text{DF}_1} \right) \left(\frac{1}{\text{DF}_2} \right) \cdots \left(\frac{1}{\text{DF}_n} \right) \tag{3.20}$$

$$C_f = C_i \left(\frac{1}{\text{DF}_\Sigma} \right) \tag{3.21}$$

$$\text{DF}_\Sigma = \frac{C_f}{C_i} \tag{3.22}$$

This approach of solving dilution or concentration problems in a single step can be used for dilutions, concentrations, and mixed procedures involving both dilutions and concentrations.

Example D7 Suppose that 0.56 g of green tea extract is diluted in boiling water to a volume of 500 mL, 25 mL of the resulting solution is mixed with 100 mL phosphate buffer, 75 mL of this solution is freeze dried, and the residue is dissolved to a final volume of 50 mL in methanol, and all of this solution is then combined with 50 mL ether. The final concentration of caffeine is analyzed by HPLC and is 12.9 µg/mL. What is the concentration of caffeine in the green tea extract (in µg/g)?

We can set the problem up like this:

Step 1: $C_i = ?$ 0.56 g (m_1) diluted to 500 mL (V_2)

Step 2: 25 mL (V_3) diluted to 125 mL $(V_4, 25 \, \text{mL} + 100 \, \text{mL})$

Step 3: 75 mL (V_5) concentrated to 50 mL (V_6)

Step 4: 50 mL (V_7) diluted to 100 mL $(V_8, 50 \, \text{mL} + 50 \, \text{mL})$,

$$C_f = \frac{12.9 \, \mu\text{g}}{\text{mL}}$$

The calculation is a back calculation of the initial concentration from the final calculation:

$$C_i = C_f \left(\frac{m \, \text{or} \, V_2}{m \, \text{or} \, V_1} \right) \left(\frac{m \, \text{or} \, V_4}{m \, \text{or} \, V_3} \right) \cdots \left(\frac{m \, \text{or} \, V_k}{m \, \text{or} \, V_{k-1}} \right)$$
$$= C_f \left(\frac{1}{\text{DF}_1} \right) \left(\frac{1}{\text{DF}_2} \right) \cdots \left(\frac{1}{\text{DF}_n} \right)$$

There are four steps, so there must be four individual DFs (this is a good way to make sure that you are doing the problem correctly; does the number of DFs you used equal the number of steps?):

$$C_i = C_f \left(\frac{m \, \text{or} \, V_2}{m \, \text{or} \, V_1} \right) \left(\frac{m \, \text{or} \, V_4}{m \, \text{or} \, V_3} \right) \left(\frac{m \, \text{or} \, V_6}{m \, \text{or} \, V_5} \right) \left(\frac{m \, \text{or} \, V_8}{m \, \text{or} \, V_7} \right)$$

$$C_i = \frac{12.9 \, \mu\text{g caffeine}}{\text{mL}} \left(\frac{500 \, \text{mL}}{0.56 \, \text{g green tea extract}} \right) \left(\frac{125 \, \text{mL}}{25 \, \text{mL}} \right) \left(\frac{50 \, \text{mL}}{75 \, \text{mL}} \right) \left(\frac{100 \, \text{mL}}{50 \, \text{mL}} \right)$$

$$= \frac{76,800 \, \mu\text{g caffeine}}{\text{g green tea extract}}$$

Another way to make sure that this calculation has been done correctly is to assign each solution a letter or number, and those should "cancel each other out" along with the units. If the units and/or solution identities don't cancel out properly, one or more concentrations, volumes, or masses have been used incorrectly. For this problem, the solutions could be identified as follows:

Step 1: $C_i = ?$ m 0.56 g (m_1) diluted to 500 mL $(V_2, \text{solution} \, A)$

Step 2: 25 mL $(V_3, \text{solution} \, A)$ diluted to 125 mL $(V_4, \text{solution} \, B, 25 \, \text{mL} + 100 \, \text{mL})$

Step 3: 75 mL $(V_5, \text{solution} \, B)$ concentrated to 50 mL $(V_6, \text{solution} \, C)$

Step 4 : 50 mL $(V_7, \text{solution}\,C)$ diluted to 100 mL

$$(V_8, \text{solution}\,D, 50\,\text{mL} + 50\,\text{mL})$$

$$C_i = \frac{12.9\,\mu g}{mL}$$

The problem would then be solved the same way with the solutions identified:

$$C_i = \frac{12.9\,\mu g\,\text{caffeine}}{mL\,\text{solution}\,D}\left(\frac{500\,mL\,\text{solution}\,A}{0.56\,g\,\text{green tea extract}}\right)$$

$$\left(\frac{125\,mL\,\text{solution}\,B}{25\,mL\,\text{solution}\,A}\right)\left(\frac{50\,mL\,\text{solution}\,C}{75\,mL\,\text{solution}\,B}\right)$$

$$\left(\frac{100\,mL\,\text{solution}\,D}{50\,mL\,\text{solution}\,C}\right) = \frac{76{,}785.7\,\mu g\,\text{caffeine}}{g\,\text{green tea extract}}$$

3.5   SPECIAL CASES

3.5.1   Extraction

There are several special cases regarding dilution that merit specific mention: extraction and homogenization/mixing. Regarding **extraction**, often, we use a solvent for which an analyte has a high affinity (e.g., lipids are more soluble in nonpolar solvents such as hexane or chloroform than the foods they are in, so these solvents "pull" the lipid out of the food) to extract that analyte from a sample. The solvent is typically immiscible with the sample as a whole. This means that the sample and the solvent do not mix and can be easily separated by centrifugation or by using a separatory funnel. Most of the analyte is transferred to the solvent, while most of the sample remains behind. For dilution and concentration purposes, we assume that 100 % of the desired analyte is quantitatively transferred from the sample into the solvent (see *Practice Problem 1*). While this assumption is never actually true, it is necessary for our purposes. Further dilutions and/or concentrations may then be performed on the extraction solvent.

In some cases, we use a single extraction step. In this situation, the sample is mixed with the solvent to achieve transfer (extraction) of the analyte from the solvent. The sample and solvent are then separated. This is different than the simple dilution problems we have done previously, in that the sample remains behind while only the analyte is transferred to the solvent. However, the principle is still the same: 100 % of the analyte that was in the sample is assumed to now reside in the solvent, and we can use the equations previously discussed (Eqs. 3.16 and 3.17).

Example E1 Suppose that peanut butter (PB) contains 0.37 g fat/g, and 1.4 g peanut butter is extracted with 100 mL hexane. What is the concentration of fat in the hexane (the assumption here is that the volume of the hexane doesn't change)?

This problem can be solved using the standard procedure:

$$C_f = C_i\left(\frac{m_i}{V_f}\right) = \frac{0.37\,g\,\text{fat}}{g\,PB}\left(\frac{1.4\,g\,PB}{100\,mL}\right) = \frac{0.00518\,g\,\text{fat}}{mL}$$

Note that this is the same as if the entire sample was diluted into the solvent, but for an extraction only a part of the sample (but presumably all of the analyte) is actually transferred.

In reality, only a fraction (generally unknown) of the analyte is transferred from the sample to the solvent. In order to overcome this, repeated extractions are often performed. These are the same as single extractions, except the extraction is repeated with fresh solvent each time and the solvent volumes from each extraction are combined into a single extract. In this case, we can treat this as a single extraction where the combined solvent volume is the final volume, and again we assume that 100 % of the analyte is transferred to the solvent.

Example E2 Suppose that the peanut butter example above is adjusted so that the peanut butter is extracted three times with 75 mL of fresh hexane, and the three extracts are combined. What is the concentration of fat in the hexane?

In this case, the final volume would be 3×75 mL = 225 mL:

$$C_f = C_i\left(\frac{m_i}{V_f}\right) = \frac{0.37\,g\,\text{fat}}{g\,PB}\left(\frac{1.4\,g\,PB}{225\,mL}\right) = \frac{0.00230\,g\,\text{fat}}{mL}$$

Extraction procedures often specify that the extraction solvent be brought up to a known final volume after extraction. This accounts for potential loss of solvent, or significant changes in volume due to extraction of sample components, during extraction and also increases the precision and accuracy of the extraction. This is handled by simply making this final volume the final volume for the dilution calculation.

Example E3 Suppose that the peanut butter example above is adjusted so that the peanut butter is extracted 2X with 100 mL hexane, and the pooled hexane extract is diluted to a final volume of 250 mL. What is the concentration of fat in the hexane?

In this case, the final volume would simply be 250 mL, as the pooled extracts (~200 mL) are further diluted with ~50 mL to an exact total volume of 250 mL. The problem would then be solved as shown above, using 250 mL as the final volume.

In summary, extraction can be thought of as a dilution (or concentration) in which the final volume is the total volume of solvent. Regardless of the number of extractions, it can be viewed as a single step as long as all the solvent extracted from the sample is kept and the final volume into which the sample is extracted is known. After extraction, dilutions or concentrations of the extract may be performed; these are handled as any other dilution or concentration.

3.5.2 Homogenization/Blending/Mixing

The other scenario that warrants consideration is **homogenization**, **blending**, and **mixing**. Often, samples need to be dispersed or blended into a solvent for sample preparation. This is typically done using a lab blender, food processor, Polytron homogenizer, stomacher, sonicator, etc. These processes cannot typically be done in volumetric glassware. Furthermore, the problem with these processes, from a calculation standpoint, is that homogenization, blending, and mixing of solids with liquids typically change the volume of a sample.

Example E4 Suppose you homogenize 2.5 g cheese (of an unknown volume) in 100 mL phosphate buffer. What is the final volume?

The final volume of the liquid will be >100 mL and is not easily measured by volumetric glassware. Therefore, in this example, the final volume is unknown. This poses a challenge for accurate dilution calculations. This process is usually overcome similarly to the extractions described above: the sample is homogenized, quantitatively transferred to a volumetric flask, and then diluted to a known final volume with the same solvent used for homogenization.

Example E5 Suppose that 22.7 g swordfish fillet is homogenized in a lab blender with 150 mL denaturing buffer. Following homogenization, the liquid is decanted into a 500 mL flask. Fresh denaturing buffer (~25 mL) is used to rinse the blender, and the rinse liquid is also decanted into the flask. The flask is brought to volume with denaturing buffer. The concentration of methylamines in the diluted homogenate is 0.92 μmol/mL. What is the concentration of methylamines in the fish?

In this case, the sum of the homogenization process is that 22.7 g fish is diluted into 500 mL final volume. Therefore, the calculation is performed as follows:

$$C_i = C_f \left(\frac{V_f}{V_i} \right) = \frac{0.92\,\mu mol\ methylamines}{mL} \left(\frac{500\,mL}{22.7\,g\ fish} \right)$$

$$= \frac{20.3\,\mu mol\ methylamines}{g\ fish}$$

In summary, mixing processes can be thought of as a dilution (or concentration) step for which the final volume is the total volume of solvent. Regardless of the number of steps involved, it can be viewed as a single step as long as all the homogenate from whole sample is kept and the final volume into which the sample is diluted is known. Keep in mind that, after mixing and quantitative dilution, subsequent dilutions or concentrations of the homogenate may be performed, and these are handled as any other dilution or concentration.

3.6 STANDARD CURVES

It is very common for analytical procedures to require a **standard curve**, which is a set of solutions containing different concentrations of the analyte. Typically, a **stock solution** containing a high concentration of the analyte is prepared, and various dilutions are performed to get the desired standard curve solutions (*see Practice Problems 3, 4, 5, and 7*). These dilutions can be done with either sequential dilutions (Fig. 3.1a) or parallel dilutions (Fig. 3.1b).

3.6.1 Sequential Versus Parallel Dilutions

In the case of **sequential dilutions**, each dilution is used to prepare the next dilution in the series:

stock solution → solution 1 → solution 2
→ solution 3, etc.

a

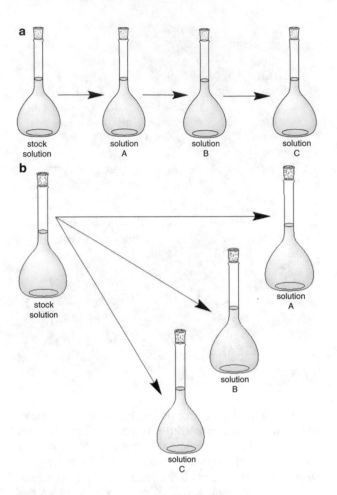

b

3.1
figure

Example standard
curve dilution schemes for sequential
dilutions (**a**) and parallel dilutions (**b**)

$$\text{Solution}\,B: C_f = C_i\left(\frac{V_i}{V_f}\right) = 0.333\,M\left(\frac{1\,\text{mL}}{3\,\text{mL}}\right)$$
$$= 0.111\,M$$

$$\text{Solution}\,C: C_f = C_i\left(\frac{V_i}{V_f}\right) = 0.111\,M\left(\frac{1\,\text{mL}}{3\,\text{mL}}\right)$$
$$= 0.0370\,M$$

$$\text{Solution}\,D: C_f = C_i\left(\frac{V_i}{V_f}\right) = 0.0370\,M\left(\frac{1\,\text{mL}}{3\,\text{mL}}\right)$$
$$= 0.0123\,M$$

For **parallel dilutions**, different DFs are applied to the stock to obtain the desired solutions:

$$\text{stock solution}\;\overset{\text{DF1}}{\rightarrow}\;\text{solution}\,1$$

$$\text{stock solution}\;\overset{\text{DF2}}{\rightarrow}\;\text{solution}\,2$$

$$\text{stock solution}\;\overset{\text{DF3}}{\rightarrow}\;\text{solution}\,3$$

Example F1 Suppose a $1\,M$ solution of citric acid is serially diluted as follows: 1 mL is diluted with 2 mL water, and the resulting solution is diluted the same way (1 mL+2 mL water) until four diluted solutions are obtained. What are the concentrations in each solution?

We can solve this problem by assigning each solution a name (Solutions A–D) and then calculating the concentration in each. For each step, the initial volume is 1 mL and the final volume is 3 mL, since 2 mL water is combined with each solution, and the initial concentration is the concentration of the previous solution (stock when making Solution A, Solution A when making Solution B, etc.):

$$\text{Solution}\,A: C_f = C_i\left(\frac{V_i}{V_f}\right) = 1\,M\left(\frac{1\,\text{mL}}{3\,\text{mL}}\right) = 0.333\,M$$

Example F2 Suppose the $1\,M$ stock solution of citric acid is diluted as follows: 1 mL of stock solution is diluted to volume in a 5 mL, 10 mL, 25 mL, and 50 mL volumetric flask. What are the citric acid concentrations in each solution? We can solve this problem by assigning each solution a name (Solutions A–D) and then calculating the concentration in each. For each dilution, the starting concentration is the stock concentration (since they were made in parallel), and the final volume is the volume of each volumetric flask:

$$\text{Solution}\,A: C_f = C_i\left(\frac{V_i}{V_f}\right) = 1\,M\left(\frac{1\,\text{mL}}{5\,\text{mL}}\right) = 0.200\,M$$

$$\text{Solution}\,B: C_f = C_i\left(\frac{V_i}{V_f}\right) = 1\,M\left(\frac{1\,\text{mL}}{10\,\text{mL}}\right) = 0.100\,M$$

$$\text{Solution}\,C: C_f = C_i\left(\frac{V_i}{V_f}\right) = 1\,M\left(\frac{1\,\text{mL}}{25\,\text{mL}}\right) = 0.040\,M$$

$$\text{Solution}\,D: C_f = C_i\left(\frac{V_i}{V_f}\right) = 1\,M\left(\frac{1\,\text{mL}}{50\,\text{mL}}\right) = 0.020\,M$$

Preparing standards in parallel is typically more accurate than in series, as each dilution is performed in a single step from the stock. This reduces error associated with each solution.

Example F3 If preparing a dilution involves an error of ~1% (a factor of 0.01), what is the error in the third solution if the solutions are prepared in parallel?

Performing three dilutions in parallel will result in each solution being off by ~1%:

$$\text{stock solution} \xrightarrow{\text{DF1}} \text{solution } 1 \left(X \pm 1\% \text{ or } 1.01X \right)$$

$$\text{stock solution} \xrightarrow{\text{DF2}} \text{solution } 2 \left(X \pm 1\% \text{ or } 1.01X \right)$$

$$\text{stock solution} \xrightarrow{\text{DF3}} \text{solution } 3 \left(X \pm 1\% \text{ or } 1.01X \right)$$

If dilutions are performed in series, the error compounds because each solution is used to prepare the next solution.

Example F4 For the same three dilutions, what is the error in the third solution if the dilutions are prepared in series?

When prepared in series, the error increases with each subsequent dilution:

$$\text{stock solution} \to \text{solution } 1 \left(\pm 1\% \text{ or } 1.01X \right)$$
$$\to \text{solution } 2 \left(\pm 1\% \text{ or } 1.01X \right) \to$$
$$\text{solution } 3 \left(\pm 1\% \text{ or } 1.01X \right)$$

$$\text{error in solution } 1 = \pm 1.01X = \pm 1\%$$

$$\text{error in solution } 2 = \pm (1.01)(1.01)X$$
$$= \pm 1.0201X = \pm 2.01\%$$

$$\text{error in solution } 3 = \pm (1.01)(1.01)(1.01)X$$
$$= \pm 1.030301X = \pm 3.0301\%$$

Parallel dilutions are often preferred over series dilutions for accuracy and improved quantification for standard curves. Regardless of how they are performed, the principles of dilution calculations apply. As with any dilution, these dilutions can be performed as "dilute to" or "dilute with" procedures, and the calculations should be performed accordingly.

3.6.2 Designing Dilution Schemes

Occasionally, you may be given the stock solution concentration and the needed concentrations of standard curve solutions and asked to design a dilution scheme. The way to do this is to calculate the needed DFs for each solution and then design a scheme to achieve the DFs.

Example F5 Suppose you have a 0.2% w/v NaCl solution and you need standard curve solutions of 0.1, 0.04, 0.02, 0.01, and 0.002% w/v. How would you make these?

First, start by calculating each DF from the stock solution:

$$\text{DF} = \frac{C_f}{C_i}$$

$$\text{DF}_A = \frac{C_f}{C_i} = \frac{0.1\%}{0.2\%} = 0.5 \left(\text{i.e., } \frac{1}{2} \right)$$

$$\text{DF}_B = \frac{C_f}{C_i} = \frac{0.04\%}{0.2\%} = 0.2 \left(\text{i.e., } \frac{1}{5} \right)$$

$$\text{DF}_C = \frac{C_f}{C_i} = \frac{0.02\%}{0.2\%} = 0.1 \left(\text{i.e., } \frac{1}{10} \right)$$

$$\text{DF}_D = \frac{C_f}{C_i} = \frac{0.01\%}{0.2\%} = 0.05 \left(\text{i.e., } \frac{1}{20} \right)$$

$$\text{DF}_E = \frac{C_f}{C_i} = \frac{0.002\%}{0.2\%} = 0.01 \left(\text{i.e., } \frac{1}{100} \right)$$

From this point, there are several options. You can dilute in series or in parallel. And, you can "dilute to" or "dilute with" to achieve the desired concentrations. For simplicity, the parallel "dilute to" example is shown in Table 3.1.

3.1 table Dilution example for a standard curve

Solution	C_i (% w/v)	Dilute (mL)	To final volume (mL)	DF	C_f (% w/v)
A	0.2	1	2	0.5	1
B	0.2	1	5	0.2	0.04
C	0.2	1	10	0.1	0.02
D	0.2	1	20	0.05	0.01
E	0.2	1	100	0.002	0.002

You should be able to design a scheme to dilute a stock solution to a wide array of diluted solutions using both parallel and series dilutions and perform either with "dilute to" or "dilute with" procedures. One consideration to keep in mind is that class A volumetric glassware comes in discreet volumes, and only those volumes can be used for dilutions where maximum accuracy and precision are desired. The most common volumes available are:

1. Class A volumetric flasks: 5, 10, 25, 50, 100, 200, 250, 500, and 1000 mL
2. Class A volumetric transfer pipettes: 1, 2, 3, 4, 5, 10, 20, 25, 50, and 100 mL

Therefore, your designed dilution scheme should use only those volumes for transferring volumes (pipettes) and bringing to volume (flasks), if possible, to optimize accuracy and precision. Some schemes may require the use of an adjustable pipettor (to transfer or add volumes such as 0.1 mL, 950 μL, etc.), particularly when small volumes are used. This is often the case when preparing standard curves (see *Practice Problems 3 and 4*).

Example F6 Suppose that you have a stock solution of 2000 ppm Fe(II) in 0.1 N HCl. You need to create 1 mL of 1000, 750, 500, 250, and 100 ppm Fe(II) by combining different volumes of the stock solution and 0.1 N HCl. What volumes should be used?

We know the starting (2000 ppm) and final concentration (1000–100 ppm) and final volume (1 mL). Therefore, we must calculate the initial volume of stock solution for each solution:

$$C_i V_i = C_f V_f \; and \, V_i = \frac{C_f V_f}{C_i}$$

$$\text{for } 1000\,\text{ppm, } V_i = \frac{C_f V_f}{C_i} = \frac{(1000\,\text{ppm})(1\,\text{mL})}{2000\,\text{ppm}}$$
$$= 0.5\,\text{mL}$$

$$\text{for } 750\,\text{ppm, } V_i = \frac{C_f V_f}{C_i} = \frac{(750\,\text{ppm})(1\,\text{mL})}{2000\,\text{ppm}}$$
$$= 0.375\,\text{mL}$$

and so forth. Once the volume of stock solution is known for each dilution, we then calculate the amount of 0.1 N HCl needed to dilute to 1 mL:

$$\text{for } 1000 \text{ ppm, } 0.5 \text{ mL} \rightarrow 1 \text{ mL} - 0.5 \text{ mL}$$
$$= 0.5 \text{ mL } 0.1\,\text{N HCl}$$

$$\text{for } 750 \text{ ppm, } 0.375 \text{ mL} \rightarrow 1 \text{ mL} - 0.375 \text{ mL}$$
$$= 0.625 \text{ mL } 0.1\,\text{N HCl}$$

and so forth. The dilution scheme would thus be as shown in Table 3.2.

3.2 table Dilution example for a standard curve

Fe(II) (ppm)	Stock (mL)	0.1 N HCL (mL)	Total volume (mL)
1000	0.5	0.5	1.0
750	0.375	0.625	1.0
500	0.25	0.75	1.0
250	0.125	0.875	1.0
100	0.05	0.95	1.0

In cases when an adjustable pipettor is used, accuracy and precisions should be optimized by frequently calibrating and maintaining the pipettors. Furthermore, using proper (and consistent) manual pipetting technique is critical in these scenarios.

3.7 UNIT CONVERSIONS

Up to this point, we have examined calculations for which the units of the dilutions and the units of the concentrations that you have been given "match up" so that no conversions are required to get the correct answer. However, this will not always be the case in actual lab practice. This will require you to perform unit conversions to get an answer that is correct and makes sense.

Example G1 Suppose that you dilute 14.4 g of yogurt to a final volume of 250 mL with water. HPLC analysis of the diluted sample indicates a riboflavin concentration of 0.0725 ng/μL. What is the riboflavin content in the yogurt?

This is a straightforward dilution problem that can be solved as follows:

$$C_i = C_f \left(\frac{V_f}{V_i} \right)$$

However, if the calculation were performed without unit conversions, the answer obtained would be as follows:

$$C_i = C_f \left(\frac{V_f}{V_i} \right) = \frac{0.0725\,\text{ng riboflavin}}{\text{μL diluted sample}}$$

$$\left(\frac{250\,\text{mL diluted sample}}{14.4\,\text{mL yogurt}} \right) = \frac{1.26\,\text{ng riboflavin} \times \text{mL}}{\text{g yogurt} \times \text{μL}}$$

Obviously, $(\text{ng} \times \text{mL})/(\text{g} \times \text{μL})$ is not an acceptable unit for expressing the riboflavin concentration in a product. Therefore, we must change either µL to mL or mL to µL for the units to be expressed correctly in the final answer. The calculation is performed correctly as follows:

$$C_i = C_f \left(\frac{V_f}{V_i} \right) = \frac{0.0725\,\text{ng riboflavin}}{\text{µL diluted sample}} \left(\frac{1000\,\text{µL}}{\text{mL}} \right)$$

$$\left(\frac{250\,\text{mL diluted sample}}{14.4\,\text{mL yogurt}} \right) = \frac{1,260\,\text{ng riboflavin}}{\text{g yogurt}}$$

If the units are not converted, the magnitude (number only) of the final answer is off by a factor of 1000 in this case, in addition to having the wrong units.

By keeping track of the units, you easily catch an incorrect value and adjust the calculation.

3.8   AVOIDING COMMON ERRORS

The most common errors associated with dilutions and corrections are:

1. Setting up the calculation incorrectly (incorrectly using the information provided)
2. Lack of needed unit conversion or incorrect unit conversion

Three strategies can be employed to avoid these mistakes or catch them if they have been made:

1. Draw a picture of the dilution or concentration scheme.
2. Perform **unit analysis** (also referred to as **dimensional analysis** or the **factor-label method**) and assign "names" to each solution in a scheme.
3. Perform the "sniff test."

3.8.1   Draw a Picture

Drawing a picture can be very useful for making sure the problem has been set up correctly. This allows you to visualize the dilution or concentration scheme, which is often helpful. Drawing a picture or diagram can help clarify the meaning of the information provided and assist in setting up the calculation.

Example H1 Suppose you are told that 1.2 mL skim milk is diluted to 100 mL with water. The picture that would be drawn might look something like Fig. 3.2a. By comparison, if you are told that 1.2 mL skim milk is diluted with 100 mL water, the picture that would be drawn might look something like Fig. 3.2b.

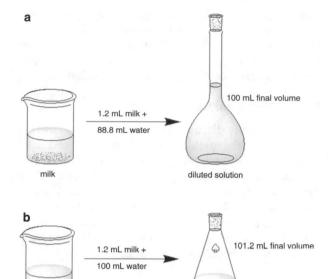

a

b

 Dilution schemes for a "dilute to" (**a**) and "dilute with" (**b**) scenario

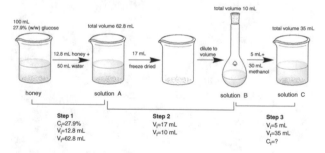

3.3 figure Diagram of a complex multistep dilution and concentration scheme

Drawing a picture can be particularly useful for visualizing complex multistep processes.

Example H2 You have 100 mL honey (27.9% glucose by weight). 12.8 mL honey is dissolved by adding 50 mL water, 17 mL of this solution is freeze dried and reconstituted to a total volume of 10 mL with water, and 5 mL of this solution is combined with 30 mL methanol. What is the % glucose by weight in the final solution?

This information can be diagramed as shown in Fig. 3.3.

Figure 3.3 shows a fairly complex scheme, and organizing the information in your head can be challenging. Drawing this scheme allows you to assign names to various solutions and organize the information so that it can easily be plugged into the calculation.

3.8.2 Unit Analysis

We have already discussed unit analysis, a key step in verifying that the calculation has been set up correctly. Along with this, assigning names to each solution in a scheme and incorporating those into the calculation can assure that each step has been set up correctly in the calculation. If the problem has been set up correctly, each solution name for the intermediate steps should be "canceled out," leaving only the final solution or sample when the calculation is complete.

Example H3 For the example in Fig. 3.3, the concentration of the final solution would be calculated as follows:

$$C_f = 27.9\% \text{ glucose} \left(\frac{12.8\,\text{mL sample}}{62.8\,\text{mL solution } A} \right)$$

$$\left(\frac{17\,\text{mL solution } A}{10\,\text{mL solution } B} \right) \left(\frac{5\,\text{mL solution } B}{35\,\text{mL solution } C} \right)$$

$$= 1.38\% \text{ glucose}$$

In this example, we see that both the units *and* the intermediate solution names cancel out. However, it would be very easy to accidentally reverse one of the dilution factors. For example, if we accidentally reversed the DF for step A → B:

$$C_f = 27.9\% \text{ glucose} \left(\frac{12.8\,\text{mL sample}}{62.8\,\text{mL solution } A} \right)$$

$$\left(\frac{10\,\text{mL solution } B}{17\,\text{mL solution } A} \right) \left(\frac{5\,\text{mL solution } B}{35\,\text{mL solution } C} \right)$$

$$= 0.478\% \text{ glucose}$$

The units still cancel out, but you would get an incorrect answer. However, by keeping track of the assigned names of each solution, you would quickly see that this step's DF is reversed because the intermediate solution names do not cancel out. Therefore, this is an easy way to quickly catch and correct errors in dilution calculations.

3.8.3 "Sniff Test"

Finally, use the "sniff test" on your calculations to detect any obvious errors. For example:

1. If the procedure is an overall dilution, then the final concentration should be lower than the initial concentration.

2. If the procedure is an overall concentration, then the final concentration should be higher than the initial concentration.

3. With very few exceptions (such as % moisture expressed on a dry weight basis for some samples), the % of an analyte in a sample will never be $\geq 100\%$.

4. Does the calculated value make sense? For example, one would not expect the following calculated values to be true:

   (a) % moisture > a few % in very dry products

   (b) % solids > a few % in dilute solutions

   (c) % fat, protein, ash, or carbohydrates that don't make sense given what is known about the product (such as vegetable oil containing 35 % moisture or 12 % fat)

   (d) Analyte levels that do not make sense given what is known about typical product composition (particularly ≥ 1 order of magnitude greater than expected)

For the sniff test, it is important that you know a little about the product you are analyzing, if possible. If the "sniff test" suggests an error, carefully reexamine how you set up the problem and how the calculation was performed.

3.9 PRACTICE PROBLEMS

(*Note: Answers to problems are in the last section of the laboratory manual.*)

The following example problems demonstrate real-world calculations involving dilutions and concentrations:

1. You randomly select a cup of applesauce (containing 113 g applesauce) from the processing line. You extract 10.3 g of the applesauce 3X with 50 mL ethyl acetate, pool the extracts, and dilute to 250 mL with ethyl acetate. You evaporate 25 mL of the extract to dryness under a stream of N_2 and redissolve the residue to 5 mL with methanol. GC analysis of the methanol indicates a methoxyfenozide concentration of 0.00334 μg/mL. What is the concentration of methoxyfenozide in the applesauce (μg/g)? What is the total amount of methoxyfenozide in the applesauce cup (μg)?

2. You are given a stock solution of (-)-epicatechin (EC, MW = 290.26 g/mol) in water (0.94 mg/mL). You prepare a series of solutions by serial dilution as follows: (1) diluting 0.5 mL of the stock solution to 10 mL with water (solution A), (2) diluting 1.5 mL of solution A with 4 mL water (solution B), and (3) diluting 3 mL solution B with 9 mL water

(solution C). What is the concentration of solution C (mg/mL and µM)? What is the overall DF? How many "fold" has the stock solution been diluted to solution C?

3. You are performing a spectrophotometric assay for riboflavin using a commercial kit. The kit comes with 2 mL of a solution of riboflavin (1.45 mg/mL). The instructions tell you to make a stock solution by diluting the riboflavin with 25 mL water. Then, the instructions tell you to make a set of standard solutions by diluting 100 µL of the stock solution with 0.5 mL (A), 0.75 mL (B), 1 mL (C), 2 mL (D), and 5 mL (E) water. What is the riboflavin concentration (mg/mL) in the stock solution and the five standards?

4. You are using a colorimetric method to measure anthocyanin pigments in raspberry juice. The method requires a total sample volume of 2 mL. You have a stock solution of 160 g/L anthocyanins, and you need solutions of 0, 0.1, 0.2, 0.3, and 0.5 mg/mL. Design a dilution scheme employing commonly available volumetric flasks, volumetric pipettes, and/or an adjustable pipettor (but do *not* use any volumes less than 0.2 mL, and the final volume of each solution must be at least 2 mL) to obtain the desired concentrations.

5. As discussed previously, samples often need to be diluted or concentrated in order to obtain analyte concentrations in the range of the standard curve. Ideally, the sample concentration would be near the middle of the standard curve values. This can be challenging, because you often do not know the approximate sample concentration prior to the analysis. For the standard curve described in *Problem 4*, the standards cover the range of 0–0.5 mg/mL. Suppose you get a sample of a new type of juice into the QA lab and you have no idea how to dilute the juice for analysis. Scientific literature suggests that this juice can have anthocyanin concentrations of anywhere from 750 to 3000 µg/mL. Based on this information, design a dilution scheme that will likely yield diluted juice samples within the range of the standard curve.

6. You are analyzing Ca content of milk using atomic absorption spectroscopy (AAS). This analysis requires dry ashing to isolate the minerals from a sample (dry ashing is essentially a concentration and extraction step: the sample is incinerated to remove all organic mineral and leave only the minerals). You dry ash a 2.8 mL sample of milk, dissolve the ash in 12 mL 1 N HCl, and dilute the solution to 50 mL with 1 N HCl. You further dilute 7 mL of this solution by adding 13 mL 1 N HCl. You then analyze 0.5 mL sample and 0.5 mL of standards (10–100 ppm Ca) by AAS. You find that the diluted sample contains 28.2 ppm Ca. What is the Ca content of the undiluted milk in ppm?

7. You are measuring caffeine (194.2 g/mol) in drinks by high-performance liquid chromatography (HPLC). You prepare standard solutions of 0–100 µM caffeine. Your method calls for dilution of the sample to 250 mL with water prior to analysis. You know that a particular drink contains 170 mg caffeine per 400 mL. How many mL should be diluted to 250 mL total to obtain a concentration in the middle of your standard curve?

Statistics for Food Analysis

Andrew P. Neilson (✉) • *Sean F. O'Keefe*
Department of Food Science and Technology,
Virginia Polytechnic Institute and State University,
Blacksburg, VA, USA
e-mail: andrewn@vt.edu; okeefes@vt.edu

S.S. Nielsen, *Food Analysis Laboratory Manual*, Food Science Text Series,
DOI 10.1007/978-3-319-44127-6_4, © Springer International Publishing 2017

4.1   INTRODUCTION

This chapter is a review of basic statistics and will demonstrate how statistics is used in the context of food analysis. It is meant to be a "survival guide" for reference. Students taking a food analysis course should have already taken an undergraduate statistics course or be taking one concurrently. A foundation for this chapter is Chap. 4, Evaluation of Analytical Data, in the *Food Analysis* textbook. The learning objectives for this chapter are to be able to

1. Calculate a mean, standard deviation, Z-score, and *t*-score for a sample dataset.
2. Determine whether a population is significantly different from a given value using a one-sample *t*-test.
3. Calculate a confidence interval for a sample mean using *t*-scores.
4. Determine if two populations are significantly different using a two-sample *t*-test.

A significant component of a typical food analysis course and the use of food analysis in a professional setting (industry, research, regulatory/government) is evaluation of analytical data. In food analysis laboratory experiments, you first collect data (values), or you are given data to work with (i.e., for laboratory exercises, homework, exams, etc.). Then, you evaluate and manipulate the data, answering questions such as these:

1. Are the values significantly different from a desired value or not?
2. Are two values significantly different from each other or not?

In your food analysis course, you will use statistics to solve problems typically encountered in the food industry, research, and regulatory or government scenarios. Data analysis concepts will be used in this chapter.

4.2   POPULATION DISTRIBUTIONS

The **distribution** of a given value in a **population** refers to how many members of a population have each value of the parameter measured. You can examine distributions by plotting a **histogram**, which is a graph with parameter values on the *x*-axis and number of members of the population having that value of the parameter (i.e., the "frequency" of each value) on the *y*-axis. Suppose there were 100 bottles of water in a population and you knew the sodium content of all of them. The histogram of this population might look like Fig. 4.1. Many populations are "normally distributed" (also known as **normal** population). Normal populations have histograms that look roughly like Fig. 4.2. Note

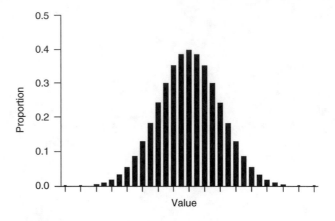

4.1 figure   Example normal distribution

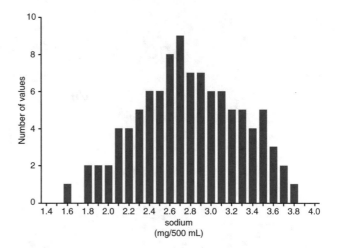

4.2 figure   Example population histogram

that the *y*-axis of a histogram could be the absolute number of population members with a given value, the proportion (fraction or decimal) of population members with a given value, or the percentage of population members with a given value. Some populations are not normally distributed, but that is a topic beyond the scope of this chapter. For the purposes of this chapter, we will assume a normal distribution. Normal population distributions are defined by two parameters (Fig. 4.3):

1. Population **mean** (μ): center of the plot
2. Population **standard deviation** (σ): a measure of the spread of the plot

For the example sodium data, you can plot the histogram of the data with the population mean (μ) and population standard deviation (σ). If you define sodium content as (x), you can calculate the mean and standard deviation of *x*:

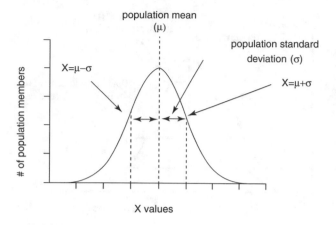

Shape of a normal population

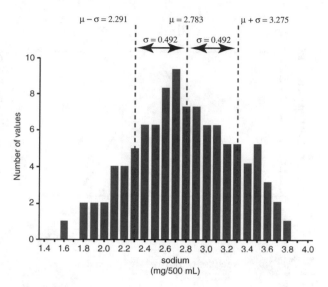

4.4 figure

Example population histogram with mean and standard deviation

Population mean of $x = \mu_x = \dfrac{\sum x_i}{n}$

$$= \dfrac{x_1 + x_2 + \ldots + x_{n-1} + x_n}{n} \quad (4.1)$$

Standard deviation of $x = \sigma_x = \sqrt{\dfrac{\sum (x_i - \bar{x})^2}{n}} \quad (4.2)$

Population mean $= \mu_x = 2.783$ standard deviation

$$= \sigma_x = \sqrt{\dfrac{\sum (x_i - 2.783)^2}{100}} = 0.492$$

You can use these values to calculate the center and spread of the data, as shown in Fig. 4.4, for example, sodium data. Normal distributions can have different shapes but are still always defined by $N(\mu, \sigma)$. The notation "$N(\mu, \sigma)$" indicates a normally distributed (N) population with center μ and standard deviation σ. The shape of the curve is dictated by the standard deviation. Figure 4.5 shows three populations, each with the same mean but different standard deviations. For a normally distributed population, if you randomly pick (or **sample**) a member of the population, the randomly selected member can have any value in the population. Since most of the population is clustered around the mean, the randomly selected value is most likely to be close to the mean. The further a value is from the mean, the less often it occurs in the population and the less likely it is to be randomly selected from the population. If we know μ and σ, we can predict the probability that any given value will be selected at random from the population, as shown in Fig. 4.6. Normal distributions have

1. 50% of values $> \mu$ and 50% $< \mu$
2. 68, 95, and 99.7% of values within ±1, 2, and 3 standard deviations (σ) of μ

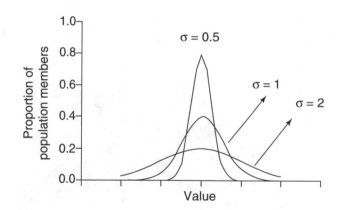

4.5 figure

Normal populations with the same mean but distinct standard deviations

Assume you randomly sample four values from the 100 water bottle population, and 1.9, 2.7, 2.9, and 2.9 mg/500 mL were selected. As seen in Fig. 4.7, three of the randomly chosen values were close to the mean, and one value was relatively far from the mean. Remember, any of the 100 values could have been selected at random. However, the values closest to the mean have the highest probability of being chosen.

4.3   Z-SCORES

Each normal population has a different μ, σ. However, this is inconvenient for statistical calculations. To make normal distributions easier to work with, you can transform, or "standardize," all normally distributed

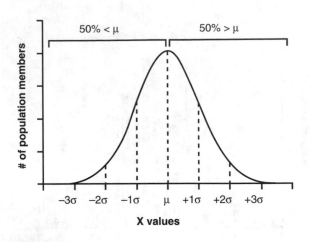

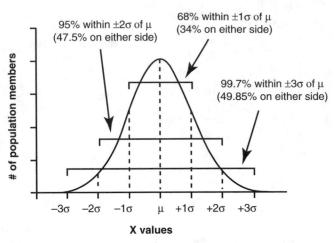

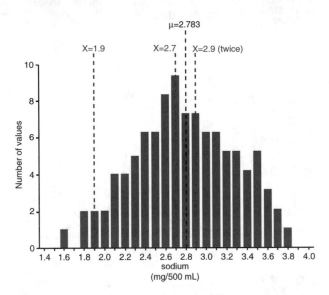

4.6 figure   Densities of normal distributions

4.7 figure   Random values from the example distribution

Z-score is the number of standard deviations that X is away from μ. The more standard deviations (i.e., multiples of σ or 1) that x (or Z) is from μ (or 0), the less likely that value is to be randomly selected from the population. You can calculate a Z-score for any x value.

> **Example C1**   For the sodium data, calculate the Z-scores for a hypothetical x value close to the mean (2.45) and one that is far from the mean (3.8).
>
> $$x = 2.45, \; Z = \frac{x - \mu}{\sigma} = \frac{2.45 - 2.783}{0.492} = \frac{-0.333}{0.492} = -0.677$$
>
> $$x = 3.8, \; Z = \frac{x - \mu}{\sigma} = \frac{3.8 - 2.783}{0.492} = \frac{1.017}{0.492} = 2.067$$
>
> Therefore, for $X \sim N(2.783, 0.492)$, x values of 2.45 and 3.8 are 0.68 standard deviations below and 2.07 standard deviations above the mean, respectively.

populations by converting the population value (x) to a standardized variable, called Z:

$$z = \frac{x - \mu}{\sigma} \qquad (4.3)$$

The variable x is normally distributed with mean μ and standard deviation σ [$x \sim N(\mu, \sigma)$]. The distribution of the **Z-scores** is $Z \sim N(0, 1)$, i.e., "the standard normal distribution" (Fig. 4.8):

1. The center of the Z-distribution is $\mu = 0$.
2. The spread of the Z-distribution is defined by $\sigma = 1$.

Each x value in a population has a corresponding Z-score. For any x value, the corresponding

The same distributions that apply to $X \sim N(\mu, \sigma)$ also apply to the $Z \sim N(1, 0)$:

1. 0% of Z values are >0 and 50% are <0.
2. 68, 95, and 99.7% of Z values are within ±1, 2, and 3 of 0.

Some other important properties of normal populations:

1. The population curve = 100% of population values
2. The area under the population curve = 100% (percent) or 1 (fraction)
3. The % area under the distribution curve between any two points = probability of randomly selecting

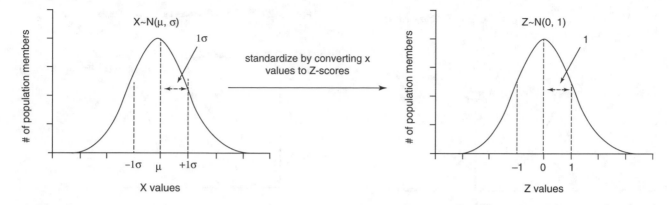

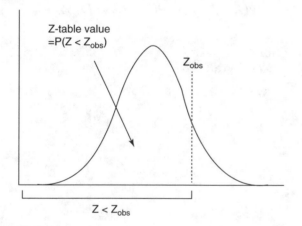

4.8 **figure** Transformation of x values to Z-scores

4.9 **figure** Meaning of Z-table values

a value in that range from the population (see below)

Based on these properties of normal populations, statisticians have developed a "Z-table" (www.normaltable.com), which contains the **probability** (P) of getting a Z-score smaller than the observed Z-score if $Z \sim N(1, 0)$ (Fig. 4.9). To use the Z-table:

1. Find your observed x value.
2. Calculate a Z-score (Z_{obs}) from the x value.
3. Find the corresponding table value $= P(Z < Z_{obs})$ if $Z \sim N(1, 0) = P(x < x_{obs})$ if $x \sim N(\mu, \sigma)$.

For ease of use, the Z-table is organized with the first two digits of Z on the left and the third digit (second decimal place: 0.00, 0.01, 0.02, etc.) across the top. Therefore, to find $P(Z < 1.08)$, you would find the row with 1.0 on the left, and then go across that row to the 0.08 column. The steps listed above can be applied to analysis of the sodium data.

Example C2 Calculate the probability of observing sodium values of *less* than 2.783 mg/500 mL (the mean) and *less* than 3.5 mg/500 mL.

Calculate Z-scores for both:

$$\text{When } x = \mu = 2.783, \ z = \frac{x - \mu}{\sigma} = \frac{2.783 - 2.783}{0.492}$$

$$= \frac{0}{0.492} = 0$$

$$\text{When } x = 3.5, \ z = \frac{x - \mu}{\sigma} = \frac{3.5 - 2.783}{0.492} = \frac{0.717}{0.492} = 1.4$$

Find table values for those Z-scores:

$$P(Z < 0) = \text{table value for } Z = 0 \to 0.50 \ (50\%)$$
$$P(Z < 1.46) = \text{table value for } Z = 1.46 \to 0.9279 \ (92.79\%)$$

For the sodium population (normally distributed, $\mu = 2.783$, $\sigma = 0.492$), the probability of observing sodium values < 2.783 and 3.5 is 50% and 92.79%, respectively.

Standardization to Z-scores eliminates the need for a different Z-score table for each population. Although the Z-table contains P values of getting a Z-score smaller than the observed Z-score, you can calculate the P value of getting a Z-score larger than the observed Z-score. The sum of the area $= 1$ (100%). Thus, the P value of getting a Z-score smaller than the observed Z-score plus the P value of getting a Z-score larger than the observed Z-score equals 100%:

$$P(Z > Z_{obs}) + P(Z < Z_{obs}) = 1(100\%) \qquad (4.4)$$

Therefore:

$$P(Z > Z_{obs}) = 1 - P(Z < Z_{obs}) \qquad (4.5)$$

The full expression then becomes:

$$P(Z > Z_{obs}) + P(Z < Z_{obs}) + P(Z = Z_{obs}) = 1(100\%) \qquad (4.6)$$

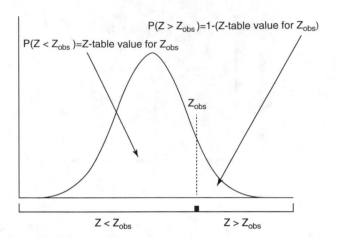

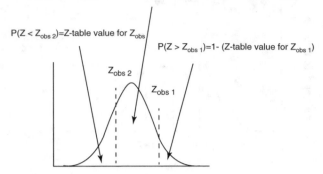

4.10 figure The probability of observing Z smaller than Z_{obs}

4.11 figure The probability of observing Z between two Z_{obs} values

$$P(Z_1 < Z < Z_2) = P(Z < Z_1) - P(Z < Z_2) \quad (4.7)$$

This equation covers all possible Z-scores. Thus, the P value of getting a Z-score larger than the observed Z-score is simply $1 - P$ value of getting a Z-score larger than the observed Z-score. Therefore, to find the P value of getting a Z-score larger than the observed Z-score (Fig. 4.10):

1. From the x value, calculate a Z-score (Z_{obs}).
2. From Z_{obs}, find table value $= P(Z < Z_{obs})$.
3. From $P(Z < Z_{obs})$, calculate $P(Z > Z_{obs}) = 1 -$ table value for $Z_{obs} = P(X > X_{obs})$ if $X \sim N(\mu, \sigma)$.

Example C3 Calculate the probability of observing sodium values of greater than 2.9 mg/500 mL, again following the steps listed previously.

Calculate the Z-score:

$$\text{When } x = 2.9, \ z = \frac{x - \mu}{\sigma} = \frac{2.9 - 2.783}{0.492} = \frac{0.117}{0.492} = 0.24$$

Find the table value for the Z-score:
$P(Z < 0.24) =$ table value for $Z = 0.24 \rightarrow 0.5948$ (59.48 %).

Convert to $P(Z > Z_{obs})$ by subtracting from 1:

$$(Z > Z_{obs}) = 1 - P(Z < Z_{obs}) = 1 - 0.5948$$
$$= 0.4052 = 40.52\%$$

So, for the sodium population, normally distributed with $\mu = 2.783$ and $\sigma = 0.492$, the probability of observing a sodium value > 2.9 is 40.52 %.

How do you calculate the probability of getting an x value (and corresponding Z-score) between two x values (with corresponding Z-scores)? The probability that a Z-score lies between two given Z-scores is simply the difference between the two table values for $P(Z < Z_{obs})$ (i.e., subtract the smaller table value from larger value) (Fig. 4.11):

Example C4 Calculate the probability of observing sodium values between 1.9 and 3.1 mg/500 mL.

Calculate the Z-scores:

$$\text{When } x = 1.9, \ z = \frac{x - m}{s} = \frac{1.9 - 2.783}{0.492} = \frac{-0.883}{0.492} = -1.79$$

$$\text{When } x = 3.1, \ z = \frac{x - \mu}{\sigma} = \frac{3.1 - 2.783}{0.492} = \frac{0.317}{0.492} = 0.64$$

Find the table values for the Z-scores:

$P(Z < 0.64) =$ table value for $Z = 0.64 \rightarrow 0.7389$
$P(Z < 0.64) =$ table value for $Z = -1.79 \rightarrow 0.0376$

Calculate the area between the two Z-scores:

$$P(Z_1 < Z < Z_2) = P(Z < Z_2) - P(Z < Z_1)$$
$$= 0.7389 - 0.0376 = 0.7013 = 70.13\%$$

So, for the sodium population which is normally distributed with $\mu = 2.783$ and $\sigma = 0.492$, the probability of observing a sodium values between 1.9 and 3.1 mg/500 mL is 70.13 %.

4.4 SAMPLE DISTRIBUTIONS

You almost never know the actual population values. This is due to several factors. First, the population is often too large to sample all members. Second, sampling often means that the product is no longer available for use. You typically sample a few members (n) and estimate population parameters from sample parameters. Population parameters are as follows:

$$\text{Population } (x) \text{ mean} = \mu_x \quad (4.8)$$

$$\text{Population}\left(x\right)\text{standard deviation} = \sigma_x \quad (4.9)$$

Sample parameters are as follows:

$$\text{Sample size} = n \quad (4.10)$$

$$\text{Mean of } \bar{x} \text{ distribution} = \mu_{\bar{x}} \approx \mu_x \quad (4.11)$$

$$\text{SD of } \bar{x} \text{ distribution} = \sigma_{\bar{x}} = \frac{\sigma_x}{\sqrt{n}} \quad (4.12)$$

Sample parameters are related to population parameters:

1. Sample mean ≈ population mean
2. Sample standard deviation < population standard deviation

This implies that sampling can be used to estimate population parameters.

> **Example D1** Determine the mean and standard deviation of the sample mean distribution if you sample five or ten members of the population.
>
> For both means:
>
> $$\mu_{\bar{x}} \approx \mu_x = 2.783$$
>
> The standard deviations of the sample means:
>
> $$\sigma_{\bar{x}} = \frac{\sigma_x}{\sqrt{n}} \quad \text{for } n = 5, \quad \sigma_{\bar{x}} = \frac{\sigma_x}{\sqrt{n}} = \frac{0.492}{\sqrt{5}} = 0.220$$
>
> $$\text{for } n = 10, \quad \sigma_{\bar{x}} = \frac{\sigma_x}{\sqrt{n}} = \frac{0.492}{\sqrt{10}} = 0.156$$

You can see that:

1. The sample mean standard deviation < population standard deviation.
2. Larger n size decreases sample mean standard deviation.

Sampling from a normal population results in a normally distributed population of all possible sample means ($\bar{x}$). The population distribution refers to the individual population values of x. The sample mean distribution refers to the values of sample means ($\bar{x}$) generated by sampling x. Because the sample mean is an *average* of the population members:

1. The average sample mean is the same as the average population member.
2. The sample mean is less widely distributed than the population (outliers in the population get diluted by taking a sample *mean*).

The larger the sample size (n), the tighter (SD) the sample mean distribution becomes:

$$\text{SD of } \bar{x} \text{ distribution} = s_{\bar{x}} = \frac{s_x}{\sqrt{n}}$$

$$\text{Therefore as } n \uparrow, s_{\bar{x}} \downarrow$$

Everything you learned about Z-scores for population values applies to sample mean values. You can transform the sample mean distribution to the Z-distribution:

$$\bar{x} \sim N\left(\mu, \frac{\sigma}{\sqrt{n}}\right) \rightarrow Z \sim N(0,1) \quad (4.13)$$

You can transform an observed sample mean into the corresponding Z-score:

$$Z = \frac{\bar{x} - \mu}{\dfrac{\sigma}{\sqrt{n}}} \quad (4.14)$$

You can use the Z-table to calculate the probability of getting a sample mean smaller or greater than an observed mean or between two observed means (same math as before).

> **Example D2** How likely would it be to have a sample mean of greater than 3.0 mg/500 mL if you sampled four bottles?
>
> Transform the $\bar{x}$ into a Z-score:
>
> $$Z = \frac{\bar{x} - \mu}{\dfrac{\sigma}{\sqrt{n}}} = \frac{3 - 2.783}{\dfrac{0.492}{\sqrt{4}}} = \frac{0.217}{0.246} = 0.88$$
>
> Find the table value for the Z-score (area to the left of Z_{obs}):
>
> $$P(Z < 0.88) = \text{table value for } Z = 0.88: 0.8106$$
>
> Find $P(Z > Z_{\text{obs}})$:
>
> $$P(Z > 0.88) = 1 - P(Z < 0.88) = 1 - 0.8106 = 0.1894$$
> $$= 18.94\%$$
>
> So, for the sodium data, the probability of sampling $n = 4$ bottles and getting a mean sodium content of >3.0 mg/500 mL is 18.9%.

4.5 CONFIDENCE INTERVALS

How confident are you that the observed sample mean is close to the actual population mean? And how close? You can calculate the probability of getting a sample mean as extreme as the one you observe, *if* you assume the population is normally distributed and the sample mean equals the population mean.

Procedure

1. Calculate Z-score for $\bar{x}$
2. Find area between $+Z$ and $-Z = P(Z < |Z_{\text{obs}}|) = C$

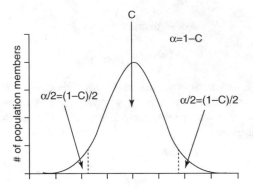

4.12 **figure** Probability of getting a sample mean as extreme as the observed mean

The area (C) is the probability of getting a sample mean as extreme as the one you observed (Fig. 4.12). You can calculate the probability of getting a sample mean more extreme than the one you observed (α) on either end, or getting a sample mean more extreme on a single end ($\alpha/2$).

$$C = P(-Z_{obs} < Z < +Z_{obs})$$
$$= P(Z < +Z_{obs}) - P(Z < -Z_{obs}) \quad (4.15)$$

$$\alpha = 1 - C \quad (4.16)$$

$$\frac{\alpha}{2} = \frac{1-C}{2} \quad (4.17)$$

Example E1 Suppose you sample $n=3$ bottles and get a mean of 2.4 mg/500 mL. Calculate the probabilities of getting a sample mean this extreme, a sample mean smaller than 1.8 mg/500 mL, and a sample mean larger than 1.8 mg/500 mL.

Calculate the Z-score for the observed population mean:

$$Z_{obs} = \frac{\bar{x} - \mu}{\frac{\sigma}{\sqrt{n}}} = \frac{2.4 - 2.783}{\frac{0.492}{\sqrt{3}}} = \frac{-0.383}{0.284} = -1.35$$

Find the $+Z$ and $-Z$ values:

$Z = -1.35$, so you are interested in the range between $Z = 1.35$ and -1.35

Find $C = P(-Z_{obs} < Z < +Z_{obs})$ from the table:

$$C = P(-Z_{obs} < Z < +Z_{obs})$$
$$= P(Z < +Z_{obs}) - P(Z < -Z_{obs})$$

$$C = P(Z < +Z_{obs}) - P(Z < -Z_{obs})$$
$$= table\,value(+Z_{obs}) - table\,value(-Z_{obs})$$

$$C = 0.9115 - 0.0885 = 0.823 = 82.3\%$$

For the sodium population which is $N(2.783, 0.492)$, there is an 82.3% chance that you would observe a sample mean within 1.35 standard deviations ($-1.35 < Z < +1.35$) on either side of the mean if you sampled three bottles from the population.

Now, find α:

$$\alpha = 1 - C = 1 - 0.823 = 0.177 = 17.7\%$$

For the sodium population which is $N(2.783, 0.492)$, there is a 17.7% chance that you would observe a sample mean > 1.35 standard deviations ($Z < -1.35$ or $Z > +1.35$, i.e., $Z > |1.35|$) on either side of the mean if you sampled three bottles from the population.

Now, find $\alpha/2$:

$$\frac{\alpha}{2} = \frac{1-C}{2} = \frac{1 - 0.823}{2} = 0.0885 = 8.85\%$$

For the sodium population which is $N(2.783, 0.492)$, there is an 8.85% chance that you would observe a sample mean > 1.35 standard deviations ($Z > +1.35$) above the mean if you sampled three bottles from the population.

Usually you do not know the actual population mean (μ) or standard deviation (σ). You sample and calculate sample mean (observed $\bar{x}$) and sample standard deviation (sample SD=$\sigma\bar{x}$) to estimate these parameters (observed $\bar{x} \approx \mu\bar{x} = \mu x$ and SD $\approx \sigma\bar{x} = \sigma x/\sqrt{n}$). It is highly unlikely that the observed sample mean is *exactly* equal to the population mean. You can calculate a range (i.e., an "interval") around our observed sample that is likely to cover the true population mean with a given level of statistical **confidence**. The **confidence interval** (CI) gives a margin of error around our sample mean:

$$CI : \bar{x} \pm margin\ of\ error \quad (4.18)$$

Suppose that you want to generate a CI that is within a specified number of standard deviations (i.e., number of Z-scores) from the mean. Using the Z-score formula:

$$Z = \frac{\bar{x} - \mu}{\frac{\sigma}{\sqrt{n}}} \rightarrow Z\left(\frac{\sigma}{\sqrt{n}}\right) = \bar{x} - \mu$$

Since it could be on either side of the mean, change to:

$$\pm Z\left(\frac{\sigma}{\sqrt{n}}\right) = \bar{x} - \mu$$

Now, you have to choose how wide the sample margin of error should be. You do this by selecting the number of standard deviations you desire to be within on each side, which gives a maximum absolute value of Z. You could decide that you want our CI to cover the true mean with

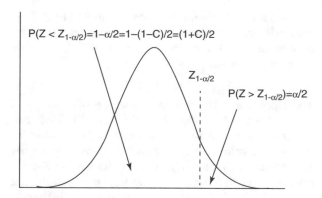

$$P(Z < Z_{1-\alpha/2}) = 1-\alpha/2 = 1-(1-C)/2 = (1+C)/2$$

$Z_{1-\alpha/2}$

$$P(Z > Z_{1-\alpha/2}) = \alpha/2$$

4.13 figure Relationship between C, $\alpha/2$, and Z-table values

a given percent confidence (C). This means that the probability that the CI does *not* cover the true mean would be $1-C$ (which was defined previously as α):

$$\alpha = 1-C \rightarrow C = 1-\alpha$$

Then, you would find the "critical" Z-scores where:

1. $P(Z_{crit} < Z < Z_{crit}) = C$ (the desired confidence level)
2. $P(Z > |Z_{crit}|) = \alpha = 1-C$ (the probability that the CI does *not* cover the true mean)

The easiest way to do this is to find $\alpha/2$ and then find the corresponding Z-score:

First, determine $\alpha/2$ from the desired confidence (C):

$$\frac{\alpha}{2} = \frac{1-C}{2} = \frac{1-0.823}{2} = 0.0885 = 8.85\%$$

$\alpha/2$ is an area to the *right* of the Z_{crit} needed to determine the area to the left of Z_{crit} (aka $Z_{1-\alpha/2}$), i.e., $P(Z < Z_{1-\alpha/2})$ (Fig. 4.13):

$$P(Z < Z_{crit}) = P\left(Z < Z_{1-\frac{\alpha}{2}}\right) = 1 - \frac{\alpha}{2} = 1 - \frac{1-C}{2} = \frac{1+C}{2}$$

$P(Z < Z_{1-\alpha/2})$ is the table value for $Z_{1-\alpha/2}$. Use this to find $Z_{1-\alpha/2}$.

Now you can calculate the "margin of error" for the CI:

$$\text{Margin of error}: \pm Z_{1-\frac{\alpha}{2}} x \frac{\sigma}{\sqrt{n}} \quad (4.19)$$

The confidence interval, with confidence level C, is:

$$\text{CI}: \bar{x} \pm \text{margin of error} \rightarrow \text{CI}: \bar{x} \pm Z_{1-\frac{\alpha}{2}} x \left(\frac{\sigma}{\sqrt{n}}\right) \quad (4.20)$$

Do not get flustered by the statistics details. You just need to be able to do the following:

1. Determine the desired C.
2. Calculate $\alpha/2$.
3. Calculate $1-\alpha/2$.
4. Use this value to find $Z_{1-\alpha/2}$.
5. Calculate the CI.
6. Get from $C \rightarrow \alpha/2 \rightarrow 1-\alpha/2 \rightarrow Z_{1-\alpha/2}$.
7. Plug these values into the formula and calculate the CI.

Typically, you use 90, 95, or 99% confidence in food analysis. We can choose any desired level of confidence that we want. However

Example E2 Suppose you sample $n=5$ cans of soda, and the mean caffeine content is 150 mg/can. You know that the population has a standard deviation (σ) of 15 mg caffeine/can. What is the 95% confidence interval for the mean caffeine content per can?

Calculate $\alpha/2$: $\dfrac{\alpha}{2} = \dfrac{1-C}{2} = \dfrac{1-0.95}{2} = \dfrac{0.05}{2} = 0.025$

Calculate $1-\alpha/2$:

$$P\left(Z < Z_{1-\frac{\alpha}{2}}\right) = 1 - \frac{\alpha}{2} = 1 - 0.025 = 0.975$$

$1-\alpha/2 = P(Z < Z_{1-\alpha/2}) = $ table value for $Z_{1-\alpha/2}$... use this value to find $Z_{1-\alpha/2}$:

$$\text{Table value} = 0.975 \rightarrow Z_{1-\frac{\alpha}{2}} = 1.96$$

Calculate the CI: $\bar{x} \pm Z_{1-\frac{\alpha}{2}} x \dfrac{\sigma}{\sqrt{n}}$

$$150\,\text{mg/can} \pm 1.96\, x \frac{15\,\text{mg/can}}{\sqrt{5}} \rightarrow$$

$$150\,\text{mg/can} \pm 13.1\,\text{mg/can}$$

Lower and upper limits $= 150\,\text{mg/can} \pm 13.1\,\text{mg/can}$
$$= 136.9 \text{ and } 163.1 \text{ mg/can}$$

Therefore, you estimate that the true population mean is somewhere between 136.9 and 163.1 mg caffeine/can with 95% confidence.

1. The more confidence you want, the broader the interval gets.
2. The less confidence you are willing to accept, the narrower the range gets.

Example E3 For the soda data, calculate the 90% CI of the sample mean.

Calculate $\alpha/2$: $\dfrac{\alpha}{2} = \dfrac{1-C}{2} = \dfrac{1-0.90}{2} = \dfrac{0.10}{2} = 0.05$

Calculate $1-\alpha/2$: $1 - \dfrac{\alpha}{2} = 1 - 0.05 = 0.95$

$$1-\alpha/2 = P(Z < Z_{1-\alpha/2}) = \text{table value for } Z_{1-\alpha/2}.$$

Use this value to find $Z_{1-\alpha/2}$:

$$\text{Table value} = 0.95 \rightarrow Z_{1-\frac{\alpha}{2}} = 1.645$$

Calculate the CI: $\bar{x} \pm Z_{1-\frac{\alpha}{2}} x \dfrac{SD}{\sqrt{n}}$

$$150\,\text{mg}/\text{can} \pm 1.645 x \frac{15\,\text{mg}/\text{can}}{\sqrt{5}} \rightarrow$$

$$150\,\text{mg}/\text{can} \pm 11.0\,\text{mg}/\text{can}$$

$$\text{Lower and upper limits} = 150\,\text{mg}/\text{can} \pm 11.0\,\text{mg}/\text{can}$$
$$= 139.0 \text{ and } 161.0\,\text{mg}/\text{can}$$

Therefore, you estimate that the true population mean is between 139.0 and 161.0 mg/can with 90% confidence. You see that our interval is tighter, but you are therefore less confident that it contains the true population mean.

4.6   *t*-SCORES

Usually you do not have large enough n and do not know σ, which are requirements for using the Z-distribution. Therefore, statisticians developed the *t*-score:

$$t = \frac{\bar{x} - \mu}{\dfrac{SD}{\sqrt{n}}} \tag{4.21}$$

Recall that the sample standard deviation (SD) is calculated as follows:

$$SD_n = \sqrt{\frac{\sum (x_i - \bar{x})^2}{n}} \text{ or } SD_{n-1} = \sqrt{\frac{\sum (x_i - \bar{x})^2}{n-1}} \tag{4.22 and 4.23}$$

You should use the "$n-1$" formula for SD when $n < 25$–30. The *t*-score is similar to the Z-score. However, there are some critical differences. The *t*-score:

1. Uses the SD of the sample (known) instead of the population σ (usually unknown)
2. Is more conservative than Z-distribution (for a given mean and SD, t is larger than Z)
3. Is typically only presented for a few selected values of $\alpha/2$ (typically 0.1, 0.05, 0.025, 0.01, and 0.005, which correspond to the commonly used confidence values of 80, 90, 95, 98, and 99%, respectively)
4. Is only listed in positive values, not negative values

Typically, you should use the *t*-distribution (and sample SD) for $n < 25$–30, and the Z-distribution and sample SD for $n > 25$–30. The *t*-distribution is presented

in a table (www.normaltable.com), similar to Z. One major difference between t and Z is that the t is based on a value called **degrees of freedom** (df). This value is based on the sample size:

$$df = n - 1 \tag{4.24}$$

Therefore, for each sample size value (n), there is a unique *t*-score. Note that the *t*-table is divided into columns for each $\alpha/2$ values and, within those columns, for each df value. Also, note that the *t*-table gives the area to the *right* of t (Fig. 4.14), whereas the Z-table gives the area to the *left* of Z. Although this is confusing at first, it actually makes t easier to use for CIs than Z. First, determine $\alpha/2$ from the desired confidence (C):

$$\frac{\alpha}{2} = \frac{1-C}{2}$$

This is the table value for t. From the table value, find $t_{\alpha/2}$, df = $n - 1$. Then, use this value to calculate the CI:

$$\text{CI} : \bar{x} \pm t_{\frac{\alpha}{2},\,df=n-1} x \frac{SD}{\sqrt{n}} \tag{4.25}$$

In a food analysis course, *t*-scores (as opposed to Z-scores) are used almost exclusively, and they have practical use in the food industry. Because t is more conservative for Z, the same confidence level will give a wider interval if calculated using t.

Example F1 Suppose that you sample $n = 7$ energy bars and obtain a mean total carbohydrate content of 47.2% with a sample standard deviation of 3.22%. Calculate the 99% confidence interval.

Use the *t*-distribution because n is relatively small.

Determine $\alpha/2$: $\dfrac{\alpha}{2} = \dfrac{1-C}{2} = \dfrac{1-0.99}{2} = \dfrac{0.01}{2} = 0.005$

Therefore, you would look in the 0.005 column on the *t*-table.

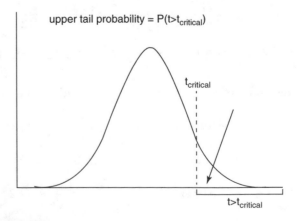

upper tail probability = P(t>t_critical)

t_critical

t>t_critical

4.14 figure   Interpretation of *t*-table values

Then, find $t_{\alpha/2}$, df $= n - 1$: df $= n-1=6-1=5$

Therefore, within the 0.005 column on the t-table, you will look at the row for df $= 5$:

$$\frac{\alpha}{2}=0.005 \text{ and df}=5 \;\rightarrow\; t_{0.005,\,df=5}=4.032$$

Calculate the CI:

$$\text{CI}:\bar{x} \pm t_{\frac{a}{2},df=n-1} \; x \frac{\text{SD}}{\sqrt{n}} \rightarrow 47.2\% \pm 4.032\, x \frac{3.22\%}{\sqrt{7}}$$

$$\rightarrow 47.2\% \pm 4.91$$

$$\text{CI}:\bar{x} \pm\; t_{\frac{a}{2},df=n-1} \; x \frac{\text{SD}}{\sqrt{n}} \rightarrow 47.2\% \pm 4.032\, x \frac{3.22\%}{\sqrt{7}}$$

$$\rightarrow 47.2\% \pm 4.91$$

Upper limit $= 47.2\% + 4.91\% = 52.1\%$

Lower limit $= 47.2\% - 4.91\% = 42.3\%$

Therefore, you estimate that the true population mean is between 42.3 and 52.5% with 99% confidence.

The take-home message is that a sample mean is a **single point estimate** that is probably not exactly correct. Therefore, a single point estimate is almost useless in food analysis because it is almost always incorrect. So, why even sample and test if it is almost always wrong?

Example F2 Suppose your company's specification says that the cheese powder to be used in your products must contain 17.0% protein by weight. You measure $n = 10$ bags of the cheese powder and calculate a mean of 16.8% by weight and a SD of 0.3%. The observed sample mean does not equal the specification value (and it is unlikely to ever fall exactly on 17.0%). Do you reject the lot because it is not exactly 17.0%? Do you keep sampling until the observed mean is 17.0%? However, suppose your company specifies that the raw ingredient must meet the specification, from $n = 10$ sampled units, using a 95% CI.

This CI is:

$$\bar{x} \pm\; t_{\frac{a}{2},df=n-1} \; x \frac{\text{SD}}{\sqrt{n}} \;\rightarrow\; 16.8\% \pm\; t_{0.025,df=4}\, x \frac{0.3\%}{\sqrt{10}}$$

$$\rightarrow\; 16.8\% \pm 2.776\, x \frac{0.3\%}{\sqrt{10}}$$

$$\text{CI}: 16.8\% \pm 0.263\% \quad \text{or} \quad 16.537\% - 17.063\%$$

Therefore, although the point estimate does not equal the specification value, you are 95% confident that the true mean lies somewhere between 16.537% and 17.063%, which contains the specification value (17.05). Therefore, based on this, you would accept the raw material.

The Example F2 demonstrates the utility of a CI. It is a range of values that have a specified likelihood of containing the true population mean, whereas the point estimate has a very small likelihood of being the true mean. You may wonder what use a CI is if it is a range of values. However, you can often get very "tight" CIs (i.e., a narrow range that is useful) by random sampling of a population and choosing an appropriate level of confidence. As discussed in the textbook Chap 4, you can use the concept of a CI to determine how large the n size should be to obtain a CI of a specific width. In addition to CIs, another technique that is often used to express the observed sample mean as a range of values is to calculate the sample mean $\pm$ SD or the mean $\pm$ SEM. The SEM, or **standard error of the mean**, is simply:

$$\text{SEM} = \frac{\text{SD}}{\sqrt{n}} \tag{4.26}$$

You will notice that the SEM term appears in the t-test and CI formulae Eqs. 4.21 and 4.25. When you express the observed sample mean $\pm$ SD or SEM, you are not giving a probability-based estimate (that would require the t-term multiplier used for t-tests and CIs) but rather simply presenting the data as an estimate of the mean based on the sample variability. The range calculated based on SEM will always be narrower, and hence less conservative, than the t-test and CI. This may not be true for the range calculated based on SD, depending on the sample size. For the cheese powder data (Example F2), you could express the estimate as:

$$\text{Mean} \pm \text{SD} = 16.8\% \pm 0.3\% = 16.5 - 17.1\%$$

$$\text{Mean} \pm \text{SEM} = 16.8\% \pm \frac{0.3\%}{\sqrt{10}} = 16.8\% \pm 0.0949$$

$$= 16.705 - 16.895\%$$

Therefore, based on these ranges, we would accept the raw material based on mean $\pm$ SD but reject it based on mean $\pm$ SEM. Therefore, it is very important to know how to use your analytical data once it is obtained. What are the specific rules you will use to evaluate data and make decisions? Typically, these are specified by company lab quality assurance/quality control manuals, industry standards, or regulatory agencies.

4.7 t-TESTS

Sometimes you need to determine if the observed sample mean indicates that the population is the same, or different from, a chosen value. You can use a procedure called a t-test if you have the sample mean ($\bar{x}$), standard deviation (SD), sample sizes (n), and a desired population mean (μ) that you want to compare the same mean to. First, determine $\alpha/2$ from the desired confidence (C) and the df value, as before:

$$\frac{\alpha}{2} = \frac{1-C}{2} \text{ and df} = n-1$$

From these values, find the table value $t_{\alpha/2,\,df=n-1}$. This is the "critical value" for a t-test, labeled as "$t_{critical}$." Then, calculate the observed t-score ("t_{obs}") for the sample mean, SD and n:

$$t = \frac{\bar{x} - \mu}{\frac{SD}{\sqrt{n}}}$$

Then, the $t_{critical}$ and t_{obs} values are compared.

$$If \; |t_{obs}| > t_{\frac{\alpha}{2},df=n-1} \rightarrow \; \text{sample mean is significantly}$$
$$\text{different from } \mu \text{ with confidence}(C)$$

$$If \; |t_{obs}| < t_{\frac{\alpha}{2},df=n-1} \rightarrow \; \text{sample mean is not significantly}$$
$$\text{different from } \mu \text{ with confidence}(C)$$

A few notes are worth mentioning:

1. The more confidence you want that you have the correct answer, the larger $t_{critcal}$ becomes, and the larger t_{obs} must be to provide evidence that the sample mean is significantly different from μ.
2. The absolute value of t_{obs} is compared vs. $t_{critical}$, as the table only lists positive t-scores.
3. You select μ to compare to the sample mean. You typically do not know the population mean, but you may have a "target value" that you are trying to reach or think the population should have, so you use that as μ.

Example G1 Suppose that your company makes a multivitamin with a label value of 35 mg vitamin E/capsule. Per FDA requirements, the label value needs to be within a certain amount of the actual value. You sample eight capsules and measure the vitamin E content. The observed sample mean is 31.7 mg/capsule, and the sample standard deviation is 3.1 mg/capsule. Can you say that the sample mean includes the label value with 99% confidence?

Determine $\alpha/2$ and df:

$$\frac{\alpha}{2} = \frac{1-C}{2} = \frac{1-0.99}{2} = \frac{0.01}{2} = 0.005 \quad \text{and}$$
$$df = n - 1 = 8 - 1 = 7$$

From these values, find the $t_{critical}$ value ($t_{\alpha/2,\,df=n-1}$):
$$t_{0.005,df=7} = 3.499$$

Calculate t_{obs} from the sample data:

$$t_{obs} = \frac{\bar{x} - \mu}{\frac{SD}{\sqrt{n}}} = \frac{31.7 - 35}{\frac{3.1}{\sqrt{8}}} = -3.01 \quad \text{and} \quad |t_{obs}| = 3.01$$

Since $|t_{obs}|$ (3.01) $< t_{\alpha/2,\,df=n-1}$ (3.499), you can say the sample mean is not significantly different from the label value (35 mg/capsule) with 99% confi-

dence. If $|t_{obs}|$ had been $> t_{\alpha/2,\,df=n-1}$ (3.499), you would say that you have evidence the sample mean was significantly different from 35 mg/capsule. Note here that you are concerned with the absolute value of t_{obs} ($|t_{obs}|$) vs. $t_{critical}$. Given your chosen confidence, you have a $t_{critical}$ value of 3.499, so as long as your $|t_{obs}|$ is <3.499 you can conclude that the observed sample mean is not significantly different from the chosen μ value. Therefore, t_{obs} could be anywhere on the interval (−3.499, +3.499) and still be considered to be not significantly different from the label value. So, the observed mean could be above or below the chosen μ, as long as it is not "too far" in either direction.

Confidence intervals and t-tests provide the same information for a chosen confidence level. Either procedure may be used to compare the sample mean and chosen μ value. Sometimes you need to determine if the means of two samples are "different" (as opposed to one sample and a chosen μ). Sample means are "point estimates." Just because they are not exactly the same does not mean that the populations are significantly different. You need a statistical basis for determining if the sample means are far enough apart such that you can say they are different with some level of confidence (or not). Given two sample means ($\bar{x}_1$ and $\bar{x}_2$), standard deviations (SD_1 and SD_2), and sample sizes (n_1 and n_2), first we decide on a confidence level (C), and we calculate $\alpha/2$ as before. Then, we determine df. For two sample means, df is calculated as follows:

$$df = n_1 + n_2 - 2 \tag{4.27}$$

Then you find the critical t-value as before, using the $\alpha/2$ and df values:

$$t_{\frac{\alpha}{2},df=n_1+n_2-2}$$

Once you have $t_{critical}$, you calculate t_{obs} as follows:

$$t_{obs} = \frac{|\bar{x}_1 - \bar{x}_2|}{\sqrt{s_p^2\left(\frac{1}{n_1} + \frac{1}{n_2}\right)}} \tag{4.28}$$

This formula differs from the one-sample t-test in that both n_1 and n_2 are used, and you employ a value known as the **pooled variance** (s_p^2) instead of the sample SD. The pooled variance is a weighted average of two sample SD values. The pooled variance is calculated as follows:

$$\text{Pooled variance} = s_p^2 = \frac{(n_1-1)\,SD_1^2 + (n_2-1)\,SD_2^2}{n_1 + n_2 - 2} \tag{4.29}$$

Variance is simply standard deviation squared. Therefore, although you will not use this value, it may be useful to understand that the "pooled standard deviation" (s_p), or the weighted average of the two sample SD values, is simply the square root of the pooled variance:

$$\text{Pooled standard deviation} = \sqrt{\text{pooled variance}}$$

$$= \sqrt{s_p^2} = \sqrt{\frac{(n_1-1)\text{SD}_1^2 + (n_2-1)\text{SD}_2^2}{n_1+n_2-2}} \qquad (4.30)$$

Once you have obtained the t_{critical} and t_{obs} values, you compare them as for the one-sample t-test:

$$t_{\text{obs}} > t_{\frac{\alpha}{2}, \text{df}=n_1+n_2-2} \rightarrow \text{means } are \text{ significantly different}$$
$$\text{with specified confidence}(C)$$

$$t_{\text{obs}} < t_{\frac{\alpha}{2}, \text{df}=n_1+n_2-2} \rightarrow \text{means } are\ not \text{ significantly different}$$
$$\text{with specified confidence}(C)$$

Example G2 Suppose your company is concerned that two production lines are producing spaghetti sauces with different acid contents. You measure the titratable acidity in samples from both lines (Line 1, 2.1; 2.0; 2.1; 2.2; 2.3; and 2.4%; Line 2, 2.7; 2.3; 2.2; 2.2; 2.4; 2.6; and 2.5%). Do the two production lines appear to be producing significantly different acidity levels with 90% confidence?

First, calculate the n, $\bar{x}$, and SD value for each sample:
  Line 1: $n=6$, $\bar{x}=2.183$ and SD$=0.147$
  Line 2: $n=7$, $\bar{x}=2.4$, and SD$=0.195$

What is the confidence level (C): 90% or 0.90?

Next, calculate $\alpha/2$: $\dfrac{\alpha}{2} = \dfrac{1-C}{2} = \dfrac{1-0.90}{2} = \dfrac{0.1}{2} = 0.05$

For the t-table, df: df $= n_1 + n_2 - 2 = 6 + 7 - 2 = 11$

Find the critical t-value on the table:
$$t_{\frac{\alpha}{2}, \text{df}=n_1+n_2-2} = t_{0.05,11} = 1.796$$

Calculate a pooled variance (s_p^2):

$$s_p^2 = \frac{(n_1-1)\text{SD}_1^2 + (n_2-1)\text{SD}_2^2}{n_1+n_2-2}$$

$$= \frac{(6-1)(0.147)^2 + (7-1)(0.195)^2}{6+7-2}$$

$$= \frac{0.1083 + 0.2286}{11} = 0.03063$$

Calculate t_{obs}:

$$t_{\text{obs}} = \frac{|\bar{x}_1 - \bar{x}_2|}{\sqrt{s_p^2\left(\dfrac{1}{n_1} + \dfrac{1}{n_2}\right)}} = \frac{|2.183 - 2.4|}{\sqrt{0.03063\left(\dfrac{1}{6} + \dfrac{1}{7}\right)}}$$

$$= \frac{|-0.217|}{0.09737} = 2.23$$

Decision:

$$t_{\text{obs}} = 2.23 \quad \text{and}$$

$$t_{\frac{\alpha}{2}, \text{df}=n_1+n_2-2} = 1.796 \rightarrow t_{\text{obs}} > t_{\frac{\alpha}{2}, \text{df}=n_1+n_2-2}$$

Thus, Lines 1 and 2 produce sauces with significantly different acidities (with 90% confidence).

4.8 PRACTICAL CONSIDERATIONS

Here are some practical considerations to help you use statistics in a food analysis course and in a career in which decisions are based on analytical data:

4.8.1 Sample Size

1. A larger sample size enables more accurate approximation of the true population values.
2. Larger sample sizes decrease the sample SD.
3. Use t for $n < 25$–30; use Z for $n > 25$–30.
4. Sample size determines df, which determines t_{critical} for a specified $\alpha/2$ value.
5. Sampling more units is often useful but not always practical or cost-effective.
6. A "happy medium" needs to be reached that provides "adequate" approximation of the mean with a fixed amount of confidence. This is done by calculating the minimum needed sample size (this is covered in the textbook chapter on sampling).

4.8.2 Confidence

1. Confidence values are chosen based on the acceptable risk to consumers (safety, poor quality, etc.), the company (quality, profit margin, formulation accuracy, etc.), or government regulatory agencies (e.g., labeling accuracy).
2. Confidence determines $\alpha/2$, which determines t_{critical} within a specified df value.
3. The more confidence is required (or desired):

(a) The wider a CI will be.
(b) The larger t_{critical} will be.
(c) The larger t_{obs} must be to be $> t_{\text{critical}}$.
(d) The larger the n size must be to show significance.

4. Increasing desired confidence increases the statistical burden required to shown significant differences; 90–99% confidence is typically used (95% is fairly standard).
5. Increasing the desired confidence is not always practical or cost-effective.

4.8.3 What Test to Use?

Table 4.1 gives "rules of thumb" for what types of tests to use for various scenarios:

4.1 table	Examples of when specific statistical tools should be used for food analysis	

Question	Test/calculation	Examples
Is an observed sample mean statistically similar to or different from a chosen value?	One-sample t-test	Is the actual value in agreement with the label value? Does the permitted level differ from the allowed level? Does the raw material meet company specifications? Is the composition of the finished product acceptable based on company specifications? Is a pipette delivering the desired volume? Is a packaging machine filling packages to the desired level?
What is the actual population value, based on the sample?	One-sample CI	What is the actual concentration of a compound of interest?
Based on samples from two populations, are these populations different?	Two-sample t-test ($\mu = 0$)	Are two lots different? Are two lines, plants, etc. producing products with different composition? Is your product the same or different from a competitor's product?

4.9   PRACTICE PROBLEMS

(*Note: Answers are at end of laboratory manual.*)

1. As QA manager for a canned soup manufacturer, you need to make sure that chicken noodle soup has the sodium content indicated on the label (343 mg/cup). You sample six cans of soup and measure the sodium content: 322.8, 320.7, 339.1, 340.9, 319.2, and 324.4 mg/cup. What is the mean of the observations (mg/cup)? What is the standard deviation of the observations (mg/cup)? Calculate the 96 % confidence interval for the true population mean, and determine both the upper and lower limits of the confidence interval. Determine if the sample mean provides strong enough evidence that the population is "out of spec" with 99 % confidence.

2. You perform moisture analysis on sweetened condensed milk using a forced-draft oven. The following data are obtained. Lot A: 86.7, 86.2, 87.9, 86.3, and 87.8 % solids; Lot B: 89.1, 88.9, 89.3, 88.8, and 89.0 %. Determine if the lots are statistically different with 95 % confidence

4.10   TERMS AND SYMBOLS

Confidence (C) Statistical probability of being correct based on chance alone.

Confidence interval (CI) A range of values (based on a point estimate plus a statistically determined margin of error) used to predict information about a population.

Degrees of freedom (df) The number of values that is free to vary independently.

Distribution All values in a population and the relative or absolute occurrence of each value.

Histogram A plot of all values in a population and the relative occurrence of each value.

Mean (μ) Average value.

Normal (N) A population distribution whose shape is defined by a mean and standard deviation.

Pooled variance (sp2) A measure of variability for two individual samples, incorporating both the sample sizes and sample standard deviations of both.

Population All individuals of interest.

Population mean (μx) The average value of all the members of a population.

Population standard deviation (σ) The average difference between the value of each population member and the population mean.

Probability (P) Statistical likelihood of an event or state occurring by chance alone.

Sample Selected members of a population, from which inferences are made about the entire population.

Sample mean (μx̄) The average value of all the members of a sample.

Sample standard deviation (SD, σx̄) The average difference between the value of each sample member and the sample mean.

Single point estimate A single value (mean or single data point) used to predict information about a population.

Standard deviation (σ) The average difference between the value of each population member and the population mean.

Standard error of the mean (SEM) A measure of variability of a sample, incorporating the sample standard deviation and the sample size.

T-score (t) A normalized value describing how many standard deviations a value is from the mean but more conservative than the Z-score.

Z-score (Z) A normalized value describing how many standard deviations a value is from the mean.

2
part

Laboratory Exercises

Nutrition Labeling Using a Computer Program

Lloyd E. Metzger (✉)
Department of Dairy Science, South Dakota State University,
Brookings, SD, USA
e-mail: lloyd.metzger@sdstate.edu

Ann M. Roland
Owl Software,
Columbia, MO, USA
e-mail: aroland@owlsoft.com

S.S. Nielsen, *Food Analysis Laboratory Manual*, Food Science Text Series,
DOI 10.1007/978-3-319-44127-6_5, © Springer International Publishing 2017

5.1 INTRODUCTION

5.1.1 Background

The 1990 Nutrition Labeling and Education Act mandated nutritional labeling of most foods. As a result, a large portion of food analysis is performed for nutritional labeling purposes. Interpretation of these regulations (21 CFR 101.9) and the appropriate usage of rounding rules, available nutrient content claims, reference amounts, and serving size can be challenging. Additionally, during the product development process, the effect of formulation changes on the nutrition label may be important. As an example, a small change in the amount of an ingredient may determine if a product can be labeled low fat. As a result, the ability to immediately approximate how a formulation change will impact the nutritional label can be valuable. In some cases, the opposite situation may occur and a concept called reverse engineering is used. In reverse engineering the information from the nutrition label is used to determine a formula for the product. Caution must be used during reverse engineering. In most cases, only an approximate formula can be obtained and additional information not provided by the nutrition label may be necessary.

The use of nutrient databases and computer programs designed for preparing and analyzing nutrition labels can be valuable in all of the situations described earlier. In this laboratory you will use a computer program to prepare a nutrition label from a product formula, determine how changes in the formula affect the nutrition label, and observe an example of reverse engineering.

5.1.2 Reading Assignment

Metzger, L.E. and Nielsen, S.S., 2017. Nutrition labeling. Ch. 3, in *Food Analysis*, 5th ed. S.S. Nielsen (Ed.), Springer, New York.

5.1.3 Objective

Prepare a nutrition label for a yogurt formula, determine how formulation changes will affect the nutrition label, and observe an example of reverse engineering.

5.1.4 Materials

TechWizard™ – Formulation and Nutrition Labeling Software for Microsoft Excel for Windows.

Owl Software. TechWizard™ Software Manual, Columbia, MO. www.owlsoft.com

5.1.5 Notes

Instructions on how to receive and install the software used for this laboratory are located online at www.owlsoft.com, under the *Academic* link located in the

heading. *It is possible that the TechWizard™ program has been updated since the publication of this laboratory manual and any changes in the procedures described below will also be found on this web page.*

*Install the software prior to the laboratory session to ensure that it works properly with your PC.

5.2 PREPARING NUTRITION LABELS FOR SAMPLE YOGURT FORMULAS

5.2.1 Procedure

1. Start the TechWizard™ program. *(Start the program, the message appears "TechWizard successfully opened the necessary files..." and click OK. Click on the Enter Program button. Select User Name "Administrator" and click OK.)*

2. Enter the Nutrition Labeling section *(From the Labeling tab select Labeling Section. You should see "TechWizard™ – Nutrition Info & Labeling with New Formats" in the top left corner of the screen.)*

3. Enter the ingredients for Formula #1 listed in Table 5.1. *(Click on the Add Ingredients button, then select an ingredient from the ingredient list window, and click the Add button. Repeat for each ingredient. After all the ingredients have been added, close the window.)*

4. Enter the amounts of the ingredients. *(Enter the percentage of each ingredient for Formula #1 in the % (wt/wt) column. Selecting the Sort button above that column and then Sort Descending will sort the ingredients by the % (wt/wt) in the formula.)*

5. Enter the serving size (common household unit and the equivalent metric quantity) and number of servings. *(First click on the Serving Size button under Common Household unit, enter 2/3 in the window, click on OK, and select cup from the units drop down list; next click on the Serving Size button under Equivalent Metric Quantity, enter 170 g in the window, click on OK, and select g from the units*

table   Sample yogurt formulas

	Formula #1 (%)	Formula #2 (%)
Milk (3.7% fat)	38.201	48.201
Milk (skim), no vit. A add	35.706	25.706
Milk (skim), condensed (35% TS)	12.888	12.888
Sweetener, sugar (Liquid)	11.905	11.905
Modified starch	0.800	0.800
Stabilizer, gelatin, dry powder, unsweetened	0.500	0.500

drop down list; and finally click on the Number of Servings button, enter 1 in the window, and click on OK.)

6. Save Formula #1. (*Click on the File tab, then Formula, and then Save Formula As. Type Student Name_Formula Name in the Formula Name window, click the Save button, then OK, and close the window.*)

7. View the nutrition label and select label options. (*Click on the View Label button, click on the Label Options button, and select Standard; select Protein -- Show ADV for voluntary nutrients since yogurt is high in protein, and enter 1 for the PDCAAS; when you have finished selecting the label options, select Apply, then OK, and then Close to view the label.*)

8. Copy and paste the nutrition label for formula #1 to a Microsoft Word file *(Click on the Copy button on the Labeling tab, select Standard Label, Enter 1 for the new label format. Click OK. When prompted that the label has been copied, click OK. Open a Word document, and type the name of the formula and paste the label. Return to the TechWizard™ program.)*

9. Edit the ingredient declarations list (*Click on the View/Edit Declaration button, and click Yes when asked – Do you wish to generate a simple formula declaration using individual ingredient declarations? – Each ingredient used in the formula can be selected in the top window and the ingredient declaration can be edited in the middle window.*)

*Note the rules for ingredient declaration are found in 21 CFR 101.4.

10. Copy and paste the ingredient declaration for Formula #1 to a Microsoft Word file. (Select *Edit Formula declaration, and click the Copy button then OK to the pop-up and close the window. The declara-*

tion can now be pasted into your Microsoft Word Document).

11. Return to the Nutrition Info & Labeling section of the TechWizard™ program. (*Go to the TechWizard™ program and click on the Return button.*)

12. Repeat the process for Formula #2 (Table 5.1). (Repeat Steps 4–10)

5.3 ADDING NEW INGREDIENTS TO A FORMULA AND DETERMINING HOW THEY INFLUENCE THE NUTRITION LABEL

Sometimes it may be necessary to add additional ingredients to a formula. As an example, let us say you decided to add an additional source of calcium to yogurt formula #1. After contacting several suppliers, you decided to add Fieldgate Natural Dairy Calcium 1000, a calcium phosphate product produced by First District Association (Litchfield, MN), to the yogurt formula. This product is a natural dairy-based whey mineral concentrate and contains 25% calcium. You want to determine how much Dairy Calcium 1000 you need to add to have 50 and 100% of the Daily Value (DV) of calcium in one serving of your yogurt. The composition of the Dairy Calcium 1000 you will add is shown in Table 5.2.

5.3.1 Procedure

1. Add the name of the new ingredient to the database. (*From the Edit Ingredient File tab, select Edit Current File; Click on Edit Ingredient File tab again in the Ingredients group; Click Add, type the ingredient name "Dairy Calcium 100_Student Name" in the Enter Ingredient Name box, and click Add. Answer Yes to the question, and click OK. Close the window.*)

2. Enter the new ingredient composition (Table 5.2). (*Select the ingredient name in the column named "Ingredients and Properties." Click Edit Selected in the Ingredients group under the Edit Ingredient File tab, the row will turn blue, and enter the amount of each component/nutrient in the appropriate column.*)

3. Edit the ingredient declaration (which will appear on the ingredient list) for the new ingredient. (*Type "Whey mineral concentrate" in the column named "Default, Spec TEXT, zzzIngredient Declaration."*)

4. Save the changes to the ingredient file. (*Click on the Finish Edit button, and answer Yes to the question. Select Close Ingred. File from the Edit Ingredient File tab.*)

5.2 **table** Composition of Fieldgate Natural Dairy Calcium 1000 (First District Association)

Component	Amount
Ash (%)	75
Calcium (mg/100 g)	25,000
Calories (cal/100 g)	40
Lactose (%)	10
Phosphorus (mg/100 g)	13,000
Protein (g/100 g)	4.0
Sugars (g/100 g)	10
Total carbohydrates (g/100 g)	10
Total solids (%)	92
Water (%)	8.0

5. **Open food analysis formula #1 in the Formula Development Section of the program.** (*Click the Formula Dev and Batching tab, and then Formula Dev. From the File menu, select Open Formula and select your Formula #1 file "Student Name_Formula Name," click on the Open button, then click on No for the first question, and click on Yes for the remaining questions*).

6. Add the new Dairy Calcium 1000 ingredient to your food analysis formula #1. (*Click on the Add Ingredients button, then select your ingredient "Dairy Calcium 1000_Student Name" from the ingredient list, click on the Add button, and close the window.*)

7. Calculate the amount of calcium (mg/100g) required to meet 50 and 100% of the DV (see example below):

$$\text{Calcium required} = \left(\frac{\text{DV for calcium}}{\text{serving size}} \right)$$
$$\times 100\,\text{g} \times \% \text{ of DV required}$$

$$\text{Calcium required for 50\% of the DV} = \left(\frac{1300\,\text{mg}}{170\,\text{g}} \right)$$
$$\times 100\,\text{g} \times 0.50$$

$$\text{Calcium required for 50\% of the DV} = 382 \left(\frac{\text{mg}}{100\,\text{g}} \right)$$

8. Enter the amount of calcium required in the formula and restrict all ingredients in the formula except skim milk and Dairy Calcium 1000. (*Find Calcium in the Property column and enter 382 in the Min and Max columns for Calcium. This lets the program know that you want to have 382 mg of calcium per 100 g. In both the Min and Max columns of the Ingredients, enter 38.201 for Milk (3.7% fat); 12.888 for Milk (skim), condensed (35% TS); 11.905 for Swtnr, sugar (Liquid); 0.800 for Modified starch; and 0.500 for Stblzr, Gelatin, dry powder, unsweetened. This lets the program adjust the amount of skim milk and Dairy Calcium 1000 (calcium phosphate) and keeps the amount of all the other ingredients constant. Click on the Formulate button, click OK.*). Record the % wt/wt for Dairy Calcium to answer Question 2 in Sect. 5.5.

9. Save the modified formula with your name. (*Click on the File tab, then Formula, and then Save Formula As. Type "Student Name_Formula #1 added calcium 50% DV" in the Formula Name window, click the Save button, then OK to close the window.*)

10. Open the new formula into the nutrition labeling section. (*Click on the Labeling tab, select*

Labeling Section, click the File tab then Open Formula, and select your formula "Student Name_ Formula #1 added calcium 50% DV," and click Open then click on No for the first question, and click on Yes for the remaining questions).

11. Make sure you have the correct serving size information (see Sect. 5.2, Step 5).

12. View and print the nutrition label for the new formula with 50% of the calcium DV. Follow the instructions described in Sect. 5.2, Steps 6–10.

13. Produce a formula and label that has 100% of the calcium DV. (*Repeat Steps 8–12 except using the calculated amount of calcium required to meet 100% of the calcium DV. You will have to perform this calculation yourself following the example in Step 7.*)

5.4 AN EXAMPLE OF REVERSE ENGINEERING IN PRODUCT DEVELOPMENT

5.4.1 Procedure

In this example the program will automatically go through the reverse engineering process. (*Start the example by selecting Cultured Products Automated Examples from the Help menu and clicking on example #4. During this example you proceed to the next step by clicking on the Next button.*) Each step is described below:

1. The information from the nutrition label for the product you want to reverse engineer is entered into the program. (*Comment: In this example serving size, calories, calories from fat, total fat, saturated fat, cholesterol, sodium, total carbohydrate, sugars, protein, vitamin A, vitamin C, calcium, and iron are entered.*)

2. The minimum and maximum levels of each nutrient are calculated on a 100-g basis. (*Comment: The program uses the rounding rules to determine the possible range of each nutrient on a 100-g basis.*)

3. The information about nutrient minimum and maximums is transferred into the Formula Development section of the program. (*Comment: The program has now converted nutrient range information into a form it can use during the formulation process.*)

4. Ingredients used in the formula are then selected based on the ingredient declaration statement on the nutrition label. (*Comment: Selecting the right ingredients can be difficult and an extensive understanding of the ingredient declaration rules is necessary. Additionally some of the required ingredients may not be in the database and will need to be added.*)

5.3 table   Recipe for chocolate chip cookies

Ingredients	Amount	Grams
Wheat flour, white, all-purpose, enriched, unbleached	2.25 cups	281.25
Swtnr, sugar, granulated	0.75 cup	
Candies, semisweet chocolate	100 g	100
Sugars, brown	0.75 cup (packed)	
Butter (salted)	1 cup	227
Egg, whole, fresh	2 units	112
Salt	0.75 tsp	

Source for ingredients: TechWizard, USDA ingredients as source
Conversion data source: USDA web page

5. Restrictions on the amount of each ingredient in the formula are imposed whenever possible. *(Comment: This is a critical step that requires knowledge about the typical levels of ingredients used in the product. Additionally, based on the order of ingredients in the ingredient declaration, approximate ranges can be determined. In this example, the amount of modified starch is limited to 0.80%, the amount of gelatin is limited to 0.50%, and the amount of culture is limited to 0.002%.)*

6. The program calculates an approximate formula. *(Comment: The program uses the information on nutrient ranges and composition of the ingredients to calculate the amount of each ingredient in the formula.)*

7. The program compares the nutrition information for the developed formula to the original nutrition label. *(Comment: This information is viewed in the Nutrition Label to Formula Spec section of the program accessed by selecting Reverse Engineering Section from the Reverse Engineering tab. Click the X to close the Cultured Products Example window when finished.)*

5.5   QUESTIONS

1. Based on the labels you produced for yogurt Formula #1 and Formula #2 in Sect. 5.2, what nutrient content claims could be made for each formula (see 21 CFR 101.13, Nutrient content claims – general principles – and 21 CFR 101-54-101.67, Specific requirements for nutrient content claims or summary tables from the FDA or other sources based on these CFR sections)?

2. How much Dairy Calcium 1000 did you have to add to the yogurt formula to have 50 and 100% of the DV of calcium in the formula?

3. If Dairy Calcium 1000 costs $2.50/lb and you are going to have 100% of the DV for calcium in your yogurt, how much extra will you have to charge for a serving of yogurt to cover the cost of this ingredient?

4. Assume you added enough Dairy Calcium 1000 to claim 100% of the DV of calcium, would you expect the added calcium to cause any texture changes in the yogurt?

5. Make a nutrition label using the chocolate chip cookie recipe and other information in Table 5.3 (for simplicity assume 0% loss of water during baking, number of servings=1, serving size=30 g). Conversion factors to get the weight of sugars and salt can be found in the US Department of Agriculture Nutrient Database for Standard Reference website: http://ndb.nal.usda.gov/. *When adding ingredients to the formula, you may have to select the tab "USDA Ingredients as Source" in the Add Ingredients window to find the ingredient you need. Enter the grams of each ingredient in the %(wt/wt) column. Select Normalize from the Labeling tab to convert the values so the total is 100. The ingredient, Candies, semisweet chocolate, does not have Added Sugars data. You can enter a temporary value to the ingredient information to calculate the Nutrition Facts. Click on Composition from the Labeling tab. In the row for that ingredient, scroll to the right until you find the column for Added Sugars with the units g. Enter the value you determine for Added Sugars (hint: look at the Sugars value) and click the Enter key. The value will be used for the label but is not saved in the ingredient file.*

RESOURCE MATERIALS

Metzger LE, Nielsen SS (2017) Nutrition labeling. Ch. 3. In: Nielsen SS edn. Food analysis, 5th edn. Springer, New York
Owl Software TechWizard™ Manual, Columbia, MO. www.owlsoft.com

Chapter 6

Accuracy and Precision Assessment

S. Suzanne Nielsen (✉)
Department of Food Science, Purdue University,
West Lafayette, IN, USA
e-mail: nielsens@purdue.edu

Charles E. Carpenter
Department of Nutrition, Dietetics and Food Sciences, Utah State University,
Logan, UT, USA
e-mail: chuck.carpenter@usu.edu

S.S. Nielsen, *Food Analysis Laboratory Manual*, Food Science Text Series,
DOI 10.1007/978-3-319-44127-6_6, © Springer International Publishing 2017

6.1   INTRODUCTION

6.1.1   Background

Volumetric glassware, mechanical pipettes, and balances are used in many analytical laboratories. If the basic skills in the use of this glassware and equipment are mastered, laboratory exercises are easier and more enjoyable, and the results obtained are more accurate and precise. Measures of accuracy and precision can be calculated based on data generated, given the glassware and equipment used, to evaluate both the skill of the user and the reliability of the instrument and glassware.

Determining mass using an analytical balance is the most basic measurement made in an analytical laboratory. Determining and comparing mass is fundamental to assays such as moisture and fat determination. Accurately weighing reagents is the first step in preparing solutions for use in various assays.

Accuracy and precision of the analytical balance are better than for any other instruments commonly used to make analytical measurements, provided the balance is properly calibrated and the laboratory personnel use proper technique. With proper calibration and technique, accuracy and precision are limited only by the readability of the balance. Repeatedly weighing a standard weight can yield valuable information about the calibration of the balance and the technician's technique.

Once the performance of the analytical balance and the technician using it has been proven acceptable, determination of mass can be used to assess the accuracy and precision of other analytical instruments. All analytical laboratories use volumetric glassware and mechanical pipettes. Mastering their use is necessary to obtain reliable analytical results. To report analytical results from the laboratory in a scientifically justifiable manner, it is necessary to understand accuracy and precision.

A procedure or measurement technique is validated by generating numbers that estimate their accuracy and precision. This laboratory includes assessment of the accuracy and precision of automatic pipettors. An example application is determining the accuracy of automatic pipettors in a research or quality assurance laboratory, to help assess their reliability and determine if repair of the pipettors is necessary. Laboratory personnel should periodically check the pipettors to determine if they accurately dispense the intended volume of water. To do this, water dispensed by the pipettor is weighed, and the weight is converted to a volume measurement using the appropriate density of water based on the temperature of the water. If replicated volume data indicate a problem with the accuracy and/or precision of the pipettor, repair is necessary before the pipettor can be reliably used again.

It is generally required that reported values minimally include the mean, a measure of precision, and the number of replicates (Smith 2017). The number of significant figures used to report the mean reflects the inherent uncertainty of the value, and it needs to be justified based on the largest uncertainty in making the measurements of relative precision of the assay. The mean value is often expressed as part of a confidence interval (CI) to indicate the range within which the true mean is expected to be found. Comparison of the mean value or the CI to a standard or true value is the first approximation of accuracy. A procedure or instrument is generally not deemed inaccurate if the CI overlaps the standard value. Additionally, a CI that is considerably greater than the readability indicates that the technician's technique needs improvement. In the case of testing the accuracy of an analytical balance with a standard weight, if the CI does not include the standard weight value, it would suggest that either the balance needs calibration or that the standard weight is not as originally issued. Accuracy is sometimes estimated by the relative error ($\%E_{rel}$) between the mean analysis value and the true value. However, $\%E_{rel}$ only reflects tendencies and in practice is often calculated even when there is no statistical justification that the mean and true value differ. Also, note that there is no consideration of the number of replicates in the calculation of $\%E_{rel}$, suggesting that the number of replicates will not affect this estimation of accuracy to any large extent. Absolute precision is reflected by the standard deviation (SD), while relative precision is calculated as the coefficient of variation (CV). Calculations of precision are largely independent of the number of replicates, except that more replicates may give a better estimate of the population variance.

Validation of a procedure or measurement technique can be performed, at the most basic level, as a single-trial validation, as is described in this laboratory that includes estimating the accuracy and precision of commonly used laboratory equipment. However, for more general acceptance of procedures, they are validated by collaborative studies involving several laboratories. Collaborative evaluations are sanctioned by groups such as AOAC International, AACC International, and the American Oil Chemists' Society (AOCS) (Nielsen 2017). Such collaborative studies are prerequisite to procedures appearing as approved methods in manuals published by these organizations.

6.1.2   Reading Assignment

Neilson A.P., Lonergan D.A., and Nielsen S.S. 2017. Laboratory standard operating procedures. Ch.1, in *Food Analysis Laboratory Manual*, 3rd ed., Nielsen S.S. (Ed.), Springer, New York.

Nielsen, S.S. 2017. Introduction to food analysis. Ch. 1, in *Food Analysis*, 5th ed. S.S, Nielsen (Ed.), Springer, New York.

Smith, J.S. 2017. Evaluation of analytical data. Ch. 5, in *Food Analysis*, 5th ed. S.S. Nielsen (Ed.), Springer, New York.

6.1.3 Objective

Familiarize, or refamiliarize, oneself with the use of balances, mechanical pipettes, and volumetric glassware, and assess accuracy and precision of data generated.

6.1.4 Principle of Method

Proper use of equipment and glassware in analytical tests helps ensure more accurate and precise results.

6.1.5 Supplies

- Beaker, 100 mL
- Beaker, 20 or 30 mL
- Beaker, 250 mL
- Buret, 25 or 50 mL
- Erlenmeyer flask, 500 mL
- Funnel, approximately 2 cm diameter (to fill buret)
- Mechanical pipettor, 1000 μL, with plastic tips
- Plastic gloves
- Ring stand and clamps (to hold buret)
- Rubber bulb or pipette pull-up
- Standard weight, 50 or 100 g
- Thermometer, to read near room temperature
- Volumetric flask, 100 mL
- 2 Volumetric pipettes, one each of 1 and 10 mL

6.1.6 Equipment

- Analytical balance
- Top loading balance

6.1.7 Notes

Before or during the laboratory exercise, the instructor is encouraged to discuss the following: (1) difference between dispensing from a volumetric pipette and a graduated pipette and (2) difference between markings on a 10-mL versus a 25- or 50-mL buret.

6.2 PROCEDURE

(Record data in tables that follow.)

1. Obtain ~400 mL deionized distilled (dd) H$_2$O in a 500-mL Erlenmeyer flask for use during this laboratory session. Check the temperature of the water with a thermometer.
2. Analytical balance and volumetric pipettes.

   (a) Tare a 100-mL beaker, deliver 10 mL of water from a volumetric pipette into the beaker, and record the weight. Repeat this procedure of taring the beaker, adding 10 mL, and recording the weight, to get six determinations on the same pipette. (Note that the total volume will be 60 mL.) (It is not necessary to empty the beaker after each pipetting.)

   (b) Repeat the procedure as outlined in Part 2a but use a 20- or 30-mL beaker and a 1.0-mL volumetric pipette. Do six determinations.

3. Analytical balance and buret.

   (a) Repeat the procedure as outlined in Part 2a, but use a 100-mL beaker and a 50-mL (or 25-mL) buret filled with water, and dispense 10 mL of water (i.e., tare a 100 mL beaker, deliver 10 mL of water from the buret into the beaker, and record the weight). (Handle the beaker wearing gloves, to keep oils from your hands off the beaker.) Repeat this procedure of taring the beaker, adding 10 mL, and recording the weight, to get six determinations on the buret. (Note that the total volume will be 60 mL.) (It is not necessary to empty the beaker after each addition.)

   (b) Repeat the procedure as outlined in Part 3a but use a 20- or 30-mL beaker and a 1.0-mL volume from the buret. Do six determinations.

4. Analytical balance and mechanical pipette. Repeat the procedure as outlined in Part 2a but use a 20- or 30-mL beaker and a 1.0-mL mechanical pipette (i.e., tare a 20- or 30-mL beaker, deliver 1 mL of water from a mechanical pipettor into the beaker, and record the weight). Repeat this procedure of taring the beaker, adding 1 mL, and recording the weight to get six determinations on the same pipettor. (Note that the total volume will be 6 mL.) (It is not necessary to empty the beaker after each pipetting.)

5. Total content (TC) versus total delivery (TD). Tare a 100-mL volumetric flask on a top loading balance. Fill the flask to the mark with water. Weigh the water in the flask. Now tare a 250-mL beaker and pour the water from the volumetric flask into the beaker. Weigh the water delivered from the volumetric flask.

6. Readability versus accuracy. Zero a top loading balance and weigh a 100-g (or 50-g) standard weight. Record the observed weight. Use gloves or finger cots as you handle the standard weight to keep oils from your hands off the weight. Repeat with the same standard weight on at least two other top loading balances, recording the observed weight and the type and model (e.g., Mettler, Sartorius) of balance used.

6.3   DATA AND CALCULATIONS

Calculate the exact volume delivered in Sect. 6.2, Parts 2–5, using each weight measurement and the known density of water (see Table 6.1). Using *volume* data, calculate the following indicators of accuracy and precision: mean, standard deviation, coefficient of variation, percent relative error, and 95% confidence interval. Use your first three measurements for $n=3$ values requested and all six measurements for $n=6$ values.

Data for Sect. 6.2, Parts 2, 3, and 4:

Rep	Volumetric pipette				Buret				Mechanical pipettor	
	1 mL		10 mL		1 mL		10 mL		1 mL	
	Wt.	Vol.	Wt.	Vol.	Wt.	Vol.	Wt.	Vol.	Wt.	Vol.
1										
2										
3										
4										
5										
6										
$n=3$										
Mean	–		–		–		–		–	
SD	–		–		–		–		–	
CV	–		–		–		–		–	
% E_{rel}	–		–		–		–		–	
$CI_{95\%}$	–		–		–		–		–	
$n=6$										
Mean	–		–		–		–		–	
SD	–		–		–		–		–	
CV	–		–		–		–		–	
% E_{rel}	–		–		–		–		–	
$CI_{95\%}$	–		–		–		–		–	

Data for Sect. 6.2, Part 5:

	Wt.	Vol.
Water in flask =		
Water in beaker =		

Data for Sect. 6.2, Part 6:

Balance	Type/model of balance	Standard weight (g)
1		
2		
3		

Viscosity and density of water at various temperatures

Temp. (°C)	Density (g/mL)	Viscosity (cps)	Temp. (°C)	Density (g/mL)	Viscosity (cps)
20	0.99823	1.002	24	0.99733	0.9111
21	0.99802	0.9779	25	0.99707	0.8904
22	0.99780	0.9548	26	0.99681	0.8705
23	0.99757	0.9325	27	0.99654	0.8513

6.4   QUESTIONS

(Questions refer to parts of Sect. 6.2)

1. Theoretically, how are standard deviation, coefficient of variation, mean, percent relative error, and 95% confidence interval affected by (1) more replicates and (2) a larger size of the measurement? Was this evident in looking at the actual results obtained using the volumetric pipettes and the buret, with $n=3$ versus $n=6$ and with 1 mL versus 10 mL? (See table below)

	Theoretical		Actual, with results obtained	
	More replicates	Larger mesurement	More replicates	Larger mesurement
Standard deviation				
Coefficient of variation				
Mean				
Percent relative error				
95% confidence interval				

2. Why are percent relative error and coefficient of variation used to compare the accuracy and precision, respectively, of the volumes from pipetting/dispensing 1 and 10 mL with the volumetric pipettes and buret in Parts 2 and 3, rather than simply the mean and standard deviation, respectively?

3. Compare and discuss the accuracy and the precision of the volumes from the 1 mL pipetted/dispensed using a volumetric pipette, buret, and

mechanical pipettor (Parts 2, 3, and 4). Are these results consistent with what would be expected?

4. If accuracy and precision using the mechanical pipettor are less than should be expected, what could you do to improve its accuracy and precision?

5. In a titration experiment using a buret, if you expect to use much less than a 10-mL volume in each titration, would you expect your accuracy and precision to be better using a 10-mL buret or a 50-mL buret? Why?

6. How do your results from Part 5 of this lab differentiate "to contain" from "to deliver"? Is a volumetric flask "to content" or "to deliver"? Which is a volumetric pipette?

7. From your results from Part 6 of this lab, would you now assume that since a balance reads to 0.01 g that it is accurate to 0.01 g?

8. What sources of error (human and instrumental) were evident or possible in Parts 2–4, and how could these be reduced or eliminated? Explain.

9. You are considering adopting a new analytical method in your lab to measure the moisture content of cereal products, how would you determine the precision of the new method and compare it to the old method? How would you determine (or estimate) the accuracy of the new method?

RESOURCE MATERIALS

Neilson AP, Lonergan DA, and Nielsen SS (2017) Laboratory standard operating procedures, Ch. 1. In: Food analysis laboratory manual, 3rd ed., Nielsen SS (ed.), Springer, New York

Nielsen SS (2017) Introduction to food analysis, Ch. 1. In: Nielsen SS (ed) Food analysis, 5th edn. Springer, New York

Smith JS (2017) Evaluation of analytical data, Ch. 4. In: Nielsen SS (ed) Food analysis, 5th edn. Springer, New York

High-Performance Liquid Chromatography

S.Suzanne Nielsen (✉)
Department of Food Science, Purdue University,
West Lafayette, IN, USA
e-mail: nielsens@purdue.edu

Stephen T. Talcott
Department of Nutrition and Food Science, Texas A&M University,
College Station, TX, USA
e-mail: stalcott@tamu.edu

S.S. Nielsen, *Food Analysis Laboratory Manual*, Food Science Text Series,
DOI 10.1007/978-3-319-44127-6_7, © Springer International Publishing 2017

7.1 INTRODUCTION

7.1.1 Background

High-performance liquid chromatography (HPLC) has many applications in food chemistry. Food components that have been analyzed with HPLC include organic acids, vitamins, amino acids, sugars, nitrosamines, certain pesticides, metabolites, fatty acids, aflatoxins, pigments, and certain food additives. Unlike gas chromatography, it is not necessary for the compound being analyzed to be volatile. It is necessary, however, for the compounds to have some solubility in the mobile phase. It is important that the solubilized samples for injection be free from all particulate matter, so centrifugation and filtration are common procedures. Also, solid-phase extraction is used commonly in sample preparation to remove interfering compounds from the sample matrix prior to HPLC analysis.

Many food-related HPLC analyses utilize reversed-phase chromatography in which the mobile phase is relatively polar, such as water, dilute buffer, methanol, or acetonitrile. The stationary phase (column packing) is relatively nonpolar, usually silica particles coated with a C_8 or C_{18} hydrocarbon. As compounds travel through the column, they partition between the hydrocarbon stationary phase and the mobile phase. The mobile phase may be constant during the chromatographic separation (i.e., isocratic) or changed stepwise or continuously (i.e., gradient). When the compounds elute separated from each other at the end of the column, they must be detected for identification and quantitation. Identification often is accomplished by comparing the volume of liquid required to elute a compound from a column (expressed as retention volume or retention time) to that of standards chromatographed under the same conditions. Quantitation generally involves comparing the peak height or area of the sample peak of interest with the peak height or area of a standard (at the same retention time). The results are usually expressed in milligrams per gram or milliliters of food sample.

7.1.2 Reading Assignment

Ismail, B.P. 2017. Basic principles of chromatography. Ch. 12, in *Food Analysis*, 5th ed. S.S. Nielsen (Ed.), Springer, New York.

Reuhs, B.L. 2017. High performance liquid chromatography. Ch. 13, in *Food Analysis*, 5th ed. S.S. Nielsen (Ed.), Springer, New York.

7.2 DETERMINATION OF CAFFEINE IN BEVERAGES BY HPLC

7.2.1 Introduction

The caffeine content of beverages can be determined readily by simple filtration of the beverage prior to separation from other beverage components using reversed-phase HPLC. An isocratic mobile phase generally provides for sufficient separation of the caffeine from other beverage components. However, separation and quantitation are much easier for soft drinks than for a beverage such as coffee, which has many more components. Commercially available caffeine can be used as an external standard to quantitate the caffeine in the beverages by peak height or area.

7.2.2 Objective

To determine the caffeine content of soft drinks by reversed-phase HPLC with ultraviolet (UV) detection, using peak height and area to determine concentrations

7.2.3 Chemicals

	CAS no.	Hazards
Acetic acid (CH_3COOH)	64-19-7	Corrosive
Caffeine	58-08-2	Harmful
Methanol, HPLC grade (CH_3OH)	67-56-1	Extremely flammable, toxic

7.2.4 Hazards, Precautions, and Waste Disposal

Adhere to normal laboratory safety procedures. Wear safety glasses at all times. Methanol waste must be handled as hazardous waste. Other waste likely may be put down the drain using a water rinse, but follow good laboratory practices outlined by environmental health and safety protocols at your institution.

7.2.5 Reagents

(** It is recommended that these solutions be prepared by the laboratory assistant before class.)

- Mobile phase**
  Deionized distilled (dd) water: HPLC-grade methanol: acetic acid, 65:35:1 (v/v/v), filtered through a Millipore filtration assembly with 0.45-µm nylon membranes and degassed

- Caffeine solutions of varying concentration for standard curve**
  Prepare a stock solution of 20 mg caffeine/100 mL dd water (0.20 mg/mL). Make standard solutions containing 0.05, 0.10, 0.15, and 0.20 mg/mL, by combining 2.5, 5.0, 7.5, and 10 mL of stock solution with 7.5, 5.0, 2.5, and 0 mL dd water, respectively.

7.2.6 Supplies

(Used by students)

- Disposable plastic syringe, 3 mL (for filtering sample)

- Hamilton glass HPLC syringe, 25 µL (for injecting sample if using manual sample loading)
- Pasteur pipettes and bulb
- Sample vials for autosampler (if using autosampler)
- Soft drinks, with caffeine
- Syringe filter assembly, (Syringe and 0.45 µm filters)
- Test tubes, e.g., 13 × 100-mm disposable culture tubes (for filtering sample)

7.2.7   Equipment

- Analytical balance
- HPLC system, with UV–vis detector
- Membrane filtering and degassing system

7.2.8   HPLC Conditions

Column	Waters µBondapak C$_{18}$ (Waters, Milford, MA) or equivalent reversed-phase column
Guard column	Waters Guard-Pak Precolumn Module with C$_{18}$ Guard-Pak inserts or equivalent
Mobile phase	dd H$_2$O: HPLC-grade methanol: acetic acid, 65:35:1 (v/v/v) (combine and then filter and degas)
Flow rate	1 mL/min
Sample loop size	10 µL
Detector	Absorbance at 254 nm or 280 nm
Sensitivity	Full-scale absorbance = 0.2
Chart speed	1 cm/min

7.2.9   Procedure

(Instructions are given for manual injection and for analysis in triplicate)

1. Filter beverage sample.
   (a) Remove plunger from a plastic 3-mL syringe, and connect syringe filter assembly (with a membrane in place) to the syringe barrel.
   (b) Use a Pasteur pipette to transfer a portion of beverage sample to the syringe barrel. Insert and depress syringe plunger to force sample through the membrane filter and into a small test tube.
2. Flush Hamilton HPLC syringe with filtered sample, and then take up 15–20 µL of filtered sample (try to avoid taking up air bubbles).

3. With HPLC injector valve in *LOAD* position, insert syringe needle into the needle port all the way.
4. Gently depress syringe plunger to completely fill the 10-µL injector loop with sample.
5. Leaving the syringe in position, *simultaneously* turn valve to *INJECT* position (mobile phase now pushes sample onto the column) and depress chart marker button on the detector (to mark start of run on chart recorder paper).
6. Remove syringe. (Leave valve in the *INJECT* position so that the loop will be continuously flushed with mobile phase, thereby preventing cross contamination.)
7. After caffeine peak has eluted, return valve to *LOAD* position in preparation for next injection.
8. Repeat Steps 3–7, injecting each caffeine standard solution in duplicate or triplicate. (Note: The laboratory assistant can inject all standard solutions prior to the laboratory session).

7.2.10   Data and Calculations

1. Chromatograms and peak area reports will be printed out for you by the laboratory assistant. Use peak areas of the standards to construct your standard curve. Use the equation of the line for your standard curve to calculate the concentration of caffeine in your sample. Use the appropriate dilution factor when calculation the caffeine concentration in your sample. Complete the tables below with all data.

Standard curve:

Caffeine conc. (mg/ml)	Rep	Peak area (cm^2)
0.05	1	
	2	
	3	
0.10	1	
	2	
	3	
0.15	1	
	2	
	3	
0.20	1	
	2	
	3	

Sample: (Complete this table for each type of sample)

Rep	Retention time (min)	Peak area (cm^2)	Conc. based on area from std. curve (mg/ml)	Conc. based on area after considering dilution factor (mg/mL)	Conc. based on area (mg/100 mL)
1					
2					
3					
$\bar{X}$					
SD =					

2. Using the mean values, calculate the concentration of caffeine in your sample expressed in terms of milligrams caffeine in a 12-oz. can (1 mL = 0.0338 oz).

7.2.11 Questions

1. Based on the triplicate values and the linearity of your standard curves, which of the two methods used to calculate concentration seemed to work best in this case? Is this what you would have expected, based on the potential sources of error for each method?
2. Why was it important to filter and degas the mobile phase and the samples?
3. How is the "reversed-phase" HPLC used here different from "normal phase" with regard to stationary and mobile phases and order of elution?
4. Mobile phase composition:
    (a) If the mobile phase composition was changed from 65:35:1 (v/v/v) to 75:25:1 (v/v/v) water to methanol to acetic acid, how would the time of elution (expressed as retention time) for caffeine be changed, and why would it be changed?
    (b) What if it was changed from 65:35:1 (water: methanol: acetic acid) to 55:45:1? How would that change the retention time and why?

7.3 SOLID-PHASE EXTRACTION AND HPLC ANALYSIS OF ANTHOCYANIDINS FROM FRUITS AND VEGETABLES

7.3.1 Introduction

Anthocyanins are naturally occurring plant pigments known for their diverse colors depending on solution pH. Analysis for anthocyanins is often difficult due to their similar molecular structure and polarity and their diversity of sugar and/or organic acid substituents. Color intensity is a common means of analyzing for anthocyanins since monomeric anthocyanins are colored bright red at low pH values from 1 to 3 (oxonium or flavylium forms) and are nearly colorless at higher pH values from 4 to 6 (carbinol or pseudobase forms). A pure anthocyanin in solution generally follows Beer's law; therefore concentration can be estimated from an extinction coefficient when an authentic standard is not available. However, many standards are commercially available with cyanidin-3-glucoside used most often for quantification purposes.

Red-fleshed fruits and vegetables contain many different anthocyanin forms due to their diverse array of esterified sugar substituents and/or acyl-linked organic acids. However, most foods contain up to six anthocyanin aglycones (without sugar or organic acid substituents, referred to as anthocyanidins) that include delphinidin, cyanidin, petunidin, pelargonidin, peonidin, and malvidin (Fig. 7.1). Sample preparation for anthocyanin analysis generally involves solid-phase extraction of these compounds from the food matrix followed by acid hydrolysis to remove sugar and/or organic acid linkages. Anthocyanidins are then easily separated by reversed-phase HPLC.

The use of solid-phase extraction (SPE) is a common chromatographic sample preparation technique used to remove interfering compounds from a biological matrix prior to HPLC analysis. This physical extraction technique is similar to an actual separation on a reversed-phase HPLC column. Although many SPE stationary phases exist, the use of reversed-phase C_{18} is commonly employed for food analysis. On a relative basis, anthocyanins are less polar than other chemical constituents in fruits and vegetables and will readily bind to a reversed-phase C_{18} SPE cartridge. Other compounds such as sugars, organic acids, water-soluble vitamins, or metal ions have little or no affinity to the cartridge. After the removal of these interferences, anthocyanins can then be efficiently eluted with alcohol, thus obtaining a semipurified extract for HPLC analysis.

Separation of compounds by HPLC involves the use of a solid support (column) over which a liquid mobile phase flows on a continuous basis. Chemical interactions with an injected sample and the stationary and mobile phases will influence rates of compound elution from a column. For compounds with similar polarities, the use of mixtures of mobile phases (gradient elution) is often employed. Reversed-phase stationary phases are most common for anthocyanin separations, and are based on column hydrophobicity of a silica-based column with varying chain lengths of n-alkanes such as C_8 or C_{18}. By setting initial chromatographic conditions to elute with a polar (water) mobile phase followed by an organic (alcohol) mobile phase, anthocyanins will generally elute in order of their polarity.

You will be analyzing anthocyanins isolated from fruits or vegetables for anthocyanidins (aglycones) following SPE and acid hydrolysis to remove sugar glycosides. After sugar hydrolysis, samples will be injected into an HPLC for compound separation. Depending on plant source, you will obtain between one and six chromatographic peaks representing common anthocyanidins found in edible plants.

7.3.2 Objective

Isolate and quantify anthocyanidin concentration from common fruits and vegetables by reversed-phase

7.1
figure Anthocyanin structures. Common substitutions on the B-ring include delphinidin (*Dp*), cyanidin (*Cy*), petunidin (*Pt*), pelargonidin (*Pg*), peonidin (*Pn*), and malvidin (*Mv*)

HPLC with Vis detection, using spectrophotometric absorbance readings and extinction coefficients of anthocyanidins to determine standard concentrations.

7.3.3   Chemicals

	CAS no.	Hazards
Hydrochloric acid (HCl)	7647-01-0	Corrosive
Methanol (CH$_3$OH)	67-56-1	Extremely flammable, toxic
o-Phosphoric acid (H$_3$PO$_4$)	7664-38-2	Corrosive

7.3.4   Hazards, Precautions, and Waste Disposal

Adhere to normal laboratory safety procedures. Wear safety glasses at all times. Use hydrochloric acid under a fume hood. Methanol waste must be handled as hazardous waste. Other waste likely may be put down the drain using a water rinse, but follow good laboratory practices outlined by your environmental health and safety protocols.

7.3.5   Reagents

(** It is recommended that these solutions be prepared by a laboratory assistant before class.)

- 4 N HCl in water (for anthocyanin hydrolysis)**
- 0.01 % HCl in water (for sample extraction)**
- 0.01 % HCl in methanol (for elution from C$_{18}$ cartridge)**
- Mobile Phase A: 100 % water (pH 2.4 with o-phosphoric acid)**
- Mobile Phase B: 60 % methanol and 40 % water (pH 2.4 with o-phosphoric acid)**

[Each mobile phase should be filtered through a 0.45-µm nylon membrane (Millipore) and degassed while stirring using either a nitrogen sparge, under vacuum (ca. 20 min/l of solvent), or by sonication.]

7.3.6   Supplies

- Beaker, Pyrex, 500 mL (for boiling water for hydrolysis)
- Blender, kitchen scale, for sample homogenization
- Disposable plastic syringe, 3–5 mL (for filtering sample)
- Filter paper (Whatman #4) and funnels
- Fruit or vegetable that contains anthocyanins (blueberries, grapes, strawberries, red cabbage, blackberries, cherries, or commercial juices that contain anthocyanins)
- Hamilton glass HPLC syringe, 25 µl (for injecting sample)
- Reversed-phase C$_{18}$ cartridge (for SPE, e.g., Waters C$_{18}$ Sep-Pak, WAT051910)
- Syringe filter (0.45-µm PTFE, polytetrafluoroethylene)
- Test tubes, screw cap, with lids (for anthocyanin hydrolysis)

7.3.7   Equipment

- Analytical balance
- Hot plate
- HPLC system, gradient, with Vis detector (520 nm)
- Membrane filtering and degassing system
- Spectrophotometer and cuvettes (1-cm path length)

7.3.8   HPLC Conditions

Column	Waters Nova-Pak C$_{18}$ (WAT044375) or equivalent reversed-phase column
Guard column	Waters Guard-Pak Precolumn Module with C$_{18}$ Guard-Pak inserts

Mobile phase	Phase A, 100% water; Phase B, 60% methanol and 40% water (both adjusted to pH 2.4 with o-phosphoric acid)
Flow rate	1 mL/min
Sample loop size	Variable: 10–100 µl
Detector	Visible at 520 nm
Gradient conditions	Linear ramp. Hold time at 100% Phase B after 15 min may vary with column length and/or column packing material

Time (min)	% Phase A	% Phase B
0	100	0
5	50	50
10	50	50
15	0	100
35	0	100 (end)
37	100	0 (equilibration)

7.3.9   Procedure

7.3.9.1 Sample Extraction

(Note to the instructor: Several different commodities can be evaluated or the experiment replicated as needed.)

1. Weigh ca. 10 g of fruit or vegetable containing anthocyanins (record exact weight) and place in blender. Add ca. 50 mL of water containing 0.01% HCl and blend thoroughly (acidified acetone, methanol, or ethanol are also suitable substitutions for water). Fruit juices that contain anthocyanins can be used without further preparation.
2. Filter homogenate through filter paper and collect aqueous filtrate.
3. Keep refrigerated until needed.

7.3.9.2 Solid-Phase Extraction

1. Condition a reversed-phase SPE cartridge by first washing with 4 mL of 100% methanol followed by 4 mL of water acidified with 0.01% HCl.
2. *Slowly* pass 1–2 mL of juice or filtrate (record exact volume) through the SPE cartridge being careful not to lose visible pigment. Anthocyanins will adhere to the SPE support and less polar compounds such as sugars, organic acids, and ascorbic acid will be removed.
3. Slowly pass an additional 4 mL of water (acidified with 0.01% HCl) through the cartridge to remove residual water-soluble components. Remove residual moisture from the cartridge by pushing air through the cartridge with an empty syringe or by flushing the cartridge with nitrogen gas until dry.
4. Elute anthocyanins with 4 mL of 0.01% HCl in methanol and collect for subsequent hydrolysis.

7.3.9.3 Acid Hydrolysis

[Note to the instructor: It is recommended that a previously extracted sample be acid hydrolyzed before class to save time].

1. Pipette 2 mL of anthocyanins, dissolved in methanol, into a screw-cap test tube and add an equal volume of aqueous 4N HCl (final acid concentration=2N) for a twofold dilution factor (see calculations below).
2. Under a fume hood, tightly cap the screw-cap vial and place in boiling water for ca. 90 min.
3. Remove test tubes and cool to room temperature before opening the vial. Filter an aliquot through a 0.45-µm PTFE syringe filter for analysis by HPLC.
4. Inject the filtered extract into the HPLC column and record peak areas for quantification of each compound (see Fig. 7.2).

7.3.10   Data and Calculations

Authentic standards for select anthocyanins can be obtained from several sources and should be used according to manufacturer's suggestions. If using anthocyanin glycosides, then the acid hydrolysis procedure should be conducted prior to HPLC analysis. Some anthocyanin suppliers include: Polyphenols Laboratories, Sandnes, Norway; INDOFINE Chemical Company, Somerville, NJ.

Cyanidin is a common anthocyanin present in large concentrations in many fruits and vegetables and will be used for sample calculations. A standard solution of cyanidin should be prepared in Mobile Phase A (water at pH 2.4) to establish a standard curve. Unless the actual concentration is known from the manufacturer, the standard should be quantified by determining its absorbance on a spectrophotometer at 520 nm against a blank of the same solvent. Using the molar extinction coefficient for cyanidin (obtained from manufacturer or expressed as cyanidin-3-glucoside equivalents, $\varepsilon=29{,}600$ for a 1-M solution and 1-cm light path), the concentration is calculated using Beer's law, $A=\varepsilon bc$, using the following calculation:

$$\text{mg / L Cyanidin} = \frac{(\text{Absorbance at } \lambda_{max})(1000)(\text{MW})}{\varepsilon}$$

where:

MW ~ 457 g/mol
$\varepsilon \sim 29{,}600$

1. Inject a series of standard concentrations into the HPLC to generate a standard curve. (*Note*: These procedures can be performed by a laboratory assistant prior to the laboratory session.)
2. Chromatograms and peak area reports will be printed out for you by the laboratory assistant. Use peak areas of the standards to construct

7.2 figure   Typical reversed-phase HPLC chromatograph of anthocyanidins (grape)

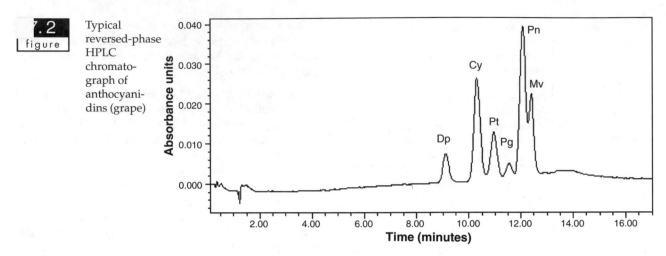

7.3 figure   Typical standard curve for cyanidin

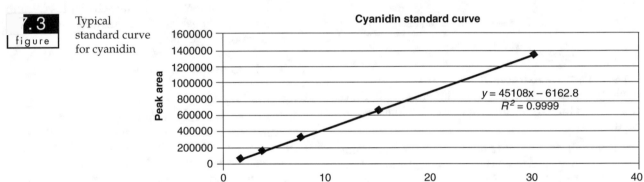

your standard curve. Use the equation of the line for your standard curve to calculate the anthocyanin concentration of your acid hydrolyzed samples (Fig 7.3). Use appropriate dilution factors.

3. Express relative concentrations of each identifiable compound as cyanidin equivalents (mg/L) based on their peak area (unless commercial standards are available for each peak in the chromatograph).

Peak	Peak area	Relative concentration (mg/L)
1		
2		
3		
4		
5		
6		

mg/L Cyanidin (in unknown)

$$= \frac{\text{Peak area} \times 2}{\text{Slope of standard curve}} \times \text{Sample dilution factors}$$

Peak area multiplied by 2 will compensate for the two-fold dilution incurred during acid hydrolysis. Sample

dilution factors are calculated based on weight of fruit/vegetable per volume of extraction solvent (sample weight + solvent volume/sample weight). Single-strength fruit juices would have a sample dilution factor of 1.

7.3.11 Questions

1. Based on chemical structure, why do anthocyanidins elute in their respective order?
2. Predict how each compound would elute from a normal-phase column.
3. If the retention time of a compound that had absolutely *no* affinity to the column was 1.5 min and the flow rate was 1 mL/min, what is the total volume of mobile phase contained in the column, tubing, and pumps? Are you surprised at this number? Why or why not?
4. What would the chromatograph look like if you injected 40 μl of a sample as compared to 20 μl?
5. What would the chromatograph look like if Mobile Phases A and B were reversed (i.e., beginning with 100% Phase B and increasing Phase A over time)?

RESOURCE MATERIALS

AOAC International (2016) Official methods of analysis, 20th edn., (On-line) Method 979.08. Benzoate, caffeine, and saccharin in soda beverages. AOAC International, Rockville, MD

Bridle P, Timberlake F (1997) Anthocyanins as natural food colours – selected aspects. Food Chem 58: 103–109

Hong V, Wrolstad RE (1990) Use of HPLC separation/photo-diode array detection for characterization of anthocyanins. J Agr Food Chem 38:708–715

Markakis P (ed) (1982) Anthocyanins as food colors. Academic Press, New York

Ismail BP (2017) Basic principles of chromatography. Ch. 12. In: Nielsen SS (ed) Food analysis, 5th edn. Springer, New York

Reuhs BL (2017) High-performance liquid chromatography. Ch. 13. In: Nielsen SS (ed) Food analysis, 5th edn. Springer, New York

Gas Chromatography

S.Suzanne Nielsen (✉)

Department of Food Science, Purdue University,
West Lafayette, IN, USA
e-mail: nielsens@purdue.edu

Michael C. Qian

Department of Food Science and Technology, Oregon State University,
Corvallis, OR, USA
e-mail: michael.qian@oregonstate.edu

S.S. Nielsen, *Food Analysis Laboratory Manual*, Food Science Text Series,
DOI 10.1007/978-3-319-44127-6_8, © Springer International Publishing 2017

8.1 INTRODUCTION

8.1.1 Background

Gas chromatography (GC) has many applications in the analysis of food products. GC has been used for the determination of fatty acids, triglycerides, cholesterol, gases, water, alcohols, pesticides, flavor compounds, and many more. While GC has been used for other food components such as sugars, oligosaccharides, amino acids, peptides, and vitamins, these substances are more suited to analysis by high performance liquid chromatography. GC is ideally suited to the analysis of volatile substances that are thermally stable. Substances such as pesticides and flavor compounds that meet these criteria can be isolated from a food and directly injected into the GC. For compounds that are thermally unstable, too low in volatility, or yield poor chromatographic separation due to polarity, a derivatization step must be done before GC analysis. The two parts of the experiment described here include the analysis of alcohols that requires no derivatization step and the analysis of fatty acids which requires derivatization. The experiments specify the use of capillary columns, but the first experiment includes conditions for a packed column.

8.1.2 Reading Assignment

Ellefson, W.C. 2017. Fat analysis. Ch. 17, in *Food Analysis*, 5th ed. S.S. Nielsen (ed.), Springer, New York.

Pike O.A., and O'Keefe, S.F. 2017. Fat characterization. Ch. 23 in *Food Analysis*, 5th ed. S.S. Nielsen (Ed.), Springer, New York.

Qian, M.C., Peterson, D.G., and Reineccius, G.A. 2017. Gas chromatography. Ch. 14, in *Food Analysis*, 5th ed. S.S. Nielsen (Ed.), Springer, New York.

<table>
<tr><td>**8.1**
table</td><td>Alcohol structure and boiling point</td></tr>
</table>

Alcohol	Structure	b.p.(°C)
Methanol	CH_3OH	64.5
Ethanol	$CH_3\text{-}CH_2OH$	78.3
n-Propanol	$CH_3\text{-}CH_2\text{-}CH_2OH$	97
Isobutyl alcohol (2-methyl-1-propanol)	$CH_3\text{-}CH\text{-}CH_2OH$ CH_3	108
Isoamyl alcohol (3-methyl-1-butanol)	$CH_3\text{-}CH\text{-}CH_2CH_2OH$ CH_3	
Active amyl alcohol (2-methyl- 1-butanol)	$CH_3\text{-}CH_2\text{-}CH\text{-}CH_2OH$ CH_3	128
Benzyl alcohol	$\bigcirc\!\!-CH_2OH$	205

8.2 DETERMINATION OF METHANOL AND HIGHER ALCOHOLS IN WINE BY GAS CHROMATOGRAPHY

8.2.1 Introduction

The quantification of higher alcohols, also known as fusel oils, in wine and distilled spirits is important because of the potential flavor impact of these compounds. These higher alcohols include n-propyl alcohol, isobutyl alcohol, and isoamyl alcohol. Some countries have regulations that specify the maximum and/or minimum amounts of total higher alcohols in certain alcoholic beverages. Table wine typically contains only low levels of higher alcohols, but dessert wines contain higher levels, especially if the wine is fortified with brandy.

Methanol is produced enzymatically during the production of wine. Pectin methyl esterase hydrolyzes the methyl ester of α-1,4-D-galacturonopyranose. The action of this enzyme, which is naturally present in grapes and may also be added during vinification, is necessary for proper clarification of the wine. White wines produced in the United States contain less methanol (4–107 mg/L) when compared with red and rosé wines (48–227 mg/L). Methanol has a lower boiling point than the higher alcohols (Table 8.1), so it is more readily volatilized and elutes earlier from a gas chromatography (GC) column.

Methanol and higher alcohols in distilled liquors are readily quantitated by gas chromatography, using an internal standard such as benzyl alcohol, 3-pentanol, or n-butyl alcohol. The method outlined below is similar to AOAC Methods 968.09 and 972.10 [Alcohols (Higher) and Ethyl Acetate in Distilled Liquors].

8.2.2 Objective

Determine the content of methanol, n-propyl alcohol, and isobutyl alcohol in wine by gas chromatography, using benzyl alcohol as the internal standard.

8.2.3 Principle of Method

Gas chromatography uses high temperatures to volatilize compounds that are separated as they pass through the stationary phase of a column and are detected for quantitation.

8.2.4 Chemicals

	CAS no.	Hazards
Benzyl alcohol	100-51-6	Harmful
Ethanol	64-17-5	Highly flammable
Isobutyl alcohol	78-83-1	Irritant
Methanol	67-56-1	Extremely flammable
n-Propyl alcohol	71-23-8	Irritant, highly flammable

8.2.5 Reagents

(**It is recommended that these solutions be prepared by the laboratory assistant before class.)

- Ethanol, 16% (vol/vol) with deionized distilled (dd) water, 500 mL**
- Ethanol, 50% (vol/vol) with dd water, 3200 mL**
- Ethanol, 95% (vol/vol) with dd water, 100 mL**
- Stock solutions**
  Prepared with known amounts of ethanol and fusel alcohols or methanol:

  1. 10.0 g of methanol and 50% (vol/vol) ethanol to 1000 mL
  2. 5.0 g of n-propyl alcohol and 50% (vol/vol) ethanol to 1000 mL
  3. 5.0 g of isobutyl alcohol and 50% (vol/vol) ethanol to 1000 mL
  4. 5.0 g of benzyl alcohol in 95% (vol/vol) ethanol to 100 mL
- Working Standard Solutions**
  Prepared from stock solutions, to contain different amounts of each of the fusel alcohols; aliquots of these are used to get standard curves. Prepare four working standards by combining:

  1. 0.5 mL of stock solutions 1, 2, and 3 with 4.5 mL of 50% (vol/vol) ethanol plus 16% (vol/vol) ethanol to 100 mL
  2. 1.0 mL of stock solutions 1, 2, and 3 with 3 mL of 50% (vol/vol) ethanol plus 16% (vol/vol) ethanol to 100 mL
  3. 1.5 mL of stock solutions 1, 2, and 3 with 1.5 mL of 50% (vol/vol) ethanol plus 16% (vol/vol) ethanol to 100 mL
  4. 2.0 mL of stock solutions 1, 2, and 3 with 16% (vol/vol) ethanol to 100 mL

(*Note*: The final concentration of ethanol in each of these working standard solutions is 18% (vol/vol) ethanol.)

8.2.6 Hazards, Precautions, and Waste Disposal

The alcohols are fire hazards; avoid open flames, breathing vapors, and contact with skin. Otherwise, adhere to normal laboratory safety procedures. Wear safety glasses at all times. Aqueous waste can go down the drain with a water flush.

8.2.7 Supplies

(Used by students)

- Mechanical pipettor, 1000 μL, with tips
- Round bottom flask, 500 mL
- Syringe (for GC)

- 6 Volumetric flasks, 100 mL
- 4 Volumetric flasks, 1000 mL

8.2.8 Equipment

- Analytical balance
- Distillation unit (heating element to fit 500-mL round bottom flask; cold water condenser)

- Gas chromatography unit:

Column	DB-Wax (30 m, 0.32 nm ID, 0.5-μm film thickness) (Agilent Technologies, Santa Clara, CA) or equivalent (capillary column) or 80/120 Carbopack BAW/5% Carbowax 20 M, 6 ft × 1/4 in OD × 2 mm ID glass column (packed column)
Injector temperature	200 °C
Column temperature	70–170 °C at 5 °C/min
Carrier gas	He at 2 mL/min (N$_2$ at 20 mL/min for packed column)
Detector	Flame ionization
Attenuation	8 (for all runs)

ID inner diameter, *OD* outer diameter, *BAW* base and acid washed

8.2.9 Procedure

(Instructions are given for single standard and sample analysis, but injections can be replicated.)

8.2.9.1 *Sample Preparation*

1. Fill a 100-mL volumetric flask to volume with the wine sample to be analyzed.
2. Pour the wine into a 500-mL round bottom flask and rinse the volumetric flask several times with dd water to complete the transfer. Add additional water if necessary to bring the volume of sample plus dd water to ca. 150 mL.
3. Distil the sample and recover the distillate in a clean 100-mL volumetric flask. Continue the distillation until the 100-mL volumetric is filled to the mark.
4. Add 1.0 mL of the stock benzyl alcohol solution to 100 mL of each working standard solution and wine sample to be analyzed.

8.2.9.2 *Analysis of Sample and Working Standard Solutions*

1. Inject 1 μL of each sample and working standard solution in separate runs on the GC column (split ratio 1:20) (For packed column, inject 5.0 μL.)
2. Obtain chromatograms and data from integration of peaks.

8.2.10 Data and Calculations

1. Calculate the concentration (mg/L) of methanol, n-propyl alcohol, and isobutyl alcohol in each of the four Working Standard Solutions (see sample calculation below).

Alcohol concentration (mg/l):

Working standard	Methanol	N-Propyl alcohol	Isobutyl alcohol
1			
2			
3			
4			

Example calculations:

Working Standard Solution #1 contains methanol + n-propyl alcohol + isobutyl alcohol, all in ethanol.

Methanol in Stock Solution #1:

$$\frac{10\,g\,methanol}{1000\,mL} = \frac{1g}{100\,mL} = \frac{0.01g}{mL}$$

Working Standard Solution #1 contains 0.5 mL of Stock Solution #1.

= 0.5 mL of 0.01 g methanol/mL
= 0.005 g methanol = 5 mg methanol

That 5 mg methanol is contained in 100 mL volume.

= 5 mg/100 mL = 50 mg/1000 mL
= 50 mg methanol/L

Repeat procedure for each alcohol in each working standard solution.

2. Calculate the peak height or peak area ratios for methanol, n-propyl alcohol, and isobutyl alcohol, compared to the internal standard, for each of the Working Standard Solutions and the wine sample. To identify which is the methanol, n-propyl alcohol, and isobutyl peak, see the example chromatogram that follows. Note that data from automatic integration of the peaks can be used for these calculations. Report the ratios in a table as shown below. Show an example calculation of concentration for each type of alcohol.

Peak hseight ratios for alcohol peaks at various concentrations of methanol, n-propyl alcohol, and isobutyl alcohol, with benzyl alcohol as internal standard:

| Alcohol Conc. (mg/l) | Peak height ratio[a] | | |
	Methanol Benzyl alcohol	n-Propyl alcohol Benzyl alcohol	Isobutyl alcohol Benzyl alcohol
25			
50			
75			
100			
150			
200			
Wine sample			

[a]Give individual values and the ratio

3. Construct standard curves for methanol, n-propyl alcohol, and isobutyl alcohol using the peak height ratios. All lines can be shown on one graph. Determine the equations for the lines.

4. Calculate the peak ratios for methanol, n-propyl alcohol, and isobutyl alcohol in the wine sample and their concentrations in mg/L.

8.2.11 Questions

1. Explain how this experiment would have differed in standard solutions used, measurements taken, and standard curves used if you had used external standards rather than an internal standard.

2. What are the advantages of using an internal standard rather than external standards for this application, and what were the appropriate criteria to use in selecting the internal standard?

8.3 PREPARATION OF FATTY ACID METHYL ESTERS (FAMEs) AND DETERMINATION OF FATTY ACID PROFILE OF OILS BY GAS CHROMATOGRAPHY

8.3.1 Introduction

Information about fatty acid profile on food is important for nutrition labeling, which involves the measurement of not only total fat but also saturated, unsaturated, and monounsaturated fat. Gas chromatography is an ideal instrument to determine (qualitatively and quantitatively) fatty acid profile or fatty acid composition of a food product. This usually involves extracting the lipids and analyzing them using capillary gas chromatography. Before such analysis, triacylglycerols and phospholipids are saponified, and the fatty acids liberated are esterified to form fatty acid methyl esters (FAMEs) so that the volatility is increased.

Two methods of sample preparation for FAME determination will be used in this experiment: (1) sodium methoxide method and (2) boron trifluoride (BF$_3$) method. In the sodium methoxide method, sodium methoxide is used as a catalyst to interesterify fatty acid. This method is applicable to saturated and unsaturated fatty acids containing from 4 to 24 carbon atoms. In the BF$_3$ method, lipids are saponified, and fatty acids are liberated and esterified in the presence of a BF$_3$ catalyst for further analysis. This method is applicable to common animal and vegetable oils and fats and fatty acids. Lipids that cannot be saponified are not derivatized and, if present in large amount, may interfere with subsequent analysis. This method is not suitable for preparation of methyl esters of fatty acids containing large amounts of epoxy, hydroperoxy, aldehyde, ketone, cyclopropyl, and cyclopentyl groups and conjugated polyunsaturated and acetylenic compounds because of partial or complete destruction of these groups.

It should be noted that AOAC Method 969.33 is used in this laboratory exercise, rather than AOAC Methods 996.06, which is the method for nutrition labeling, with a focus on *trans* fats. Compared to AOAC Method 969.33, Method 996.06 used a longer and more expensive capillary column, requires a longer analysis time per sample, and involves more complicated calculations.

8.3.2 Objective

Utilize two methods to prepare methyl esters from fatty acids in food oils and then determine the fatty acid profile and their concentration in the oils by gas chromatography.

8.3.3 Chemicals

	CAS no.	Hazards
Boron trifluoride (BF$_3$)	7637-07-2	Toxic, highly flammable
Hexane	110-54-3	Harmful, highly flammable, dangerous for the environment
Methanol	67-56-1	Extremely flammable
Sodium chloride (NaCl)	7647-14-5	Irritant
Sodium hydroxide (NaOH)	1310-73-2	Corrosive
Sodium sulfate (Na$_2$SO$_4$)	7757-82-6	Harmful
Sodium methoxide	124-41-4	Toxic, highly flammable

8.3.4 Reagents and Samples

- Boron trifluoride (BF$_3$) – in methanol, 12–14 % solution
- Hexane (GC grade. If fatty acids contain 20 C atoms or more, heptane is recommended.)
- Methanolic sodium hydroxide 0.5 N (Dissolve 2 g of NaOH in 100 mL of methanol.)
- Oils: pure olive oil, safflower oil, salmon oil
- Reference standard [Gas–liquid chromatography (GLC)-60 Reference Standard FAME 25 mg is dissolved in 10 mL hexane, (Table 8.2) (Nu-Chek Prep, Inc. Elysian, MN)]
- Sodium methoxide, 0.5 *M* solution in methanol (Aldrich)
- Sodium chloride, saturated
- Sodium sulfate, anhydrous granular

8.3.5 Hazards, Precautions, and Waste Disposal

Do all work with the boron trifluoride in the hood; avoid contact with the skin, eyes, and respiratory tract. Wash all glassware in contact with boron trifluoride immediately after use. Otherwise, adhere to normal laboratory safety procedures. Wear safety glasses at all times. Boron trifluoride, hexane, and sodium methoxide must be disposed of as hazardous wastes. Other wastes likely may be put down the drain using a water rinse, but follow good laboratory practices outlined by environmental health and safety protocols at your institution.

8.3.6 Supplies

(Used by students)

- Boiling flask, 100 mL, with water-cooled condenser for saponification and esterification

8.2 table FAME GLC-60 reference standard

No.	Chain	Item	Weight %
1	C4:0	Methyl butyrate	4.0
2	C6:0	Methyl caproate	2.0
3	C8:0	Methyl caprylate	1.0
4	C10:0	Methyl caprate	3.0
5	C12:0	Methyl laurate	4.0
6	C14:0	Methyl myristate	10.0
7	C14:1	Methyl myristoleate	2.0
8	C16:0	Methyl palmitate	25.0
9	C16:1	Methyl palmitoleate	5.0
10	C18:0	Methyl stearate	10.0
11	C18:1	Methyl oleate	25.0
12	C18:2	Methyl linoleate	3.0
13	C18:3	Methyl linolenate	4.0
14	C20:0	Methyl arachidate	2.0

- Pasteur pipette
- Syringe
- Vials or sample bottle with tight-seal cap

8.3.7 Equipment

- Analytical balance
- Centrifuge
- Vortex mixer
- Gas chromatography unit (with running conditions):

Instrument	Gas chromatograph (Agilent 6890 or similar)
Detector	Flame ionization detector
Capillary column	DB-Wax or equivalent
Length	30 m
Internal diameter (ID)	0.32 mm
Df	1.0 μm
Carrier gas	He
Makeup gas	Nitrogen
Sample injection	1 μl
Split ratio	1:20
Flow rate	2 mL/min (measured at room temperature)
Injector temperature	250 °C
Detector temperature	250 °C
Temperature program	
Initial oven temperature	100 °C
Initial time	2 min
Rate	5 °C/min
Final temperature	230 °C
Final time	10 min

8.3.8 Procedure

(Instructions are given for single sample preparation and injection, but injections of samples and standards can be replicated.)

8.3.8.1 *Preparation of Methyl Esters*

Method A. Preparation of Methyl Esters by Boron Trifluoride (Adapted from AOAC Method 969.33)

Notes Methyl ester should be analyzed as soon possible or sealed in an ampule and stored in a freezer. You might also add equivalent 0.005 % 2, 6-di-*tert*-butyl-4-methylphenol (BHT). Sample size needs to be known to determine the size of the flask and the amount of reagents, according to Table 8.3.

1. Add 500 mg sample (see Table 8.3) to 100-mL boiling flask. Add 8 mL methanolic NaOH solution and boiling chip.

8.3
table

Determination of flask size and amount of reagent from approximate sample size

Sample (mg)	Flask (mL)	0.5 N NaOH (mL)	BF₃ reagent (mL)
100–250	50	4	5
250–500	50	6	7
500–750	100	8	9
750–1000	100	10	12

2. Attach condenser and reflux until fat globules disappear (about 5–10 min).
3. Add 9 mL BF₃ solution through condenser and continue boiling for 2 min.
4. Add 5 mL hexane through condenser and boil for 1 more min.
5. Remove the boiling flask and add ca. 15 mL saturated NaCl solution.
6. Stopper flask and shake vigorously for 15 s while solution is still tepid.
7. Add additional saturated NaCl solution to float hexane solution into neck of flask.
8. Transfer 1 mL upper hexane solution into a small bottle and add anhydrous Na₂SO₄ to remove H₂O.

Method B Preparation of Methyl Esters by Sodium Methoxide Method

1. Using a Pasteur pipette to transfer, weigh 100 mg (±5 mg) of sample oil to the nearest 0.1 mg into a vial or small bottle with a tight-sealing cap.
2. Add 5 mL of hexane to the vial and vortex briefly to dissolve lipid.
3. Add 250 μl of sodium methoxide reagent, cap the vial tightly, and vortex for 1 min, pausing every 10 s to allow the vortex to collapse.
4. Add 5 mL of saturated NaCl solution to the vial, cap the vial, and shake vigorously for 15 s. Let stand for 10 min.
5. Remove the hexane layer and transfer to a vial containing a small amount of Na₂SO₄. Do not transfer any interfacial precipitate (if present) or any aqueous phase.
6. Allow the hexane phase containing the methyl esters to be in contact with Na₂SO₄ for at least 15 min prior to analysis.
7. Transfer the hexane phase to a vial or small bottle for subsequent GC analysis. (Hexane solution can be stored in the freezer.)

8.3.8.2 *Injection of Standards and Samples into GC*

1. Rinse the syringe three times with hexane and three times with the reference standard mixture (25 mg of 20A GLC Reference Standard FAME dissolved in 10 mL hexane). Inject 1 μl of standard solution, remove syringe from injection port, and then press start button. Rinse syringe again three times with solvent. Use the chromatogram obtained as described below.
2. Rinse the syringe three times with hexane and three times with the sample solution prepared by Method A. Inject 1 μl of sample solution, remove syringe from injection port, and then press start button. Rinse syringe again three times with solvent. Use the chromatogram obtained as described below.
3. Repeat Step 3 for sample solution prepared by Method B.

8.3.9 Data and Calculations

1. Report retention times and relative peak areas for the peaks in the chromatogram from the FAME reference standard mixture. Use this information to identify the 14 peaks in the chromatogram.

Peak	Retention time	Peak area	Identity of peak
1			
2			
3			
4			
5			
6			
7			
8			
9			
10			
11			
12			
13			
14			

2. Using the retention times for peaks in the chromatogram from the FAME reference standard mixture, and your knowledge of the profile of the oil, identify the peaks in the chromatograms for each type of oil analyzed, [Cite your source(s) of information on the fatty acid profile of each oil.] Report results for samples from both methods of derivatization.

Results from chromatograms using *boron trifluoride method* to prepare methyl esters:

Peak	Safflower oil Retention time	Identity	Pure olive oil Retention time	Identity	Salmon oil Retention time	Identity
1						
2						
3						
4						
5						
6						
7						
8						
9						
10						
11						
12						
13						
14						

Results from chromatograms using *sodium methoxide method* to prepare methyl esters:

Peak	Safflower oil Retention time	Identity	Pure olive oil Retention time	Identity	Salmon oil Retention time	Identity
1						
2						
3						
4						
5						
6						
7						
8						
9						
10						
11						
12						
13						
14						

3. For the one oil analyzed by your group, prepare a table (with appropriate units) comparing your experimentally determined fatty acid profile to that found in your cited literature source.

	Quantity determined		
	Quantity in literature	Boron trifluoride method	Sodium methoxide method
C4:0			
C6:0			
C8:0			
C10:0			

	Quantity determined		
	Quantity in literature	Boron trifluoride method	Sodium methoxide method
C12:0			
C14:0			
C14:1			
C16:0			
C16:1			
C18:0			
C18:1			
C18:2			
C18:3			
C20:0			

Type of oil tested:

8.3.10 Questions

1. Comment on the similarities and differences in the fatty acid profiles in question #3 of Data and Calculations, comparing experimental data to literature reports. From the results, compare and decide which method of esterification to obtain FAMEs was better for your sample.

2. The approach taken in this lab provides a fatty acid profile for the oils analyzed. This is sufficient for most analytical questions regarding fatty acids. However, determining the fatty acid profile is not quite the same thing as quantifying the fatty acids in the oil. (Imagine that you wanted to use the results of your GC analysis to calculate the amount of mono- and polyunsaturated fatty acids as grams per a specified serving size of the oil). To make this procedure sufficiently quantitative for a purpose like that just described, an internal standard must be used.

   (a) Why is the fatty acid profiling method used in this lab inadequate to quantify the fatty acids?
   (b) What are the characteristics required of a suitable internal standard for FAME quantification by GC and how does this overcome the problem(s) identified in (a)?
   (c) Would the internal standard be added to the reference standard mixture and the sample or only to one of these?
   (d) When would the internal standard be added?

RESOURCE MATERIALS

Amerine MA, Ough CS (1980) Methods for analysis of musts and wine. Wiley, New York.

AOAC International (2016) Official methods of analysis, 20th edn., (On-line). Methods 968.09, 969.33, 972.10, 996.06. AOAC International, Rockville, MD

Martin GE, Burggraff JM, Randohl DH, Buscemi PC (1981) Gas–liquid chromatographic determination of congeners in alcoholic products with confirmation by gas chromatography/mass spectrometry. J Assoc Anal Chem 64:186

Ellefson WC (2017) Fat analysis. Ch. 17. In: Nielsen SS (ed) Food analysis, 5th edn. Springer, New York

Pike OA, O'Keefe SF (2017) Fat characterization. Ch. 23. In: Nielsen SS (ed) Food analysis, 5th edn. Springer, New York

Qian MC, Peterson DG, Reineccius GA (2017) Gas chromatography. Ch. 14. In: Nielsen SS (ed) Food analysis, 5th edn. Springer, New York

Mass Spectrometry with High-Performance Liquid Chromatography

Baraem P. Ismail

Department of Food Science and Nutrition, University of Minnesota,
St. Paul, MN, USA
e-mail: bismailm@umn.edu

S.S. Nielsen, *Food Analysis Laboratory Manual*, Food Science Text Series,
DOI 10.1007/978-3-319-44127-6_9, © Springer International Publishing 2017

9.1 INTRODUCTION

9.1.1 Background

Mass spectrometry (MS) is an analytical technique that provides information about the molecular weight and chemical characteristics of a compound. This technique is commonly linked to high-performance liquid chromatography (LC–MS), gas chromatography (GC–MS), inductively coupled plasma (ICP–MS), and other methods.

A mass spectrometer has three essential functions: sample ionization, ion separation, and ion detection. Components in a sample can be ionized by various ionization techniques; regardless of the source, the charged particles that are generated are separated based on their mass-to-charge ratio. Separation occurs in the mass analyzer which sorts the different ion masses by applying electric and magnetic fields. These ions are then detected which provide data on the abundance of each ion fragment present.

For LC–MS analysis, the compounds are first separated on a LC system. After detection in the LC system (by ultraviolet–visible spectroscopy (UV–VIS)), the eluent enters the ion source of a mass spectrometer where the compounds get ionized and subsequently separated based on "mass-to-charge" (m/z) ratio in the analyzer. The separated ions are then detected using an electron multiplier. Ionization using electrospray ionization (ESI) often produces precursor ions with some fragmentation. The ion which undergoes fragmentation is the precursor ion, and the ions that are produced upon fragmentation of the precursor ion are called product ions. To obtain further structural information about the compound, tandem mass spectrometry (MS/MS) can be employed. After ionization, the ions of interest are separated and fragmented by collision-induced dissociation (CID) by using an inert gas (helium, argon, etc.) as the collision gas. The energy given to the collision gas is varied depending on the desired extent of fragmentation, which will provide valuable structural information. The mass spectrum obtained by tandem MS contains only the product ions. A case study is presented to better understand the advantages of MS in tandem.

9.1.2 Reading Assignment

Reuhs, B.L. 2017. High-performance liquid chromatography. Ch. 13 in *Food Analysis*, 5th ed. S.S. Nielsen (Ed.), Springer, New York.

Smith, J.S., and Thakur, R.A. 2017. Mass spectrometry. Ch. 11 in *Food Analysis*, 5th ed. S.S. Nielsen (Ed.), Springer, New York.

9.1.3 Objectives

Determine and identify various phytochemicals (mainly isoflavones, Table 9.1 and Fig. 9.1) in soy flour, and understand how to analyze a mass spectrum to obtain mass and structural information of a compound.

9.1.4 Chemicals

	CAS No.	Hazards
HPLC grade acetonitrile	1334547-72-6	Highly flammable, acute toxicity, irritant
HPLC grade methanol	67-56-1	Highly flammable, acute toxicity, serious health hazard

9.1.5 Hazards, Precautions, and Waste Disposal

Adhere to normal laboratory safety procedures. Wear safety glasses at all times. Methanol and acetonitrile waste must be handled as hazardous waste. Other wastes likely may be put down the drain using a water rinse, but follow good laboratory practices outlined by environmental health and safety protocols at your institution.

9.1.6 Reagents

- Deionized distilled water
- HPLC grade water
- Methanol/water mixture, 80/20

9.1.7 Supplies

- Aluminum foil
- Automatic pipettors, 1000 μL, 200 μL, 100 μL, and 10 μL
- Autosampler vials, 1.5 mL, with caps
- 3 Beakers, 125 mL, each containing either 80/20 methanol, deionized distilled water, or acetonitrile
- 1 Beaker, 125 mL, for balancing the centrifuge tubes
- Buchner funnel
- 2 Centrifuge tubes, 50 mL centrifuge tubes
- Defatted soy flour
- Erlenmeyer flask, 25 mL (rinsed prior with 80/20 methanol)
- Erlenmeyer flask, 10 mL (rinsed prior with 80/20 methanol)
- 3 Graduated cylinders, 10 mL
- Filter paper, Whatman 42, 90 mm diameter
- Glass rods
- Markers
- Parafilm
- Rotary evaporator (rotovap) collection flasks (125 mL or smaller) (rinsed prior with 80/20 methanol)
- Sidearm flask, 125 mL, with a nozzle to plug the vacuum pump (rinsed prior with 80/20 methanol)
- Stir bars and stirrer
- Syringes, 3 mL
- Syringe filters, non-sterile, nylon; pore size 0.45 μm; diameter 25 mm

9.1 table Twelve known isoflavones found in soybean and their molecular weight (MW)

Isoflavone	MW	Isoflavone	MW	Isoflavone	MW
Daidzein	254	Genistein	270	Glycitein	284
Daidzin	416	Genistin	432	Glycitin	446
Acetyldaidzin	458	Acetylgenistin	474	Acetylglycitin	488
Malonyldaidzin	502	Malonylgenistin	518	Malonylglycitin	532

a

Aglycone

Non-conjugated glucoside

6″-O-Acetylglucoside

6″-O-Malonylglucoside

b

4″-O-Malonylglucoside

9.1 figure (**a**) Structures and numbering of the 12 known isoflavones categorized as aglycone, nonconjugated glucoside, acetylglucoside, and malonylglucoside. R1 can be -H in the case of daidzin and genistin or -OCH₃ in the case of glycitin, while R2 can be -H in the case of daidzin and glycitin or -OH in the case of genistin. (**b**) Structures and numbering system of 4″-O-malonylglucosides (malonylglucoside isomers)

9.2				

table Example combinations of ionization sources and mass analyzers coupled with chromatographic separation

Name	Ionization source	Mass analyzer	Tandem MS	Separation mode
LTQ Orbitrap	Electrospray ionization (ESI)	Linear ion trap and orbital trap	Yes	Liquid chromatography
5500 QTRAP®	ESI	Triple quadrupole with Q3 linear trap	Yes	Liquid chromatography
4000 QTRAP®	ESI	Triple quadrupole with Q3 linear trap	Yes	Liquid chromatography
LTQ	ESI	Linear ion trap	Yes	Liquid chromatography
LCT Premier	ESI	Time of flight (TOF)	No	Liquid chromatography
4800 AB Sciex	Matrix-associated laser desorption ionization (MALDI)	TOF–TOF	Yes	Liquid chromatography
Pegasus 4D	Electron impact (EI)	TOF	No	Gas chromatography
Agilent 6130	Dual ESI/atmospheric pressure chemical ionization (APCI)	Quadrupole	Yes	Liquid chromatography
LCQ (2)	ESI	Ion trap	Yes	Liquid chromatography
LCQ(1),	Dual ESI/APCI	Ion trap	Yes	Liquid chromatography
Biflex III	MALDI	TOF	No	Liquid chromatography
QSTAR XL	ESI	Quadrupole TOF	Yes	Liquid chromatography
Saturn 3	EI/chemical ionization (CI)	Ion trap	Yes	Gas chromatography

- Tape
- Vacuum pump
- Vortex

9.1.8 Equipment

- Analytical balance
- Centrifuge, for 15 mL tubes
- High-performance liquid chromatography system
- Mass-selective detector (quadrupole), equipped with electrospray ionization source
- Rotary evaporator

Various combinations of ionization sources and mass analyzers can be used for LC–MS and GC–MS analysis (see Table 9.2). This laboratory exercise as describes utilizes a LC–MS system coupled with electrospray ionization (ESI) and a quadrupole mass analyzer.

9.2 PROCEDURE

9.2.1 Sample Preparation

1. Weigh 0.05 gm of defatted soy into a labeled 25 mL Erlenmeyer flask.
2. Add 10 mL of HPLC grade acetonitrile to the sample and put in a stir bar.
3. Stir on a magnetic stir plate until the sample is thoroughly mixed (stirrer speed = 7, time = 5 min).
4. Add 9 mL DDW and stir for 2 h (stirrer speed = 7). Occasionally, scrape the surface of the plastic bottle with a glass rod to make the sample sticking to the sides go into the solution.

(There will be a sample on the stirrer prepared by the TA and ready for you to take to the next step.)

5. Quantitatively transfer the sample solution from the Erlenmeyer flask to a labeled centrifuge tube. Rinse the flask three times with 1 mL acetonitrile and add the rinse to the centrifuge tube. Use the scale to balance the tube with a tube containing water.
6. Centrifuge the solution in a centrifuge. Example conditions with a Marathon 3200 centrifuge are:

   Speed = 4000 rpm
   Time = 15 min
   Temperature = 15 °C

7. Take the centrifuge tube carefully out of the rotor so that the precipitate does not get disturbed.
8. Filter the permeate using a Buchner funnel, labeled 125 mL filter flask with a nozzle to plug the vacuum pump, Whatman 42 filter paper, and vacuum pump. The setup should look like in Fig. 9.2.
9. Quantitatively transfer the filtered solution into a labeled rotary evaporator (rotovap) collection flask. Again rinse three times with 1 mL acetonitrile and add the rinse to the rotovap collection flask.
10. Evaporate to dryness using a rotovap. Example conditions with a Buchi R-II Rotavapor are:

    Cooler temperature = 4 °C.
    Water bath temperature = 38 °C.
    Rotovap RPM = 120.
    Time = 15 min (may take up to 30 min or so).

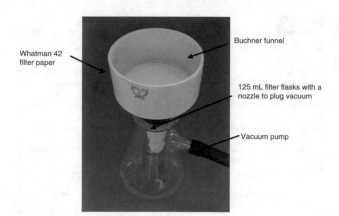

Whatman 42 filter paper

Buchner funnel

125 mL filter flasks with a nozzle to plug vacuum

Vacuum pump

9.2
figure

Setup to filter sample permeate

Pressure: since it is a two-step evaporation, evaporating of acetonitrile will be done at 115 mbar, and evaporation of water will be done at 30 mbar.

Note: If the available pump is not adequate enough to use for evaporating water from the sample, shell freezing is used to freeze the sample so it can be dried using a lyophilizer (freeze dryer).

11. There will be a sample already dried out for you and ready for the next step. Add 10 mL of 80/20 methanol/water and quantitatively transfer the solution into a labeled 25 or 10 mL Erlenmeyer flask. This time do not rinse; just transfer using a pipette.
12. Vortex for 2 min (Speed = 5).
13. Attach a 0.45 μm filter onto the end of a 3 mL syringe.
14. Pour some of the sample into the filter assembly.
15. Push sample through the filter into an autosampler vial.
16. Label the sample vial (include group name, date, sample name).
17. Seal the flask with parafilm, wrap with silver foil, and label adequately.

9.2.2   LC–MS Procedure

A Shimadzu LC-10 AD HPLC equipped with two solvent pumps and a CT-10A column oven will be used. The chromatographic column is a YMC AM-303 (ODS, 250 mm × 4.6 mm i.d.) comprising C18 reversed-phase packing (5 μm average pore size), equipped with a C18 guard column (4 mm × 20 mm i.d.).

The chosen flow rate is 1 mL/min and oven temp of 45 °C. A linear HPLC gradient will be used: Solvent A will be 0.1 % (v/v) formic acid, and solvent B will be acetonitrile. The gradient time program is described in Table 9.3.

9.3
table   HPLC gradient time program

Solvent B concentration (5 %)	Time (min)
11	0 (Initial)
14	30
14	35
30	40
30	50
11	52
11	60

9.2.3   MS Conditions

Mass spectrometry is performed on a Waters ZQ quadrupole instrument. Positive ionization will be employed. Masses are scanned from 200 to 600 m/z over 60 min, at a scan rate of one scan/s.

Eight selected ions of the following m/z are monitored, [M+H⁺]: 255, 271, 417, 433, 459, 475, 503, and 519.

The injection volume is 20 μL; the split ratio is 1:3 (so ¼ of the flow goes into the mass spectrometer, ¾ flow through the UV detector). The source temperature is 150 °C; the desolvation temperature is 450 °C. The desolvation gas flow is 600 L/h; the cone gas flow is 75 L/h. The cone voltage is 30 eV; the capillary voltage is 3 kV.

1. Turn on the gas flow by pressing the button "API gas" on the tune page.
2. Click the button "Press to Operate" on the tune page.
3. Turn on the oven and the degasser.
4. Load the HPLC conditions on the inlet page and turn on the flow.
5. Wait until the HPLC column is conditioned and the oven has reached the desired temperature.
6. Enter the file name, MS file, inlet file, and MS tune file name, and save the data set.
7. Before you inject sample, make sure that the mass spectrometer is in the "Load" mode.
8. Press the "Start" button, inject the sample, click on "Start" again, and press the "Inject" button on the mass spectrometer.

9.3   DATA AND CALCULATIONS

Analyze the total ion chromatogram and mass spectra of the selected isoflavones provided to you at the end of the lab exercise. To help you understand how to analyze a mass spectrum to obtain mass and structural information of a compound, see the Case Study, Sect. 9.5. Use the resource materials provided to answer the general questions below, and analyze the ion chromatogram and mass spectra of the isoflavones.

9.4   QUESTIONS

9.4.1   General Questions

1. What is a molecular ion? Does electron impact (EI) ionization produce a single precursor ion? How about ESI? Answer by explaining the difference between soft ionization to hard ionization.
2. List the ionization interfaces that are capable of delivering soft ionization.
3. Which ion travels farthest in unit time (low m/z or high m/z) in time-of-flight mass analyzer? How does the length of time-of-flight mass analyzer affect the resolution?
4. Explain the different between a quadrupole and an ion trap.
5. What more information can MS/MS (i.e., MS in tandem) provide versus just one MS?

9.4.2   Questions Specific to Isoflavone Analysis

1. Discuss the peaks and how you identify them.
2. Indicate if there were any co-elutions and how one can determine that.
3. Discuss the presence of isomers, if any, and how useful the MS was in identifying their presence, pointing to advantages and limitations.
4. Were you able to see precursor ions? Were you able to see product ions? Use a table format (Table 9.4) to list the precursor ions identified and their product ions as in the example table below.

9.5   CASE STUDY

Malonylgenistin is a type of isoflavone abundantly found in soybeans. Epidemiological studies have shown the association of isoflavones with many health benefits such as reducing the risk of cancer, alleviation of postmenopausal symptoms, etc. Processing of soybeans into various soy products results in the conversion of malonylgenistin to various compounds. Monitoring the conversions of malonylgenistin to different compounds is important to determine if the resultant compounds have any biological significance.

9.4
table   Identified isoflavones, precursor, and product ions m/z values

Retention Time	Isoflavone	Precursor ion m/z	List of product ions (name and m/z)

Recent studies have shown the formation of a new derivative upon heat treatment of malonylgenistin. The wavescans obtained using a photodiode array detector (PDA) of the new derivative and malonylgenistin were similar. Hence, LC/UV failed to provide any information about the structural differences between the two compounds. MS analysis of the two compounds revealed that the compounds had same mass (m/z = 519 Dalton) indicating that the new derivative is an isomer of malonylgenistin (Fig. 9.3). Tandem mass spectrometry was employed to probe structural differences between the two isomers. There were two noticeable differences in the product ion spectra (Fig. 9.4).

1. Isomer fragmented more at a lower collision level as compared to malonylgenistin.
2. Malonylgenistin product ion spectra had an ion with m/z = 433, which was absent in the isomer product ion spectra.

These differences highlight the structural differences between the two compounds. Nuclear magnetic resonance has to be employed to obtain further structural details.

RESOURCE MATERIALS

Smith JS, Thakur RA (2017) Mass spectrometry. Ch. 11. In: Nielsen SS (ed) Food analysis, 5th edn. Springer, New York

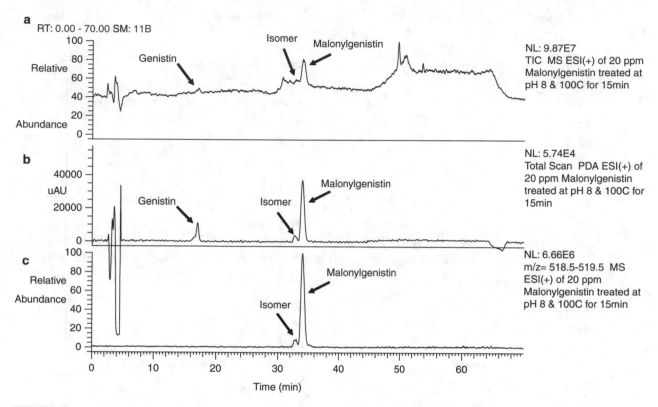

a
RT: 0.00 - 70.00 SM: 11B

NL: 9.87E7
TIC MS ESI(+) of 20 ppm Malonylgenistin treated at pH 8 & 100C for 15min

b

NL: 5.74E4
Total Scan PDA ESI(+) of 20 ppm Malonylgenistin treated at pH 8 & 100C for 15min

c

NL: 6.66E6
m/z= 518.5-519.5 MS ESI(+) of 20 ppm Malonylgenistin treated at pH 8 & 100C for 15min

9.3 figure   ESI–MS analysis malonylgenistin and its isomer: (**a**) total ion chromatogram, (**b**) PDA view of malonylgenistin solution, and (**c**) reconstructed single ion chromatogram of m/z 519 ion (protonated molecule of a malonylgenistin)

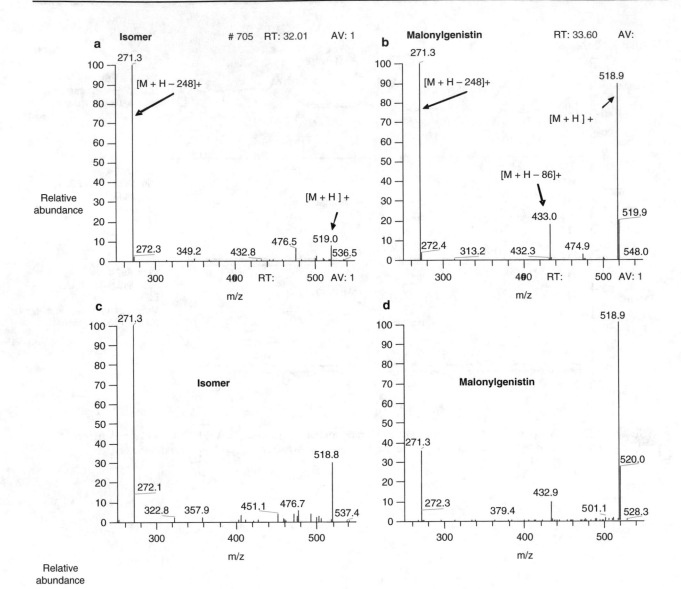

ESI–MS/MS analysis of the protonated forms of isomer and malonylgenistin at various collision levels: (**a**) isomer at 20 %, (**b**) malonylgenistin at 20 %, (**c**) isomer at 17 %, and (**d**) malonylgenistin at 17 % collision

Moisture Content Determination

S.Suzanne Nielsen

Department of Food Science, Purdue University,
West Lafayette, IN, USA
e-mail: nielsens@purdue.edu

S.S. Nielsen, *Food Analysis Laboratory Manual*, Food Science Text Series,
DOI 10.1007/978-3-319-44127-6_10, © Springer International Publishing 2017

10.1 INTRODUCTION

10.1.1 Background

The moisture (or total solids) content of foods is important to food manufacturers for a variety of reasons. Moisture is an important factor in food quality, preservation, and resistance to deterioration. Determination of moisture content also is necessary to calculate the content of other food constituents on a uniform basis (i.e., dry weight basis). The dry matter that remains after moisture analysis is commonly referred to as total solids.

While moisture content is not given on a nutrition label, it must be determined to calculate total carbohydrate content. Moisture content of foods can be determined by a variety of methods, but obtaining accurate and precise data is commonly a challenge. The various methods of analysis have different applications, advantages, and disadvantages (see Sect. 10.1.2). If the ash content also is to be determined, it is often convenient to combine the moisture and ash determinations. In this experiment several methods to determine the moisture content of foods will be used and the results compared. Note that in some cases, the method is not ideal for the sample type (e.g., moisture content of corn syrup or basil by forced draft oven), but the intent is to compare results to those obtained using more appropriate methods (i.e., vacuum oven for corn syrup, toluene distillation for basil). Summarized below are the food samples proposed for analysis and the methods used. However, note that other types of food samples could be analyzed, and groups of students could analyze different types of food samples. It is recommended that all analyses be performed in triplicate, as time permits.

	Corn syrup	Corn flour	Milk (liquid)	Nonfat dry milk	Basil
Forced draft oven	X	X	X	X	X
Vacuum oven	X				
Microwave drying	X		X		
Rapid moisture analyzer		X			
Toluene distillation				X	X
Karl Fischer method		X		X	
Near-infrared analyzer		X			

10.1.2 Reading Assignment

Mauer, L.J., and Bradley, R.L., Jr. 2017. Moisture and total solids analysis, Ch. 15, in *Food Analysis*, 5th ed. S.S. Nielsen (Ed.), Springer, New York.

10.1.3 Overall Objective

The objective of this experiment is to determine and compare the moisture contents of foods by various methods of analysis.

10.2 FORCED DRAFT OVEN

10.2.1 Objective

Determine the moisture content of corn syrup and corn flour using a forced draft oven method.

10.2.2 Principle of Method

The sample is heated under specified conditions, and the loss of weight is used to calculate the moisture content of the sample.

10.2.3 Supplies

(*Note*: Samples for moisture analysis could be the same products tested also for ash, fat, and protein analysis, by procedures in respective chapters. Content of ash, fat, and protein could then be expressed on both a wet weight basis and dry weight basis.)

- Beaker, 20–30 mL (for corn syrup)
- Basil (fresh), 15 g (ground)
- Beaker, 25–50 mL (to pour corn syrup into pans)
- Corn flour, 10 g
- Corn syrup, 15 g
- 3 Crucibles (preheated at 550 °C for 24 h)
- 2 Desiccators (with dried desiccant)
- Liquid milk, 20 mL
- Nonfat dry milk (NFDM), 10 g
- Plastic gloves (or tongs)
- 2 Spatulas
- 5 Trays (to hold/transfer samples)
- 2 Volumetric pipettes, 5 mL
- 6 Weighing pans – disposable aluminum open pans (for use with corn syrup) (predried at 100 °C for 24 h)
- 6 Weighing pans – metal pans with lids [for use with corn flour and nonfat dry milk (NFDM)] (predried at 100 °C for 24 h)

10.2.4 Equipment

- Analytical balance, 0.1 mg sensitivity
- Forced draft oven
- Hot plate

10.2.5 Note

Glass microfiber filters (e.g., GF/A, Whatman, Newton, MA), predried for 1 h at 100 °C, can be used to cover samples to prevent splattering in the forced draft oven and the vacuum oven. Instructors may want to have students compare results with and without these fiberglass covers.

10.2.6 Cautions and Hazards

Be sure to label all containers used with complete information, or record container information linker to each sample. Use gloves or tongs when handling sample pans and crucibles. These pans and crucibles have been dried and stored in desiccators prior to weighing. They will pick up moisture by sitting on the counter, so remove them from the desiccator only just before use. Open desiccators slowly to avoid damage and danger from broken glass.

10.2.7 Procedure

Instructions are given for analysis in triplicate.

10.2.7.1 *Moisture in Corn Syrup*
1. Label dried pans (disposable aluminum open pans) and weigh accurately.
2. Place 5 g of sample in the pan and weigh accurately. (Because corn syrup is very hygroscopic, work quickly, using a plastic transfer pipette, as you weigh the corn syrup.)
3. Place in a forced draft oven at 98–100 °C for 24 h.
4. Store in a desiccator until samples are weighed.
5. Calculate percentage moisture (wt/wt) as described below.

10.2.7.2 *Moisture in Corn Flour (Method 44–15A of AACC International, One-Stage Procedure)*
1. Weigh accurately the dried pan with lid. (Note identifier number on pan and lid.)
2. Place 2–3 g of sample in the pan and weigh accurately.
3. Place in a forced draft oven at 130 °C for 1 h. Be sure metal covers are ajar, to allow water loss.
4. Remove from oven, realign covers to close, cool, and store in desiccator until samples are weighed.
5. Calculate percentage moisture (wt/wt) as described below.

10.2.7.3 *Moisture in Liquid Milk (AOAC Method 990.19, 990.20)*
1. Label and weigh accurately the predried crucibles (550 °C for 24 h). (Note identified number on crucible.)
2. Place 5 g of sample in the crucible and weigh accurately.
3. Evaporate a majority of water on a hot plate; do not dry the sample completely. (Gently heat the milk in the crucibles. Wear gloves as you handle the crucibles, swirling the milk to coat the sides of the crucible. Try to avoid development of a film on the surface, until most of the water has been evaporated.)
4. Place in a forced draft oven at 100 °C for 3 h.
5. Store in a desiccator until samples are weighed.
6. Calculate percentage moisture (wt/wt) as described below.

Note: Ash content of this milk sample could be determined by placing the milk sample, dried at 100 °C for 3 h, in a muffle furnace at 550 °C for 18–24 h. After cooling in a desiccator, the crucibles containing ashed milk would be weighed and the ash content calculated. See Ash Analysis laboratory in Chap. 11.

10.2.7.4 *Moisture of Nonfat Dry Milk*
1. Weigh accurately the dried pan with lid. (Note identifier number on pan and lid.)
2. Place 3 g of sample in the pan and weigh accurately.
3. Place pan in a forced draft oven at 100 °C for 24 h.
4. Store in a desiccator until samples are weighed.
5. Calculate percentage moisture (wt/wt) as described below.

10.2.7.5 *Moisture in Fresh Basil*
1. Label dried pans (disposable aluminum open pans) and weigh accurately.
2. Place 3 g of ground sample in the pan and weigh accurately.
3. Place in a forced draft oven at 98–100 °C for 24 h.
4. Store in a desiccator until samples are weighed.
5. Calculate percentage moisture (wt/wt) as described below.

10.2.8 Data and Calculations

Calculate percentage moisture (wt/wt):

$$\% \text{ moisture} = \frac{\text{wt of H}_2\text{O in sample}}{\text{wt of wet sample}} \times 100$$

$$\% \text{moisture} = \frac{\left(\begin{array}{c}\text{wt of wet}\\\text{sample+pan}\end{array}\right) - \left(\begin{array}{c}\text{wt of dried}\\\text{sample+pan}\end{array}\right)}{\left(\begin{array}{c}\text{wt of wet}\\\text{sample+pan}\end{array}\right) - \left(\text{wt of pan}\right)} \times 100$$

Sample	Rep.	Pan (g)	Pan + wet sample (g)	Pan + dried sample (g)	Wet sample (g)	Water (g)	Moisture content (%)
Corn syrup	1						
	2						
	3						
							$\bar{X} =$
							SD =
Corn flour	1						
	2						
	3						
							$\bar{X} =$
							SD =
Liquid milk	1						
	2						
	3						
							$\bar{X} =$
							SD =
Nonfat dry milk	1						
	2						
	3						
							$\bar{X} =$
							SD =
Fresh basil	1						
	2						
	3						
							$\bar{X} =$
							SD =

10.3 VACUUM OVEN

10.3.1 Objective

Determine the moisture content of corn syrup by the vacuum oven method, with and without the addition of sand to the sample.

10.3.2 Principle

The sample is heated under conditions of reduced pressure to remove water, and the loss of weight is used to calculate the moisture content of the sample.

10.3.3 Supplies

- Corn syrup, 30 g
- Desiccator (with dried desiccant)
- 3 Glass stirring rods (ca. 2–3 cm long, predried at 100 °C for 3 h)
- Plastic gloves (or tongs)
- Pipette bulb or pump
- Sand, 30 g (predried at 100 °C for 24 h)
- 2 Spatulas
- Volumetric pipette, 5 mL
- 6 Weighing pans – disposable aluminum open pans (predried at 100 °C for 3 h)

10.3.4 Equipment

- Analytical balance, 0.1 mg sensitivity
- Vacuum oven (capable of pulling vacuum to <100 mm of mercury)

10.3.5 Cautions and Hazards

See same information in Sect. 10.2.6.

10.3.6 Procedure

10.3.6.1 *Moisture of Corn Syrup, Without the Use of Drying Sand*

1. Label weighing pans (i.e., etch identifier into tab of disposable aluminum pan) and weigh accurately.
2. Place 5 g of sample in the weighing pan and weigh accurately.
3. Dry at 70 °C and a vacuum of at least 26 in. for 24 h, but pull and release the vacuum slowly. (Note that samples without drying sand will

bubble up and mix with adjoining samples if pans are too close together.) Bleed dried air into the oven as vacuum is released.

4. Store in a desiccator until samples are cooled to ambient temperature. Weigh.

10.3.6.2 *Moisture of Corn Syrup, with the Use of Drying Sand*

1. Label weighing pan, add 10 g dried sand and stirring rod, then weigh accurately.
2. Add 5 g of sample and weigh accurately. Add 5 mL of deionized distilled (dd) water. Mix with

stirring rod being careful not to spill any of the samples. Leave the stirring bar in the pan.

3. Dry at 70 °C and a vacuum of <100 mm mercury for 24 h. Bleed dried air into the oven as vacuum is released.
4. Store in a desiccator until samples are cooled to ambient temperature. Weigh.

10.3.7 Data and Calculations

Calculate percentage moisture (wt/wt) as in Sect. 10.2.8.

Sample	Rep.	Pan (g)	Pan + wet sample (g)	Pan + dried sample (g)	Wet sample (g)	Water (g)	Moisture content (%)
Corn syrup without sand	1						
	2						
	3						
							$\bar{X} =$
							SD =
Corn syrup with sand	1						
	2						
	3						
							$\bar{X} =$
							SD =

10.4 MICROWAVE DRYING OVEN

10.4.1 Objective

Determine the moisture content of corn syrup and milk (liquid) using a microwave drying oven.

10.4.2 Principle

The sample is heated using microwave energy, and the loss of weight is used to calculate the moisture content of the sample.

10.4.3 Supplies

- Corn syrup, 4 g
- Glass stirring rod (to spread corn syrup)
- Milk (liquid), 4 g
- 6 Paper pads (for use in microwave oven)
- Pasteur pipette and bulb (to spread milk sample)
- Plastic gloves

10.4.4 Equipment

- Microwave drying oven (e.g., from CEM Corporation, Matthew, NC)

10.4.5 Procedure

Follow instructions from the manufacturer for the use of the microwave drying oven, regarding the following:

- Turning on instrument and warming up
- Loading method for specific application (i.e., sets time, power, etc.)
- Taring instrument
- Testing sample
- Obtaining results

10.4.6 Data and Calculations

Sample	Rep.	Moisture content (%)	Water/dry matter (g/g)
Corn syrup	1		
	2		
	3		
		$\bar{X} =$	$\bar{X} =$
		SD =	SD =
Milk (liquid)	1		
	2		
	3		
		$\bar{X} =$	$\bar{X} =$
		SD =	SD =

10.5 RAPID MOISTURE ANALYZER

10.5.1 Objective

Determine the moisture content of corn flour using a rapid moisture analyzer.

10.5.2 Principle

The sample placed on a digital balance is heated under controlled high-heat conditions, and the instrument automatically measures the loss of weight to calculate the percentage moisture or solids.

10.5.3 Supplies

- Corn flour, 10 g
- Plastic gloves
- Spatula

10.5.4 Equipment

- Rapid moisture analyzer (e.g., from Computrac®, Arizona Instrument LLC., Chandler, AZ)

10.5.5 Procedure

Follow instructions from the manufacturer for the use of the rapid moisture analyzer, regarding the following:

- Turning on instrument and warming up
- Select test material
- Taring instrument
- Testing sample
- Obtaining results

10.5.6 Data and Calculations

Sample	Moisture content (%)			
	1	2	3	Mean
Corn flour				

10.6 TOLUENE DISTILLATION

10.6.1 Objective

Determine the moisture content of basil by the toluene distillation method.

10.6.2 Principle

The moisture in the sample is codistilled with toluene, which is immiscible in water. The mixture that distills off is collected, and the volume of water removed is measured.

10.6.3 Chemicals

	CAS no.	Hazards
Toluene	108-88-3	Harmful, highly flammable

10.6.4 Hazards, Cautions, and Waste Disposal

Toluene is highly flammable and is harmful if inhaled. Use adequate ventilation. Wear safety glasses and gloves at all times. For disposal of toluene waste, follow good laboratory practices outlined by environmental health and safety protocols at your institution.

10.6.5 Supplies

- Fresh basil, 40–50 g
- NFDM, 40–50 g
- Toluene, ACS grade

10.6.6 Equipment

- Analytical balance, 0.1 mg sensitivity.
- Glass distillation apparatus with ground glass joints: (1) boiling flask, 250 mL or 300 mL, round-bottom, short-neck flask with a TS 24/40 joint; (2) West condenser with drip tip, 400 mm in length with a TS 24/40 joint; (3) Bidwell-Sterling trap, TS 24/40 joint, 3-mL capacity graduated in 0.1-mL intervals.
- Heat source, capable of refluxing toluene in the apparatus above (e.g., heating mantle connected to voltage controller). No open flame!
- Nylon bristle buret brush, ½ in. in diameter, and a wire loop. (It should be long enough to extend through the condenser, ca. 450 mm. Flatten the loop on the buret brush and use this brush, inverted, as a wire to dislodge moisture drops in the moisture trap.)

10.6.7 Procedure

1. Grind the fresh basil with a small tabletop food grinder. Pulse grind the sample in 5–10 s intervals. Avoid long pulses and excessive grinding to prevent frictional heat.
2. Weigh approximately 40 g of sample (basil or NFDM) accurately (amount chosen to yield 2–5 mL water).
3. Transfer sample quantitatively to distilling flask. Add sufficient toluene to cover the sample completely (not less than 75 mL).
4. Assemble the apparatus appropriate. Fill the trap with toluene by pouring it through the condenser until it just fills the trap and begins to flow into the flask. Insert a loose nonabsorbing

cotton plug into the top of the condenser to prevent condensation of atmospheric moisture in the condenser.

5. Bring to boil and reflux at about two drops per second until most of the water has been collected in the trap and then increase the reflux rate to ca. four drops per second.

6. Continue refluxing until two consecutive readings 15 min apart show no change. Dislodge any water held up in the condenser with a brush or wire loop. Rinse the condenser carefully with ca. 5 mL toluene. Dislodge any moisture droplets adhering to the Bidwell-Sterling trap or toluene trapped under the collected moisture. For this, use the wire. Rinse wire with a small amount (10 mL) of toluene before removing from the apparatus.

7. Continue refluxing for 3–5 min, remove the heat, and cool the trap to 20 °C in a suitable water bath.

8. Calculate the moisture content of the sample:

$$\% \,\text{Moisture} = \left[\text{vol. of water (mL)} / \text{wt of sample (g)} \right] \times 100$$

10.6.8 Notes

1. Flask, condenser, and receiver must be scrupulously clean and dry. For example, the apparatus, including the condenser, could be cleaned with potassium dichromate-sulfuric acid cleaning solution, rinsed with water, rinsed with $0.05\,N$ potassium hydroxide solution, rinsed with alcohol, and then allowed to drain for 10 min. This procedure will minimize the adherence of water droplets to the surfaces of the condenser and the Bidwell-Sterling trap.

2. A correction blank for toluene must be conducted periodically by adding 2–3 mL of distilled water to 100 mL of toluene in the distillation flask and then following the procedure in Steps 2–6 in Sect. 10.6.7.

10.6.9 Data and Calculations

Wt. sample (g)	Vol. water removed (mL)	Moisture content (%)

10.7 KARL FISCHER METHOD

10.7.1 Objective

Determine the moisture content of NFDM and corn flour by the Karl Fischer (KF) method.

10.7.2 Principle

When the sample is titrated with the KF reagent, which contains iodine and sulfur dioxide, the iodine is reduced by sulfur dioxide in the presence of water from the sample. The water reacts stoichiometrically with the KF reagent. The volume of KF reagent required to reach the endpoint of the titration (visual, conductometric, or coulometric) is directly related to the amount of water in the sample.

10.7.3 Chemicals

	CAS no.	Hazards
Karl Fischer reagent		Toxic
2-Methoxyethanol	109-86-4	
Pyridine	110-86-1	
Sulfur dioxide	7446-09-5	
Iodine	7553-56-2	Harmful, dangerous to the environment
Methanol, anhydrous	67-56-1	Extremely flammable
Sodium tartrate dihydrate ($Na_2C_4H_4O_6 \cdot 2H_2O$)	868-18-8	

10.7.4 Reagents

- KF reagent
- Methanol, anhydrous
- Sodium tartrate dihydrate, 1 g, dried at 150 °C for 2 h

10.7.5 Hazards, Cautions, and Waste Disposal

Use the anhydrous methanol in an operating hood, since the vapors are harmful and are toxic. Otherwise, adhere to normal laboratory safety procedures. Use appropriate eye and skin protection. The KF reagent and anhydrous methanol should be disposed of as hazardous wastes.

10.7.6 Supplies

- Corn flour
- Graduated cylinder, 50 mL
- NFDM
- 2 Spatulas
- Weighing paper

10.7.7 Equipment

- Analytical balance, 0.1 g sensitivity
- KF titration unit, nonautomated unit (e.g., from Barnsted Themaline, Berkeley, CA, Aquametry Apparatus) or automated unit

10.7.8 Procedure

Instructions are given as for a nonautomated unit and for analysis in triplicate. If using an automated unit, follow instructions of the manufacturer.

10.7.8.1 *Apparatus Setup*

Assemble titration apparatus and follow instructions of the manufacturer. The titration apparatus includes the following: buret, reservoir for reagent, magnetic stirring device, reaction/titration vessel, electrodes, and circuitry for dead stop endpoint determination. Note that the reaction/titration vessel of the KF apparatus (and the anhydrous methanol within the vessel) must be changed after analyzing several samples (exact number depends on type of sample). Remember that this entire apparatus is very fragile. To prevent contamination from atmospheric moisture, all openings must be closed and protected with drying tubes.

10.7.8.2 *Standardizing Karl Fischer Reagent*

The KF reagent is standardized to determine its water equivalence. Normally, this needs to be done only once a day, or when changing the KF reagent supply.

1. Add approximately 50 mL of anhydrous methanol to reaction vessel through the sample port.
2. Put the magnetic stir bar in the vessel and turn on the magnetic stirrer.
3. Remove the caps (if any) from drying tube. Turn the buret stopcock to the *filling position*. Hold one finger on the air-release hold in the rubber bulb and pump the bulb to fill the buret. Close the stopcock when the KF reagent reaches the desired level (at position 0.00 mL) in the buret.
4. Titrate the water in the solvent (anhydrous methanol) by adding enough KF reagent to just change the color of the solution from clear or yellow to dark brown. This is known as the KF endpoint. Note and record the conductance meter reading. (You may titrate to any point in the brown KF zone on the meter, but make sure that you always titrate to that same endpoint for all subsequent samples in the series.) Allow the solution to stabilize at the endpoint on the meter for at least 1 min before proceeding to the next step.
5. Weigh, to the nearest milligram, approximately 0.3 g of sodium tartrate dihydrate, previously dried at 150 °C for 2 h.
6. Fill the buret with the KF reagent and then titrate the water in the sodium tartrate dihydrate sample as in Sect. 10.7.8.2, Step 4. Record the volume (mL) of KF reagent used.

7. Calculate the KF reagent water (moisture) equivalence (KFR$_{eq}$) in mg H$_2$O/mL:

$$\text{KFR}_{eq} = \frac{36\,\text{g}/\text{mol} \times S \times 1000}{230.08\,\text{g}/\text{mol} \times A}$$

where:

   S = weight of sodium tartrate dihydrate (g)
   A = mL of KF reagent required for titration of sodium tartrate dihydrate

10.7.8.3 *Titration of Sample*

1. Prepare samples for analysis and place in reaction vessel as described below.

   If samples are in powder form:

- Use an analytical balance to weigh out approximately 0.3 g of sample, and record the exact sample weight (S) to the nearest milligram.
- Remove the conductance meter from the reaction vessel, and then transfer your sample to the reaction vessel through the sample port immediately. (Use an extra piece of weighing paper to form a cone-shaped funnel in the sample port, and then pour your sample through the funnel into the reaction vessel.)
- Put the conductance meter and stopper back in the reaction vessel. The color of the solution in the vessel should change to light yellow, and the meter will register below the KF zone on the meter.

   If any samples analyzed are in liquid form:

- Use a 1-mL syringe to draw up about 0.1 mL of sample. Weigh the syringe with sample on an analytical balance and record the exact weight (S_1) to the nearest milligram.
- Inject 1–2 drops of sample into the reaction vessel through the sample port and then weigh the syringe again (S_0), to the nearest milligram.
- Sample weight (S) is the difference of S_1 and S_0.

$$S = S_1 - S_0$$

1. Put the stopper back in the sample port of the reaction vessel. The color of the solution in the vessel should change to light yellow, and the meter will register below the KF zone on the meter.
2. Fill the buret, then titrate the water in the sample as in Sect. 10.7.8.2, Step 4. Record the volume (mL) of KF reagent used.
3. To titrate another sample, repeat Steps 5–7 above (Sect. 10.7.8.2) with the new sample. After

titrating several samples (exact number depends on the nature of the sample), it is necessary to start with fresh methanol in a clean reaction vessel. Record the volume (mL) of KF reagent used for each titration.

10.7.9 Data and Calculations

Calculate the moisture content of the sample as follows:

$$\% H_2O = \frac{KFR_{eq} \times K_s}{S} \times 100$$

where:

KFR_{eq} = water equivalence of KF reagent (mg H_2O/mL)

K_s = mL of KF reagent required for titration of sample

S = weight of sample (mg)

Karl Fischer reagent water equivalence (KFR_{eq}):

Rep.	Wt. sodium tartrate dihydrate (g)	Buret start (mL)	Buret end (mL)	Volume titrant (mL)	Calculated KFR_{eq}
1					
2					
3					
					$\bar{X} =$

Moisture content of samples by Karl Fischer method:

Sample Rep	Wt. sample (g)	Buret start (mL)	Buret end (mL)	Volume titrant (mL)	Moisture content (%)

10.8 NEAR-INFRARED ANALYZER

10.8.1 Objective

Determine the moisture content of corn flour using a near-infrared (NIR) analyzer.

10.8.2 Principle

Specific frequencies of infrared radiation are absorbed by the functional groups characteristic of water (i.e., the –OH stretch of the water molecule). The concentration of moisture in the sample is determined by measuring the energy that is reflected or transmitted by the sample, which is inversely proportional to the energy absorbed.

10.8.3 Supplies

- Corn flour
- Pans and sample preparation tools for near-infrared analyzer

10.8.4 Equipment

- Near-infrared analyzer

10.8.5 Procedure

Follow instructions from the manufacturer for the use of the near-infrared analyzer, regarding the following:

- Turning on instrument and warming up
- Calibrating instrument
- Testing sample
- Obtaining results

10.8.6 Data and Calculations

Corn flour % moisture			
1	2	3	Mean

10.9 QUESTIONS

1. In separate tables, summarize the results from the various methods used to determine the moisture content of each type of food sample analyzed: (a) corn syrup, (b) liquid milk, (c) corn flour, (d) NFDM, and (e) basil. Include in each table the following for each method: (a) data from individual determinations, (b) mean value, (c) standard deviation, (d) observed appearance of samples, (e) relative advantages of method, and (f) relative disadvantages of method.

2. Calculate the moisture content of the liquid milk samples as determined by the forced draft oven and microwave drying oven methods in terms of g H_2O/g dry matter and include this in a table of results.

Method	Liquid milk moisture content	
	Mean % moisture	Mean g water/g dry matter
Forced draft oven		
Microwave drying oven		

3. For the liquid milk sample analyzed for moisture content using a forced draft oven, why was the milk sample partially evaporated on a hot plate before being dried in the oven?

4. Of the various methods used to measure the moisture content of corn syrup, based on concerns for accuracy and precision, what method would you choose if you needed to measure moisture content again? Explain your answer.

5. What is the difference between moisture content and water activity measurements?

6. What method would you use to measure the moisture content of cornflakes for a) rapid quality control and b) a research project? Explain your answers. For each method, what would you have to do to the cornflakes before measuring the moisture content?

7. Explain the theory/principles involved in predicting the concentrations of various constituents in a food sample by NIR analysis. Why do we say "predict" and not "measure"? What assumptions are being made?

8. Your quality control lab has been using a hot air oven method to make moisture determinations on various products produced in your plant. You have been asked to evaluate the feasibility of switching to new methods (the specific one would depend on the product) for measuring moisture content.

(a) Describe how you would evaluate the accuracy and precision of any new method.

(b) What common problems or disadvantages with the hot air oven method would you seek to reduce or eliminate using any new method?

(c) You are considering the use of a toluene distillation procedure or Karl Fischer titration method for some of your products that are very low in moisture. What are the advantages of each of these methods over the hot air oven method in the proposed use? What disadvantages or potential problems might you encounter with the other two methods?

Acknowledgments This experiment was developed in part with materials provided by Dr. Charles E. Carpenter, Department of Nutrition, Dietetics and Food Sciences, Utah State University, Logan, UT, and by Dr. Joseph Montecalvo, Jr., Department of Food Science and Nutrition, California Polytechnic State University, San Luis Obispo, CA. Arizona Instrument Corp., Tempe, AZ, is acknowledged for its partial contribution of a Computrac moisture analyzer for use in developing a section of this laboratory exercise.

RESOURCE MATERIALS

AACC International (2010) Approved methods of analysis, 11th edn. (On-line). AACC International, St. Paul, MN

AOAC International (2016) Official methods of analysis, 20th edn. (On-line). AOAC International, Rockville, MD

Mauer LJ, Bradley RL Jr (2017) Moisture and total solids analysis, Ch. 15. In: Nielsen SS (ed) Food analysis, 5th edn. Springer, New York

Wehr HM, Frank JF (eds.) (2004) Standard methods for the examination of dairy products, 17th edn. American Public Health Association, Washington, DC

Ash Content Determination

Baraem P. Ismail

Department of Food Science and Nutrition, University of Minnesota,
St. Paul, MN, USA
e-mail: bismailm@umn.edu

S.S. Nielsen, *Food Analysis Laboratory Manual*, Food Science Text Series,
DOI 10.1007/978-3-319-44127-6_11, © Springer International Publishing 2017

11.1   INTRODUCTION

11.1.1   Background

Ash refers to the inorganic residue remaining after either ignition or complete oxidation of organic matter in a food sample. The inorganic residue consists mainly of the minerals present in the food sample. Determining the ash content is part of the proximate analysis for nutritional evaluation. Also, ashing is the first step in the preparation of a sample for specific elemental analysis. Two major types of ashing procedures are commonly used, dry ashing and wet ashing. Dry ashing is heating food at elevated temperatures (500–600 °C) in a muffle furnace. Water and volatiles will evaporate, and organic matter will burn in the presence of oxygen and convert to CO_2 and oxides of N_2. In contrast, wet ashing is based on oxidizing organic matter using acids and oxidizing agents or their combination. Minerals are thus solubilized without oxidation. Food with high-moisture content, such as vegetables, is often dried prior to ashing. Food with high-fat content, such as meat, may need to be dried and their fat extracted prior to ashing. The ash content can be expressed on a wet basis or a dry basis.

11.1.2   Reading Assignment

Harris, G.K., and Marshall, M. R. 2017. Ash analysis. Ch. 16, in *Food Analysis*, 5th ed. S.S. Nielsen (Ed.), Springer, New York.

11.1.3   Objective

Determine the ash content of a variety of food products by the dry ashing technique and express on a wet weight basis and dry weight basis.

11.1.4   Principle of Method

Organic materials are incinerated at elevated temperatures (550 °C) in a muffle furnace, and inorganic matter (ash) remains. Ash content is measured by weight of inorganic matter remaining.

11.1.5   Chemical

	CAS No.	Hazards
Hydrochloric acid (HCl)	7647-01-0	Corrosive

11.1.6   Hazards, Cautions, and Waste Disposal

Concentrated hydrochloric acid is corrosive; avoid breathing vapors and contact with skin and clothes. The muffle furnace is extremely hot. Use gloves and tongs when handling crucibles. The crucibles have been dried and stored in desiccators prior to weighing. They will pick up moisture by sitting on the counter, so remove them from the desiccators only just before use. Open desiccators slowly to avoid damage and danger from broken glass.

11.1.7   Supplies

- Ashed crucibles (numbered) (prewashed with $0.2N$ HCl, heated in a muffle furnace at 550 °C for 24 h, and stored in a desiccator prior to use)
- Variety of food products, e.g., Cheddar cheese, Parmesan cheese, pasteurized processed cheese, dry baby cereal (rice), whole wheat flour, all-purpose flour, quinoa, and dry breakfast cereal

11.1.8   Equipment

- Analytical balance
- Desiccator
- Electric muffle furnace

11.2   PROCEDURE

Note: Food products such as cheese will need to be dried before ashing (i.e., also determine the moisture content). For dry food products such as those listed above, drying is not needed before ashing. However, moisture content must be determined to calculate ash content on a dry weight basis. Follow standard procedures such as those described in the moisture determination experiment to obtain the moisture content of all samples to be ashed.

1. Remove ashed crucibles from the desiccator and record weight and number of crucible in the table.
2. Accurately weigh ca. 2 g of sample (note that cheese samples are pre-dried and placed in desiccators) into the crucible, and record weight on the spreadsheet. Prepare triplicate samples for each type of food product analyzed.
3. Place crucibles in muffle oven at 550 °C for 24 h.
4. Turn off the muffle furnace and allow it to cool (might take a few hours).
5. Remove crucibles from the muffle furnace and place into a desiccator to cool (note that this may need to be done by a teaching assistant). Return the following day to weigh the ashed sample and record weight of crucible plus ashed sample in the table.

11.3   DATA AND CALCULATIONS

$$\text{Weight of ash} = (\text{weight of crucible and ash})$$
$$- \text{weight of crucible}$$

$$\% \, Ash = \left(\text{weight of ash} / \text{original sample weight}\right) \times 100$$

Report the average ash %, standard deviation, and coefficient of variation for the food product analyzed. Also calculate average ash % on a dry basis, using the average % moisture value determined in the moisture analysis experiment.

Note: For cheese samples, the % ash obtained is on a dry weight basis (dwb), since samples to be ashed need to be pre-dried. Also, calculate % ash on a wet weight basis (wwb), using the moisture % obtained in the moisture analysis experiment.

Converting wet basis to dry basis:

$$\% \text{ash on dry basis} = \% \text{ash on wet basis} \times 100$$
$$/ \left(100 - \% \text{moisture content}\right)$$

Converting dry basis to wet basis:

$$\% \text{ash on wet basis} = \% \text{ash on dry basis} \times$$
$$\left(100 - \% \text{moisture content}\right) / 100$$

Rep	Crucible number	Crucible wt. (g)	Crucible + un-ashed sample (g)	Un-ashed sample (g)	Crucible + ashed sample (g)	Ash (g)	Moisture (%) (previously determined)	Ash (% wwb)	Ash (% dwb)
Sample A									
1									
2									
3									
								$\bar{X} =$	$\bar{X} =$
								SD =	SD =
								CV =	CV =
Sample B									
1									
2									
3									
								$\bar{X} =$	$\bar{X} =$
								SD =	SD =
								CV =	CV =

11.4 QUESTIONS

1. For this laboratory we used a dry ashing technique, in what instance would you want to use a wet ashing technique?
2. What are the advantages and disadvantages to using a dry ashing technique?
3. What are the disadvantages of using a wet ashing technique?
4. Why was it necessary to pre-acid wash, pre-ash, and dry in a desiccator the crucibles prior to use?
5. Were your results comparable to USDA reported values? Explain any discrepancies.

RESOURCES

Harris GK, Marshall MR (2017) Ash analysis, Ch. 16. In: Nielsen SS (ed) Food Analysis, 5th edn. Springer, New York

Fat Content Determination

S. Suzanne Nielsen (✉)
*Department of Food Science, Purdue University,
West Lafayette, IN, USA
e-mail: nielsens@purdue.edu*

Charles E. Carpenter
*Department of Nutrition, Dietetics and Food Sciences, Utah State University,
Logan, UT, USA
e-mail: chuck.carpenter@usu.edu*

S.S. Nielsen, *Food Analysis Laboratory Manual*, Food Science Text Series,
DOI 10.1007/978-3-319-44127-6_12, © Springer International Publishing 2017

12.1 INTRODUCTION

12.1.1 Background

The term "lipid" refers to a group of compounds that are sparingly soluble in water but show variable solubility in a number of organic solvents (e.g., ethyl ether, petroleum ether, acetone, ethanol, methanol, benzene). The lipid content of a food determined by extraction with one solvent may be quite different from the lipid content as determined with another solvent of different polarity. Fat content is determined often by solvent extraction methods (e.g., Soxhlet, Goldfish, Mojonnier), but it also can be determined by nonsolvent wet extraction methods (e.g., Babcock, Gerber) and by instrumental methods that rely on the physical and chemical properties of lipids (e.g., infrared, density, X-ray absorption). The method of choice depends on a variety of factors, including the nature of the sample (e.g., dry versus moist), the purpose of the analysis (e.g., official nutrition labeling or rapid quality control), and instrumentation available (e.g., Babcock uses simple glassware and equipment; infrared requires an expensive instrument).

This experiment includes the Soxhlet, Goldfish, Mojonnier, and Babcock methods. If samples analyzed by these methods can be tested by an instrumental method for which equipment is available in your laboratory, data from the analyses can be compared. Low-moisture snack foods are suggested for analysis and comparison by the Soxhlet and Goldfish methods, and milk by the Mojonnier and Babcock methods. However, other appropriate foods could be substituted, and results compared between methods. Also, the experiment specifies the use of petroleum ether as the solvent for the Soxhlet and Goldfish methods, for reasons of cost and safety. Anhydrous ethyl ether could be used for both extraction methods, but appropriate precautions must be taken in handling the solvent, plus the snack foods would need to be dried before extraction. Using low-moisture content snack foods (versus dried samples) admittedly introduces a low bias with petroleum ether extraction, and requires that separate samples be analyzed for moisture content, to correct for moisture in the fat content calculations. In this laboratory exercise, the moisture content data are used to express the fat content on a wet and dry weight basis.

12.1.2 Reading Assignment

Ellefson, W.C. 2017. Fat analysis. Ch. 17, in *Food Analysis*, 5rd ed. S.S. Nielsen (Ed.), Springer, New York.

12.1.3 Objective

Determine the lipid contents of various snack foods by the Soxhlet and Goldfish methods, and determine the lipid content of milk by the Mojonnier and Babcock methods.

12.2 SOXHLET METHOD

12.2.1 Principle of Method

Fat is extracted, semicontinuously, with an organic solvent. Solvent is heated and volatilized, then is condensed above the sample. Solvent drips onto the sample and soaks it to extract the fat. At 15–20 min intervals, the solvent is siphoned to the heating flask, to start the process again. Fat content is measured by the weight loss of sample or weight of fat removed.

12.2.2 Chemicals

	CAS no.	Hazards
Petroleum ether	8032-32-4	Harmful, highly flammable, dangerous for environment
(or ethyl ether)	60-29-7	Harmful, extremely flammable

12.2.3 Hazards, Precautions, and Waste Disposal

Petroleum ether and ethyl ether are fire hazards; avoid open flames, breathing vapors, and contact with skin. Ether is extremely flammable, is hygroscopic, and may form explosive peroxides. Otherwise, adhere to normal laboratory safety procedures. Wear gloves and safety glasses at all times. Petroleum ether and ether liquid wastes must be disposed of in designated hazardous waste receptacles.

12.2.4 Supplies

- 3 Aluminum weighing pans, predried in 70 °C vacuum oven for 24 h
- Beaker, 250 mL
- Cellulose extraction thimbles, predried in 70 °C vacuum oven for 24 h
- Desiccator
- Glass boiling beads
- Glass wool, predried in 70 °C vacuum oven for 24 h
- Graduated cylinder, 500 mL
- Mortar and pestle
- Plastic gloves
- Snack foods (need to be fairly dry and able to be ground with a mortar and pestle)
- Spatula
- Tape (to label beaker)
- Tongs
- Weighing pan (to hold 30-g snack food)

12.2.5 Equipment

- Analytical balance
- Soxhlet extractor, with glassware
- Vacuum oven

12.2.6 Procedure

(Instructions are given for analysis in triplicate.)

1. Record the fat content of your snack food product as reported on the package label. Also record serving size so you can calculate g fat/100-g product.
2. Slightly grind ~30-g sample with mortar and pestle (excessive grinding will lead to greater loss of fat in mortar).
3. Wearing plastic gloves, remove three predried cellulose extraction thimbles from the desiccator. Label the thimbles on the outside with your initials and a number (use a lead pencil), then weigh accurately on an analytical balance.
4. Place ~2–3 g of sample in the thimble. Reweigh. Place a small plug of dried glass wool in each thimble. Reweigh.
5. Place the three samples in a Soxhlet extractor. Put ~350-mL petroleum ether in the flask, add several glass boiling beads, and extract for 6 h or longer. Place a 250-mL beaker labeled with your name below your samples on the Soxhlet extraction unit. Samples in thimbles will be placed in the beaker after extraction and before drying.
6. Remove thimbles from the Soxhlet extractor using tongs, air dry overnight in a hood, then dry in a vacuum oven at 70 °C, 25 in. mercury, for 24 h. Cool dried samples in a desiccator then reweigh.
7. Correct for moisture content of product as follows:

    (a) Using the remainder of the ground sample and three dried, labeled, and weighed aluminum sample pans, prepare triplicate 2–3-g samples for moisture analysis.
    (b) Dry sample at 70 °C, 25 in. mercury, for 24 h in a vacuum oven.
    (c) Reweigh after drying, and calculate moisture content of the sample.

12.2.7 Data and Calculations

Using the weights recorded in the tables below, calculate the percent fat (wt/wt) on a wet weight basis as determined by the Soxhlet extraction method. If the fat content of the food you analyzed was given on the label, report this theoretical value.

Name of snack food
Label g fat/serving
Label serving size (g)
Label g fat/100-g product

Data from Soxhlet extraction:

Rep	Thimble (g)	Wet sample + thimble (g)	Wet sample (g)	Wet sample + thimble + glass wool (g)	Dry, extracted sample + thimble + glass wool (g)	Fat + moisture (g)	Fat + moisture (%)	% Fat (wwb)	% Fat (dwb)
1									
2									
3									
								$\bar{X}$ =	
								SD =	

$$\%\left(\text{Fat + Moisture}\right) = \frac{\left[\left(\text{Wet sample + thimble + glass wool, g}\right) - \left(\text{Dry, extracted sample + thimble + glass wool, g}\right)\right]}{\left(\text{Wet sample, g}\right)} \times 100$$

Data from moisture analysis:

Rep	Pan (g)	Pan + wet sample (g)	Pan + dried sample (g)	% Moisture
1				
2				
3				
			$\bar{X}$ =	
			SD =	

Calculation of % moisture:

$$\frac{\left(\text{Wt of wet sample} + \text{Pan}\right) - \left(\text{Wt of dried sample} + \text{Pan}\right)}{\left(\text{Wt of wet sample} + \text{Pan}\right) - \left(\text{Wt of pan}\right)} \times 100$$

Calculation of % fat, wet weight basis:

$$\% \text{Fat} \left(\text{wt} / \text{wt, wet weight basis}\right) =$$

$$\left(\% \text{Fat} + \% \text{Moisture}\right) - \left(\% \text{Moisture}\right)$$

(*Note:* Use average % moisture in this calculation.)

Calculate % fat on a dry weight basis:

$$\% \text{ Fat on dry basis} = \left(\% \text{ Fat on wet basis} \times 100\right) / \left(100 - \% \text{ Moisture content}\right)$$

12.2.8 Questions

1. The Soxhlet extraction procedure utilized petroleum ether. What were the advantages of using it rather than ethyl ether?
2. What were the advantages of using the Soxhlet extraction method rather than the Goldfish extraction method?
3. If the fat content measured here differed from that reported on the nutrition label, how might this be explained?

12.3 GOLDFISH METHOD

12.3.1 Principle

Fat is extracted, continuously, with an organic solvent. Solvent is heated and volatilized, then is condensed above the sample. Solvent continuously drips through the sample to extract the fat. Fat content is measured by the weight loss of sample or weight of fat removed.

12.3.2 Chemicals

Same as for Sect. 12.2.2.

12.3.3 Hazards, Precautions, and Waste Disposal

Same as for Sect. 12.2.2.

12.3.4 Supplies

Same as for Sect. 12.2.2.

12.3.5 Equipment

- Analytical balance
- Goldfish extraction apparatus
- Hot plate
- Vacuum oven

12.3.6 Procedure

(Instructions are given for analysis in triplicate.)
Note: Analyze samples in triplicate.

1. Follow Steps 1–4 in Sect. 12.2.6.
2. Place the thimble in the Goldfish condenser bracket. Push the thimble up so that only about 1–2 cm is below the bracket. Fill the reclaiming beaker with petroleum ether (50 mL) and transfer to beaker. Seal beaker to apparatus using gasket and metal ring. Start the water flow through the condenser. Raise the hot plate up to the beaker, turn on, and start the ether boiling. Extract for 4 h at a condensation rate of 5–6 drops per second.
3. Follow Steps 6 and 7 in Sect. 12.2.6.

12.3.7 Data and Calculations

Record the data and do the calculations for the Goldfish method using the same tables and equations given for the Soxhlet method in Sect. 12.2.7. Calculate the percent fat (wt/wt) on a wet weight basis. If the fat content of the food you analyzed was given on the label, report this theoretical value:

Name of snack food
Label g fat/serving
Label serving size (g)
Label g fat/100-g product

12.3.8 Questions

1. What would be the advantages and disadvantages of using ethyl ether rather than petroleum ether in a solvent extraction method, such as the Goldfish method?
2. What were the advantages of using the Goldfish extraction method rather than the Soxhlet extraction method?
3. If the fat content measured here differed from that reported on the nutrition label, how might this be explained?

12.4 MOJONNIER METHOD

12.4.1 Principle

Fat is extracted with a mixture of ethyl ether and petroleum ether. The extract containing the fat is dried and expressed as percent fat by weight.

The assay uses not only ethyl ether and petroleum ether, but also ammonia and ethanol. Ammonia dissolves the casein and neutralizes the acidity of the product to reduce its viscosity. Ethanol prevents gelation of the milk and ether, and aids in the separation of the ether-water phase. Ethyl ether and petroleum ether serve as lipid solvents, and petroleum ether decreases the solubility of water in the ether phase.

12.4.2 Chemicals

	CAS no.	Hazards
Ammonium hydroxide	1336-21-6	Corrosive, dangerous for the environment
Ethanol	64-17-5	Highly flammable
Petroleum ether	8032-32-4	Harmful, highly flammable, dangerous for environment
(Ethyl ether)	60-29-7	Harmful, extremely flammable

12.4.3 Hazards, Precautions, and Waste Disposal

Ethanol, ethyl ether, and petroleum ether are fire hazards; avoid open flames, breathing vapors, and contact with skin. Ether is extremely flammable, is hygroscopic, and may form explosive peroxides. Ammonia is a corrosive; avoid contact and breathing vapors. Otherwise, adhere to normal laboratory safety procedures. Wear gloves and safety glasses at all times. Petroleum ether and ether liquid wastes must be disposed of in designated hazardous waste receptacles. The aqueous waste can go down the drain with a water rinse.

12.4.4 Supplies

- Milk, whole and 2 % fat
- Mojonnier extraction flasks, with stoppers
- Mojonnier fat dishes with lids
- Plastic gloves
- Tongs

12.4.5 Equipment

- Analytical balance
- Hot plate
- Mojonnier apparatus (with centrifuge, vacuum oven, and cooling desiccator)

12.4.6 Notes

Reagents must be added to the extraction flask in the following order: water, ammonia, alcohol, ethyl ether, and petroleum ether. The burets on the dispensing cans or tilting pipets are graduated for measuring the proper amount. Make triplicate determinations on both the sample and reagent blanks. The procedure given here is for fresh milk. Other samples may need to be diluted with distilled water in Step 2 and require different quantities of reagents in subsequent steps. Consult the instruction manual or AOAC International *Official Methods of Analysis* for samples other than fresh milk.

12.4.7 Procedure

(Instructions are given for analysis in triplicate.)

1. Turn on power unit and temperature controls for oven and hot plate on the fat side of the Mojonnier unit.
2. Warm milk samples to room temperature and mix well.
3. When oven is at 135 °C, heat cleaned fat dishes in oven under a vacuum of 20-in. mercury for 5 min. Handle dishes from this point on with tongs or gloves. Use three dishes for each type of milk samples, and two dishes for the reagent blank.
4. Cool dishes in cooling desiccator for 7 min.
5. Weigh dishes, record weight of each dish and its identity, and place dishes in desiccator until use.
6. Weigh samples accurately (ca. 10 g) into Mojonnier flasks. If weighing rack is used, fill curved pipettes and place in rack on the balance. Weigh each sample by difference.
7. Add chemicals for the first extraction in the order and amounts given below. After each addition of chemicals, stopper the flask and shake by inverting for 20 s.

	First extraction		Second extraction	
Chemicals	Step	Amount (mL)	Step	Amount (mL)
Ammonia	1	1.5	–	None
Ethanol	2	10	1	5
Ethyl ether	3	25	2	15
Petroleum ether	4	25	3	15

8. Place the extraction flasks in the holder of the centrifuge. Place both flask holders in the centrifuge. Operate the centrifuge to run at 30 turns in 30 s, to give a speed of 600 rpm (revolutions per minute). [In lieu of centrifuging, the flasks can be allowed to stand 30 min until a clear separation line forms, or three drops of phenolphthalein indicator (0.5% w/v ethanol) can be added during the first extraction to aid in determining the interface.]
9. Carefully pour off the ether solution of each sample into a previously dried, weighed, and cooled fat dish. Most or all of the ether layer should be poured into the dish, but none of the remaining liquid must be poured into the dish.
10. Place dishes with ether extract on hot plate under glass hood of Mojonnier unit, with power unit running. (If this hot plate is not available, use a hot plate placed in a hood, with the hot plate at 100 °C.)

11. Repeat the extraction procedure a second time for the samples in the Mojonnier flasks, following the sequence and amount given in the table above. Again, after each addition of chemicals, stopper the flask and shake by inverting for 20 s. Centrifuge the flasks again, as described above. Distilled water may be added now to the flask to bring the dividing line between ether and water layers to the center of the neck of flask. If this is done, repeat the centrifugation.

12. Pour ether extract into the respective fat dish (i.e., the ether for a specific sample should be poured into the same fat dish used for that sample from the first extraction), taking care to remove all the ether but none of the other liquid in the flask.

13. With lids ajar, complete the evaporation of ether, either very carefully on the hot plate (this can be problematic and a fire hazard) or open in a hood. In using a hot plate, the ether should boil slowly; not fast enough to cause splattering. If the plate appears to be too hot and boiling is too fast, only part of the dish should be placed on the hot plate. If instead using an operating hood, leave collection containers with lids ajar to have them evaporated by the next day.

14. When all the ether has been evaporated from the dishes, place the dishes in the vacuum oven 70–75 °C for 10 min with a vacuum of at least 20 in.

15. Cool the dishes in the desiccator for 7 min.

16. Accurately weigh each dish with fat. Record weight.

12.4.8 Data and Calculations

Calculate the fat content of each sample. Subtract the average weight of the reagent blank from the weight of each fat residue in the calculation.

12.4.9 Questions

1. List possible causes for high and low results in a Mojonnier fat test.
2. How would you expect the elimination of alcohol from the Mojonnier procedure to affect the results? Why?
3. How would you propose to modify the Mojonnier procedure to test a solid, nondairy product? Explain your answer.

12.5 BABCOCK METHOD

12.5.1 Principle

Sulfuric acid is added to a known amount of milk sample in a Babcock bottle. The acid digests the protein, generates heat, and releases the fat. Centrifugation and hot water addition isolate the fat into the graduated neck of the bottle. The Babcock fat test uses a volumetric measurement to express the percent of fat in milk or meat by weight.

12.5.2 Note

The fat column in the Babcock test should be at 57–60 °C when read. The specific gravity of liquid fat at that temperature is approximately 0.90 g per mL. The calibration on the graduated column of the test bottle reflects this fact and enables one to make a volumetric measurement, which expresses the fat content as percent by weight.

12.5.3 Chemicals

	CAS No.	Hazards
Glymol (red reader)	8042-47-5	Toxic, irritant
Sulfuric acid	7664-93-9	Corrosive

Rep	Milk start (g)	Milk end (g)	Milk tested (g)	Dish (g)	Dish + fat (g)	Calculated % fat
Reagent blanks:						
1						–
2						–
Samples:						$\bar{X} =$
1						
2						
3						
						$\bar{X} =$
						SD =

$$\% \text{ Fat} = 100 \times \left\{ \left[(\text{wt dish} + \text{fat}) - (\text{wt dish}) \right] - (\text{avg wt blank residue}) \right\} / \text{wt sample}$$

12.5.4 Hazards, Precautions, and Waste Disposal

Concentrated sulfuric acid is extremely corrosive; avoid contact with skin and clothes and breathing vapors. Wear gloves and safety glasses at all times. Otherwise, adhere to normal laboratory safety procedures. Sulfuric acid and glymol wastes must be disposed in a designated hazardous waste receptacle.

For safety and accuracy reasons, dispense the concentrated sulfuric acid from a bottle fitted with a repipettor (i.e., automatic bottle dispenser). Fit the dispenser with a thin, semirigid tube to dispense directly and deep into the Babcock bottle while mixing contents. Set the bottle with dispenser on a tray to collect spills. Wear corrosive- and heat-resistant gloves when mixing the sulfuric acid with samples.

12.5.5 Supplies

- 3 Babcock bottles
- Babcock caliber (or shrimp divider)
- Measuring pipette, 10 mL
- Pipette bulb or pump
- Plastic gloves
- Standard milk pipette (17.6 mL)
- Thermometer

12.5.6 Equipment

- Babcock centrifuge
- Water bath

12.5.7 Procedure

(Instructions are given for analysis in triplicate.)

1. Adjust milk sample to ca. 38 °C and mix until homogenous. Using a standard milk pipette, pipette 17.6 mL of milk into each of three Babcock bottles. After the pipette has emptied, blow out the last drops of milk from the pipette tip into the bottle. Allow milk samples to adjust to ca. 22 °C.
2. Dispense ca. 17.5 mL of sulfuric acid (specific gravity 1.82–1.83) and carefully add into the test bottle, with mixing during and between additions, taking care to wash all traces of milk into the bulb of the bottle. Time for complete acid addition should not exceed 20 s. Mix the milk and acid thoroughly. Be careful not to get any of the mixture into the column of the bottle while shaking. Heat generated behind any such lodged mixture may cause a violent expulsion from the bottle.
3. Place bottles in centrifuge heated to 60 °C. Be sure bottles are counterbalanced. Position bottles so that bottlenecks will not be broken in horizontal configuration. Be sure that the heater of the centrifuge is on.
4. Centrifuge the bottles for 5 min after reaching the proper speed (speed will vary depending upon the diameter of the centrifuge head).
5. Stop the centrifuge and add soft hot water (60 °C) until the liquid level is within 0.6 cm of the neck of the bottle. Carefully permit the water to flow down the side of the bottle. Again, centrifuge the bottles for 2 min.
6. Stop the centrifuge and add enough soft hot water (60 °C) to bring the liquid column near the top graduation of the scale. Again, centrifuge the bottles for 1 min.
7. Remove the bottles from the centrifuge and place in a heated (55–60 °C, preferably 57 °C) water bath deep enough to permit the fat column to be below the water level of the water bath. Allow bottles to remain at least 5 min before reading.
8. Remove the samples from the water bath one at a time, and quickly dry the outside of the bottle. Add glymol (red reader) to the top of fat layer. Immediately use a divider or caliper to measure the fat column to the nearest 0.05%, holding the bottle in a vertical position at eye level. Measure from the highest point of the upper meniscus to the bottom of the lower meniscus.
9. Reject all tests in which the fat column is milky or shows the presence of curd or charred matter, or in which the reading is indistinct or uncertain. The fat should be clear and sparkling, the upper and lower meniscus clearly defined, and the water below the fat column should be clear.
10. Record the readings of each test and determine the mean % fat and the standard deviation.

12.5.8 Data and Calculations

Rep	Measured % fat
1	
2	
3	
	$\bar{X} =$
	SD =

12.5.9 Questions

1. What are the possible causes of charred particles in the fat column of the Babcock bottle?
2. What are the possible causes of undigested curd in the Babcock fat test?
3. Why is sulfuric acid preferred over other acids for use in the Babcock fat test?

RESOURCE MATERIALS

AOAC International (2016) Official methods of analysis, 20th edn, (On-line). AOAC International, Rockville, MD

Ellefson WC (2017) Fat analysis. Ch. 17. In: Nielsen SS (ed) Food analysis, 5th edn. Springer, New York

Wehr HM, and Frank JF (eds) (2004) Standard methods for the examination of dairy products. 17th edn. American Public Health Administration, Washington, DC

Protein Nitrogen Determination

S.Suzanne Nielsen

Department of Food Science, Purdue University,
West Lafayette, IN, USA
e-mail: nielsens@purdue.edu

S.S. Nielsen, *Food Analysis Laboratory Manual*, Food Science Text Series,
DOI 10.1007/978-3-319-44127-6_13, © Springer International Publishing 2017

13.1 INTRODUCTION

13.1.1 Background

The protein content of foods can be determined by numerous methods. The Kjeldahl method and the nitrogen combustion (Dumas) method for protein analysis are based on nitrogen determination. Both methods are official for the purposes of nutrition labeling of foods. While the Kjeldahl method has been used widely for over 100 years, the recent availability of automated instrumentation for the Dumas method in many cases is replacing use of the Kjeldahl method.

13.1.2 Reading Assignment

Chang, S.K.C., Zhang, Y. 2017. Protein analysis. Ch. 18, in *Food Analysis*, 5th ed. S.S. Nielsen (Ed.), Springer, New York.

13.1.3 Notes

Both the Kjeldahl and nitrogen combustion methods can be done without automated instrumentation, but they are commonly done with automated instruments. The descriptions below are based on the availability of such automated instrumentation. If protein content of samples analyzed by Kjeldahl and/or nitrogen combustion has been estimated in a previous experiment by near infrared analysis, values can be compared between methods.

13.2 KJELDAHL NITROGEN METHOD

13.2.1 Objective

Determine the protein content of corn flour using the Kjeldahl method.

13.2.2 Principle of Method

The Kjeldahl procedure measures the nitrogen content of a sample. The protein content then can be calculated assuming a ratio of protein to nitrogen for the specific food being analyzed. The Kjeldahl procedure can be basically divided into three parts: (1) digestion, (2) distillation, and (3) titration. In the digestion step, organic nitrogen is converted to an ammonium in the presence of a catalyst at approximately 370 °C. In the distillation step, the digested sample is made alkaline with NaOH, and the nitrogen is distilled off as NH_3. This NH_3 is "trapped" in a boric acid solution. The amount of ammonia nitrogen in this solution is quantified by titration with a standard HCl solution.

A reagent blank is carried through the analysis and the volume of HCl titrant required for this blank is subtracted from each determination.

13.2.3 Chemicals

	CAS No.	Hazards
Boric acid (H_3BO_3)	10043-35-3	
Bromocresol green	76-60-8	
Ethanol, 95%	64-17-5	Highly flammable
Hydrochloric acid, conc. (HCl)	7647-01-0	Corrosive
Methyl red	493-52-7	
Sodium hydroxide (NaOH)	1310-73-2	Corrosive
Sulfuric acid, conc. (H_2SO_4)	7664-93-9	Corrosive
Kjeldahl digestion tablets		Irritant
Potassium sulfate (K_2SO_4)	7778-80-5	
Cupric sulfate	7758-98-7	
Titanium dioxide (TiO_2)	13463-67-7	
Tris (hydroxymethyl) aminomethane (THAM)	77-86-1	Irritant

13.2.4 Reagents

(** It is recommended that these solutions be prepared by the laboratory assistant before class.)

- Sulfuric acid (concentrated, N-free)
- Catalyst/salt mixture (Kjeldahl digestion tablets)
    Contains potassium sulfate, cupric sulfate, and titanium dioxide
    Note: There are several types of Kjeldahl digestion tablets that contain somewhat different chemicals.
- Sodium hydroxide solution, 50%, w/v, NaOH in deionized distilled (dd) water **
    Dissolve 2000 g sodium hydroxide (NaOH) pellets in ~3.5 L dd water. Cool. Add dd water to make up to 4.0 L.
- Boric acid solution **
    In a 4-L flask, dissolve 160-g boric acid in ca. 2 L boiled, and still very hot, dd water. Mix and then add an additional 1.5 L of boiled, hot dd water. Cool to room temperature under tap water (caution: glassware may break due to sudden cooling) or leave overnight. When using the rapid procedure, the flask must be shaken occasionally to prevent recrystallization of the boric acid. Add 40 mL of bromocresol green solution (100-mg bromocresol green/100-mL ethanol) and 28 mL of methyl red solution (100-mg methyl red/100-mL ethanol). Dilute to 4 L of water and mix carefully. Transfer 25 mL of the boric acid solution to a receiver flask and distill a digested blank (a digested catalyst/salt/acid mixture). The contents of the flask should then be a neutral gray. If not, titrate with $0.1 N$ NaOH solution

until this color is obtained. Calculate the amount of NaOH solution necessary to adjust the boric acid solution in the 4-L flask with the formula:

$$\text{mL } 0.1\,N \text{ NaOH} = \frac{(\text{mL titer}) \times (4000\text{ mL})}{(25\text{ mL})}$$

Add the calculated amount of $0.1\,N$ NaOH solution to the boric acid solution. Mix well. Verify the adjustment results by distilling a new blank sample. Place adjusted solution into a bottle equipped with a 50-mL repipettor.

- Standardized HCl solution**

  Dilute 3.33-mL conc. HCl to 4 L with dd water. Empty old HCl solution from the titrator reservoir and rinse three times with a small portion of the new HCl solution. Fill the titrator with the new HCl solution to be standardized. Using a volumetric pipet, dispense 10-mL aliquots of the THAM solution prepared as described below into three Erlenmeyer flasks (50 mL). Add 3–5 drops indicator (3 parts 0.1 % bromocresol green in ethanol to 1 part of 0.2 % methyl red in ethanol) to each flask and swirl. Titrate each solution with the HCl solution to a light pink endpoint. Note the acid volume used and calculate the normality as described below.

  Calculation to standardize HCl solution:

  $$\text{Normality} = \frac{\text{mL THAM} \times \text{THAM Normality}}{\text{average acid volume (AAV)}}$$
  $$= \frac{20\text{ mL} \times 0.01\,N}{\text{AAV}}$$

  Write the normality of the standardized HCl solution on the stock container.

- Tris (hydroxymethyl) aminomethane (THAM) Solution – (0.01 N) **

  Place 2 g of THAM in a crucible. Leave in a drying oven (95 °C) overnight. Let cool in a desiccator. In a 1-L volumetric flask, dissolve 1.2114 g of oven dried THAM in distilled water. Dilute to volume.

13.2.5 Hazards, Cautions, and Waste Disposal

Concentrated sulfuric acid is extremely corrosive; avoid breathing vapors and contact with skin and clothes. Concentrated sodium hydroxide is a corrosive. Wear corrosion-resistant gloves and safety glasses at all times. Perform the digestions in an operating hood with an aspirating fume trap attached to the digestion unit. Allow samples to cool in the hood before removing the aspirating fume trap from the digestion unit. Otherwise, adhere to normal laboratory safety procedures. The waste of combined sulfu-

ric acid and sodium hydroxide has been largely neutralized (check pH to ensure it is pH 3–9), so it can be discarded down the drain with a water rinse. However, for disposing any chemical wastes, follow good laboratory practices outlined by environmental health and safety protocols at your institution.

For safety and accuracy reasons, dispense the concentrated sulfuric acid from a bottle fitted with a repipettor (i.e., automatic dispenser). Fit the dispenser with a thin, semirigid tube to dispense directly into the Kjeldahl tube. Set the bottle with dispenser on a tray to collect spills.

13.2.6 Supplies

(Used by students)

- Corn flour (not dried)
- 5 Digestion tubes
- 5 Erlenmeyer flasks, 250 mL
- Spatula
- Weighing paper

13.2.7 Equipment

- Analytical balance
- Automatic titrator
- Kjeldahl digestion and distillation system

13.2.8 Procedure

(Instructions are given for analysis in triplicate. Follow manufacturer's instructions for specific Kjeldahl digestion and distillation system used. Some instructions given here may be specific for one type of Kjeldahl system.)

13.2.8.1 *Digestion*

1. Turn on digestion block and heat to appropriate temperature.
2. Accurately weigh approximately 0.1 g of corn flour. Record the weight. Place corn flour in digestion tube. Repeat for two more samples.
3. Add one catalyst tablet and appropriate volume (e.g., 7 mL) of concentrated sulfuric acid to each tube with corn flour. Prepare duplicate blanks: one catalyst tablet + volume of sulfuric acid used in the sample + weigh paper (if weigh paper was added with the corn flour samples).
4. Place rack of digestion tubes on digestion block. Cover digestion block with exhaust system turned on.
5. Let samples digest until digestion is complete. The samples should be clear (but neon green), with no charred material remaining.
6. Take samples off the digestion block and allow to cool with the exhaust system still turned on.
7. Carefully dilute digest with an appropriate volume of dd water. Swirl each tube.

13.2.8.2 Distillation

1. Follow appropriate procedure to start up distillation system.
2. Dispense appropriate volume of boric acid solution into the receiving flask. Place receiving flask on distillation system. Make sure the tube coming from the distillation of the sample is submerged in the boric acid solution.
3. Put sample tube from Sect. 13.2.8.1 in place, making sure it is seated securely, and proceed with the distillation until completed. In this distillation process, a set volume of NaOH solution will be delivered to the tube and, a steam generator will distill the sample for a set period of time.
4. Upon completing distillation of one sample, proceed with a new sample tube and receiving flask.
5. After completing distillation of all samples, follow manufacturer's instructions to shut down the distillation unit.

13.2.8.3 Titration

1. Record the normality of the standardized HCl solution, as determined by the teaching assistant.
2. If using an automated pH meter titration system, follow manufacturer's instructions to calibrate the instrument. Put a magnetic stir bar in the receiver flask from Sect. 13.2.8.2 and place it on a stir plate. Keep the solution stirring briskly while titrating, but do not let the stir bar hit the electrode. Titrate each sample and blank to an endpoint pH of 4.2. Record volume of HCl titrant used.
3. If using a colorimetric endpoint, put a magnetic stir bar in the receiver flask, place it on a stir plate, and keep the solution stirring briskly while titrating. Titrate each sample and blank with the standardized HCl solution to the first faint gray color. Record volume of HCl titrant used.

13.2.9 Data and Calculations

Calculate the percent nitrogen and the percent protein for each of your duplicate or triplicate corn flour samples, then determine average values. The corn flour sample you analyzed was not a dried sample. Report percent protein results on a wet weight basis (wwb) and on a dry weight basis (dwb). Assume a moisture content of 10 % (or use the actual moisture content if previously determined on this corn flour sample). Use 6.25 for the nitrogen to protein conversion factor.

$$\% N = \text{Normality HCl} * \times \frac{\text{corrected acid vol.(mL)}**}{\text{g of sample}} \times \frac{14 \, g \, N}{mol} \times 100$$

* Normality is in mol/1000 mL.
** Corrected acid vol = (mL std. acid for sample) − (mL std. for blank)

$$\% \text{Protein} = \% N \times \text{Protein Factor}$$

$$\frac{\% \text{ Protein, wwb}}{[\% \text{ Solids (wet)}/100 \, \%]} = \% \text{ Protein, dwb}$$

	Rep	Sample wt. (g)	Vol. HCl titrant (mL)	% Nitrogen	% Protein, wwb	% Protein, dwb
Blank	1	–		–	–	–
	2	–		–	–	–
			$\bar{X} =$			
Sample	1					
	2					
	3				$\bar{X} =$	$\bar{X} =$
					SD =	SD =

13.2.10 Questions

1. If the alkali pump timer on the distillation system was set to deliver 25 mL of 50 % NaOH and 7 mL of concentrated H_2SO_4 was used to digest the sample, how many milliliters of the 50 % NaOH is actually required to neutralize the amount of sulfuric acid used in the digestion? How would your results have been changed if the alkali pump timer malfunctioned and delivered only 15 mL of the 50 % NaOH? (Molarity of conc. $H_2SO_4 = 18$.)
2. Could phenolphthalein be used as an indicator in the Kjeldahl titration? Why or why not?
3. Describe the function of the following chemicals used in this determination:

   (a) Catalyst pellet
   (b) Borate
   (c) H_2SO_4
   (d) NaOH

4. Why was it not necessary to standardize the boric acid solution?
5. Explain how the factor used to calculate the percent protein for your product was obtained, and why the protein factors for some other cereal grains (e.g., wheat, oats) differ from that for corn.
6. For each of the disadvantages of the Kjeldahl method, give another protein analysis method that overcomes (at least partially) that disadvantage.

13.3 NITROGEN COMBUSTION METHOD

13.3.1 Objective

Determine the protein content of corn flour using the nitrogen combustion method.

13.3.2 Principle of Method

The nitrogen combustion method measures the nitrogen content of a sample. The protein content then is calculated assuming a ratio of protein to nitrogen for the specific food being analyzed. In the assay, the sample is combusted at a high temperature (900–1110 °C) to release nitrogen gas and other products (i.e., water, other gases). The other products are removed, and the nitrogen is quantitated by gas chromatography using a thermal conductivity detector.

13.3.3 Chemicals

	CAS No.	Hazards
Ethylenediaminetetraacetic acid (EDTA), disodium salt ($Na_2EDTA \cdot 2H_2O$)	60-00-4	Irritant

(The other chemicals used are specific to each manufacturer for the columns within the instrument.)

13.3.4 Hazards, Cautions, and Waste Disposal

During operation, the front panel of the instrument gets very hot. Check instructions of manufacturer for any other hazards, especially those associated with maintenance of instrument.

13.3.5 Supplies

(Used by students)

　Corn flour
　Sample cup

13.3.6 Equipment

Nitrogen combustion unit

13.3.7 Procedure

Follow manufacturer's instructions for start-up, analyzing samples, and shutdown.

Weigh appropriate amount of sample into a tared sample cup on an analytical balance. (Sample weight will be coordinated with sample number in autosampler, if autosampler is used.) Remove sample from balance and prepare for insertion following manufacturer's instructions. If an autosampler is used, the weighed sample must be placed into autosampler in the appropriate slot for the sample number. Repeat this procedure for EDTA standard. Sample and standard should be run in duplicate or triplicate.

13.3.8 Data and Calculations

Record the percent nitrogen content for each of your duplicate or triplicate corn flour samples. Calculate protein content from percent nitrogen data, and determine average percent protein. The corn flour sample you analyzed was not a dried sample. Report percent protein results on a wet weight basis (wwb) and on a dry weight basis (dwb). Assume a moisture content of 10 % (or use the actual moisture content if previously determined on this corn flour sample). Use 6.25 for the nitrogen to protein conversion factor.

Sample	% Nitrogen	% Protein, wwb	% Protein, dwb
1			
2			
3			
		$\bar{X} =$	$\bar{X} =$
		SD =	SD =

13.3.9 Questions

1. What are the advantages of the nitrogen combustion method compared to the Kjeldahl method?
2. Explain why ethylenediaminetetraacetic acid (EDTA) can be used as a standard to check the calibration of the nitrogen analyzer.
3. If you analyzed the corn flour sample by both the Kjeldahl and nitrogen combustion methods, compare the results. What might explain any differences?

RESOURCE MATERIALS

Chang SKC, Zhang Y (2017) Protein analysis. Ch. 18. In: Nielsen SS (ed) Food analysis, 5th edn. Springer, New York

AOAC International (2016) Official methods of analysis, 20th edn, (On-line). Method 960.52 (Micro-Kjeldahl method) and Method 992.23 (Generic combustion method). AOAC International, Rockville, MD

Total Carbohydrate by Phenol-Sulfuric Acid Method

S. Suzanne Nielsen

Department of Food Science, Purdue University,
West Lafayette, IN, USA
e-mail: nielsens@purdue.edu

S.S. Nielsen, *Food Analysis Laboratory Manual*, Food Science Text Series,
DOI 10.1007/978-3-319-44127-6_14, © Springer International Publishing 2017

14.1 INTRODUCTION

14.1.1 Background

The phenol-sulfuric acid method is a simple and rapid colorimetric method to determine total carbohydrates in a sample. The method detects virtually all classes of carbohydrates, including mono-, di-, oligo-, and polysaccharides. Although the method detects almost all carbohydrates, the absorptivity of the different carbohydrates varies. Thus, unless a sample is known to contain only one carbohydrate, the results must be expressed arbitrarily in terms of one carbohydrate.

In this method, the concentrated sulfuric acid breaks down any polysaccharides, oligosaccharides, and disaccharides to monosaccharides. Pentoses (5-carbon compounds) then are dehydrated to furfural, and hexoses (6-carbon compounds) to hydroxymethyl furfural. These compounds then react with phenol to produce a yellow-gold color. For products that are very high in xylose (a pentose), such as wheat bran or corn bran, xylose should be used to construct the standard curve for the assay and measure the absorption at 480 nm. For products that are high in hexose sugars, glucose is commonly used to create the standard curve, and the absorption is measured at 490 nm. The color for this reaction is stable for several hours, and the accuracy of the method is within ±2% under proper conditions.

Carbohydrates are the major source of calories in soft drinks, beer, and fruit juices, supplying 4 Cal/gram carbohydrate. In this experiment, you will create a standard curve with a glucose standard solution, use it to determine the carbohydrate concentration of soft drinks and beer, then calculate the caloric content of those beverages.

14.1.2 Reading Assignment

BeMiller, J.N. 2017. Carbohydrate analysis. Ch. 19, in *Food Analysis*, 5th ed. S.S. Nielsen (Ed.), Springer, New York.

14.1.3 Objective

Determine the total carbohydrate content of soft drinks and beers.

14.1.4 Principle of Method

Carbohydrates (simple sugars, oligosaccharides, polysaccharides, and their derivatives) react in the presence of strong acid and heat to generate furan derivatives that condense with phenol to form stable yellow-gold compounds that can be measured spectrophotometrically.

14.1.5 Chemicals

	CAS No.	Hazards
D-Glucose ($C_6H_{12}O_6$)	50-99-7	
Phenol (C_6H_6O)	108-95-2	Toxic
Sulfuric acid (H_2SO_4)	7664-93-9	Corrosive

14.1.6 Reagents

(** It is recommended that these solutions be prepared by the laboratory assistant before class.)

- Glucose std. solution, 100 mg/L **
- Phenol, 80% wt/wt in H_2O, 1 mL **
  Prepare by adding 20-g deionized distilled (dd) water to 80 g of redistilled reagent grade phenol (crystals).
- Sulfuric acid, concentrated

14.1.7 Hazards, Cautions, and Waste Disposal

Use concentrated H_2SO_4 and the 80% phenol solution with caution. Wear gloves and safety glasses at all times, and use good lab technique. The concentrated H_2SO_4 is very corrosive (e.g., to clothes, shoes, skin). The phenol is toxic and must be discarded as hazardous waste. Other waste not containing phenol likely may be put down the drain using a water rinse, but follow good laboratory practices outlined by environmental health and safety protocols at your institution.

14.1.8 Supplies

(Used by students)

- Beer (lite and regular, of same brand)
- Bottle to collect waste
- Cuvettes (tubes) for spectrophotometer
- Erlenmeyer flask, 100 mL, for dd water
- 2 Erlenmeyer flasks, 500 mL, for beverages
- Gloves
- Mechanical, adjustable volume pipettors, 1000 μL and 100 μL (or 200 μL), with plastic tips
- Pasteur pipettes and bulb
- Parafilm®
- Pipette bulb or pump
- Repipettor (for fast delivery of 5-mL conc. H_2SO_4)
- Soft drinks (clear-colored, diet and regular, of same brand)
- 20 test tubes, 16–20-mm internal diameter
- Test tube rack
- 4 Volumetric flasks, 100 mL *or* 2 Volumetric flasks, 1000 mL
- Volumetric pipette, 5 mL
- 2 Volumetric pipettes, 10 mL

14.1.9 Equipment

- Spectrophotometer
- Vortex mixer
- Water bath, maintained at 25 °C

14.2 PROCEDURE

(Instructions are given for analysis in duplicate.)

1. Standard curve tubes: Using the glucose standard solution (100 mg glucose/L) and dd water as indicated in the table below, pipette aliquots of the glucose standard into clean test tubes (duplicates for each concentration) such that the tubes contain 0–100 μL of glucose (use 1000-μl mechanical pipettor to pipette samples), in a total volume of 2 mL. These tubes will be used to create a standard curve, with values of 0–100-μg glucose/2 mL. The 0-μg glucose/2 mL sample will be used to prepare the reagent blank.

	μg Glucose/2 mL					
	0	20	40	60	80	100
mL glucose stock solution	0	0.2	0.4	0.6	0.8	1.0
mL dd water	2.0	1.8	1.6	1.4	1.2	1.0

2. Record caloric content from label: You will analyze for total carbohydrate content: (1) a regular and diet soft drink of the same brand, and/or (2) a regular and lite beer of the same brand. Before you proceed with the sample preparation and analysis, record the caloric content on the nutrition label of the samples you will analyze.
3. Decarbonate the beverages: With the beverages at room temperature, pour approximately 100 mL into a 500-mL Erlenmeyer flask. Shake gently at first (try not to foam the sample if it is beer) and continue gentle shaking until no observable carbon dioxide bubbles appear. If there is any noticeable suspended material in the beverage, filter the sample before analysis.
4. Sample tubes: So the sample tested will contain 20–100-μg glucose/2 mL, the dilution procedure and volumes to be assayed are given below.

	Dilution	Volume assayed (mL)
Soft drink		
Regular	1:2000	1
Diet	0	1
Beer		
Regular	1:2000	1
Lite	1:1000	1

Recommended dilution scheme for 1:2000 dilution:

(a) Pipette 5 mL of beverage into a 100-mL volumetric flask, and dilute to volume with dd water. Seal flask with Parafilm® and mix well (this is a 1:20 dilution). Then, pipette 1.0 mL of this 1:20 diluted beverage into another 100-mL volumetric flask. Dilute to volume with dd water. Seal flask with Parafilm® and mix well.

OR

(b) Pipette 1.0 mL of beverage into a 1000-mL volumetric flask, and dilute to volume with dd water. Seal flask with Parafilm® and mix well. Then, in a test tube, combine 1 mL of the 1:1000 diluted beverage and 1 mL dd water. Mix well.

Recommended dilution scheme for 1:1000 dilution:

(a) Pipette 10 mL of beverage into a 100-mL volumetric flask, and dilute to volume with dd water. Seal flask with Parafilm® and mix well (this is a 1:10 dilution). Then, pipette 1.0 mL of this 1:10 diluted beverage into another 100-mL volumetric flask. Dilute to volume with dd water. Seal flask with Parafilm® and mix well.

OR

(b) Pipette 1.0 mL of beverage into a 1000-mL volumetric flask, and dilute to volume with dd water. Seal flask with Parafilm® and mix well.

5. After dilution as indicated, pipette 1.0 mL of sample into a test tube and add 1.0 mL of dd water. Analyze each diluted sample in duplicate.
6. Phenol addition: To each tube from Parts 1 and 4 containing a total volume of 2 mL, add 0.05-mL 80 % phenol (use 100 or 200-μl mechanical pipettor). Mix on a Vortex test tube mixer.
7. H_2SO_4 addition: To each tube from Part 6, add 5.0-mL H_2SO_4. The sulfuric acid reagent should be added rapidly to the test tube. Direct the stream of acid against the liquid surface rather than against the side of the test tube in order to obtain good mixing. (These reactions are driven by the heat produced upon the addition of H_2SO_4 to an aqueous sample. Thus, the rate of addition of sulfuric acid must be standardized.) Mix on a Vortex test tube mixer. Let tubes stand for 10 min and then place in a 25 °C bath for 10 min (i.e., to cool them to room temperature). Vortex the test tubes again before reading the absorbance.
8. Reading absorbance: Wear gloves to pour samples from test tubes into cuvettes. Do not

rinse cuvettes with water between samples. Zero the spectrophotometer with the standard curve sample that contains 0-µg glucose/2 mL (i.e., blank). Retain this blank sample in one cuvette for later use. Read absorbances of all other samples at 490 nm. Read your standard curve tubes from low to high concentration (i.e., 20 µg/2 mL up to 100 µg/2 mL), and then read your beverage samples. To be sure that the outside of the cuvettes are free of moisture and smudges, wipe the outside of the cuvette with a clean paper wipe prior to inserting it into the spectrophotometer for a reading.

9. Absorbance spectra: Use one of the duplicate tubes from a standard curve sample with an absorbance reading of 0.5-0.8. Determine the absorbance spectra from 450–550 nm by reading the tube at 10-nm intervals. Zero the spectrophotometer with the blank at each 10-nm interval.

14.3 DATA AND CALCULATIONS

1. Summarize your procedures and results for all standards and samples in the tables immediately below. Use the data for the standard curve samples in the first table to calculate the equation for the line, which is used to calculate the concentrations in the original samples reported in the second table.

Standard Curve:

Sample identity	A_{490} msmt 1	A_{490} msmt 2	Avg.
Blank			
Std. 20 µg			
Std. 40 µg			
Std. 60 µg			
Std. 80 µg			
Std. 100 µg			

Samples:

Sample identity	A_{490}	ug glucose/ 2 mL	Dilution scheme	Glucose equivalent µg/mL, original sample	Glucose equivalent g/L, original sample
Soft drink, reg.					
Soft drink, reg.					
Soft drink, diet					

Sample identity	A_{490}	ug glucose/ 2 mL	Dilution scheme	Glucose equivalent µg/mL, original sample	Glucose equivalent g/L, original sample
Soft drink, diet					
Beer, reg.					
Beer, reg.					
Beer, lite					
Beer, lite					

Sample calculation for soft drink, regular:

Equation of the line: $y = 0.011x + 0.1027$

$$y = 0.648$$

$$x = 49.57\,\mu g\,/\,2\,mL$$

$$C_i = C_f (V_2 / V_1)(V_4 / V_3)$$

(See Chap. 3 in this laboratory manual, C_i = initial concentration; C_f = final concentration)

$$C_i = (49.57\,\mu g\,glucose\,/\,2\,mL) \times (2000\,mL\,/\,1\,mL)$$
$$\times (2\,mL\,/\,1\,mL) = 99\,140\,\mu g\,/\,mL$$
$$= 99.14\,mg\,/\,mL$$
$$= 99.14\,g\,/\,L$$

2. Construct a standard curve for your total carbohydrate determinations, expressed in terms of glucose (A_{490} versus µg glucose/2 mL). Determine the equation of the line for the standard curve.

3. Calculate the concentration of glucose in your soft drink samples and beer samples, in terms of (a) grams/liter and (b) g/12 fl. oz. (Note: 29.56 mL/fl. oz.)

4. Calculate the caloric content (based only on carbohydrate content) of your soft drink samples and beer samples in term of Cal/12 fl. oz.

Sample	g Glucose/ 12 fl. oz.	Measured Cal/12 fl. oz.	Nutrition label Cal/12 fl. oz.
Soft drink			
Regular			
Diet			
Beer			
Regular			
Lite			

5. Plot the absorbance spectra obtained by measuring the absorbance between 450 and 550 nm.

wave-length	450	460	470	480	490	500	510	520	530	540	550
Abs.											

14.4 QUESTIONS

1. What are the advantages, disadvantages, and sources of error for this method to determine total carbohydrates?
2. Your lab technician performed the phenol-H_2SO_4 analysis on food samples for total carbohydrates but the results showed low precision, and the values seemed a little high. The technician had used new test tubes (they had never been used, and were taken right from the cardboard box). What most likely caused these results? Why? Describe what happened.
3. If you started with a glucose standard solution of 10-g glucose/liter, what dilution of this solution would be necessary such that you could pipette 0.20, 0.40, 0.60, 0.80, and 1.0 mL of the diluted glucose standard solution into test tubes and add water to 2 mL for the standard curve tubes (20–100 µg/2 mL)? Show all calculations.
4. If you had not been told to do a 2000-fold dilution of a soft drink sample, and if you know the approximate carbohydrate content of regular soft drinks (US Department of Agriculture Nutrient Database for Standard Reference indicates ca. 3 g carbohydrate/fl. oz.), how could you have calculated the 2000-fold dilution was appropriate if you wanted to use 1 mL of diluted soft drink in the assay. Show all calculations.
5. How does your calculated value compare to the caloric content on the food label? Do the rounding rules for Calories explain any differences? (See Metzger and Nielsen, 2017, Table 3.3). Does the alcohol content (assume 4–5% alcohol at 7 Cal/g) of beer explain any differences?
6. Was it best to have read the absorbance for the standard curve and other samples at 490 nm? Explain why a wavelength in this region is appropriate for this reaction.

Acknowledgment This laboratory was developed with input from Dr Joseph Montecalvo, Jr., Department of Food Science & Nutrition, California Polytechnic State University, San Luis Obispo, California.

RESOURCE MATERIALS

BeMiller JN (2017) Carbohydrate analysis, Ch. 19. In: Nielsen SS (ed) Food analysis, 5th edn. Springer, New York

Dubois M, Gilles KA, Hamilton JK, Rebers PA, Smith F (1956) Colorimetric method for determination of sugars and related substances. Anal Chem 28: 350–356

Metzger LE, Nielsen SS (2017) Nutrition labeling. Ch. 3. In: Nielsen SS edn. Food analysis, 5th edn. Springer, New York

Vitamin C Determination by Indophenol Method

S. Suzanne Nielsen

Department of Food Science, Purdue University,
West Lafayette, IN, USA
e-mail: nielsens@purdue.edu

S.S. Nielsen, *Food Analysis Laboratory Manual*, Food Science Text Series,
DOI 10.1007/978-3-319-44127-6_15, © Springer International Publishing 2017

15.1   INTRODUCTION

15.1.1   Background

Vitamin C is an essential nutrient in the diet, but is easily reduced or destroyed by exposure to heat and oxygen during processing, packaging, and storage of food. The instability of vitamin C makes it more difficult to ensure an accurate listing of vitamin C content on the nutrition label.

The official method of analysis for vitamin C determination of juices is the 2,6-dichloroindophenol titrimetric method (AOAC Method 967.21). While this method is not official for other types of food products, it is sometimes used as a rapid, quality control test for a variety of food products, rather than the more time-consuming microfluorometric method (AOAC Method 984.26). The procedure outlined below is from AOAC Method 967.21.

15.1.2   Reading Assignment

AOAC International. 2016. *Official Methods of Analysis*, 20th ed., (On-line). Method 976.21, AOAC International, Rockville, MD.

Pegg, R.B., and Eitenmiller, R.R. 2017. Vitamin analysis. Ch. 20, in *Food Analysis*, 5th ed. S.S. Nielsen (Ed.), Springer, New York.

15.1.3   Objective

Determine the vitamin C content of various orange juice products using the indicator dye 2,6-dichloroindophenol in a titration method.

15.1.4   Principle of Method

Ascorbic acid reduces the indicator dye to a colorless solution. At the endpoint of titrating an ascorbic acid-containing sample with dye, excess unreduced dye is a rose-pink color in the acid solution. The titer of the dye can be determined using a standard ascorbic acid solution. Food samples in solution then can be titrated with the dye and the volume for the titration used to calculate the ascorbic acid content.

15.1.5   Chemicals

	CAS no.	Hazards
Acetic acid (CH_3COOH)	64-19-7	Corrosive
Ascorbic acid	50-81-7	
2,6-Dichloroindophenol (DCIP) (sodium salt)	620-45-1	
Metaphosphoric acid (HPO_3)	37267-86-0	Corrosive
Sodium bicarbonate ($NaHCO_3$)	144-55-8	

15.1.6   Reagents

(**It is recommended that samples be prepared by the laboratory assistant before class.)

- Ascorbic acid standard solution (prepare only at time of use)

    Accurately weigh (on an analytical balance) approximately 50 mg ascorbic acid [preferably US Pharmacopeia (USP) Ascorbic Acid Reference Standard]. Record this weight. Transfer to a 50-mL volumetric flask. Dilute to volume *immediately before use* with the metaphosphoric acid-acetic acid solution (see below for preparation of this solution).
- Indophenol solution – dye

    To 50 mL deionized distilled (dd) water in a 150-mL beaker, add and stir to dissolve 42 mg sodium bicarbonate, and then add and stir to dissolve 50 mg 2,6-dichloroindophenol sodium salt. Dilute mixture to 200 mL with dd water. Filter through fluted filter paper into an amber bottle. Close the bottle with a stopper or lid and store refrigerated until used.
- Metaphosphoric acid-acetic acid solution

    To a 250-mL beaker, add 100 mL dd water then 20 mL acetic acid. Add and stir to dissolve 7.5 g metaphosphoric acid. Dilute mixture to 250 mL with distilled water. Filter through fluted filter paper into a bottle. Close the bottle with a stopper or lid and store refrigerated until used.
- Orange juice samples**

    Use products processed and packaged in various ways (e.g., canned, reconstituted frozen concentrate, fresh squeezed, not-from-concentrate). Filter juices through cheesecloth to avoid problems with pulp when pipetting. Record from the nutrition label for each product the percent of the Daily Value for vitamin C.

15.1.7   Hazards, Precautions, and Waste Disposal

Preparation of reagents involves corrosives. Use appropriate eye and skin protection. Otherwise, adhere to normal laboratory safety procedures. Waste likely may be put down the drain using a water rinse, but follow good laboratory practices outlined by environmental health and safety protocols at your institution.

15.1.8   Supplies

(Used by students)

- Beaker, 250 mL
- 2 Bottles, glass, 200–250 mL, one amber and one clear, both with lids or stoppers
- Buret, 50 or 25 mL

- 9 Erlenmeyer flasks, 50 mL (or 125 mL)
- Fluted filter paper, two pieces
- Funnel, approx. 6–9 cm diameter (to hold filter paper)
- Funnel, approx. 2–3 cm diameter (to fill buret)
- 2 Glass stirring rods
- Graduated cylinder, 25 mL
- Graduated cylinder, 100 mL
- Pipette bulb or pump
- Ring stand
- 3 Spatulas
- Volumetric flask, 50 mL
- Volumetric flask, 200 mL
- Volumetric flask, 250 mL
- 2 Volumetric pipettes, 2 mL
- Volumetric pipette, 5 mL
- Volumetric pipette, 7 mL
- Volumetric pipette, 10 or 20 mL
- Weighing boats or paper

15.1.9 Equipment

- Analytical balance

15.1.10 Notes

The instructor may want to assign one or two types of orange juice samples to each student (or lab group) for analysis, rather than having all students analyze all types of orange juice samples. Quantities of supplies and reagents specified are adequate for each student (or lab group) to standardize the dye and analyze one type of orange juice sample in triplicate.

15.2 PROCEDURE

(Instructions are given for analysis in triplicate.)

15.2.1 Standardization of Dye

1. Pipette 5 mL metaphosphoric acid-acetic acid solution into each of three 50-mL Erlenmeyer flasks.
2. Add 2.0 mL ascorbic acid standard solution to each flask.
3. Using a funnel, fill the buret with the indophenol solution (dye) and record the initial buret reading.
4. Place the Erlenmeyer flask under the tip of the buret. Slowly add indophenol solution to standard ascorbic acid solution until a light but distinct rose-pink color persists for >5 s (takes about 15–17 mL). Swirl the flask as you add the indophenol solution.
5. Note final buret reading and calculate the volume of dye used.
6. Repeat Steps 3–5 for the other two standard samples. Record the initial and final buret

readings and calculate the volume of dye used for each sample.
7. Prepare blanks: Pipette 7.0 mL metaphosphoric acid-acetic acid solution into each of three 50-mL Erlenmeyer flasks. Add to each flask a volume of distilled water approximately equal to the volume of dye used above (i.e., average volume of dye used to titrate three standard samples).
8. Titrate the blanks in the same way as Steps 3–5 above. Record initial and final buret readings for each titration of the blank, and then calculate the volume of dye used.

15.2.2 Analysis of Juice Samples

1. Pipet into each of three 50-mL Erlenmeyer flasks 5 mL metaphosphoric acid-acetic acid solution and 2 mL orange juice.
2. Titrate each sample with the indophenol dye solution (as you did in Sect. 15.2.1, Steps 3–5) until a light but distinct rose-pink color persists for >5 s.
3. Record the initial and final readings and calculate the difference to determine the amount of dye used for each titration.

15.3 DATA AND CALCULATIONS

15.3.1 Data

	Rep	Buret start (mL)	Buret end (mL)	Vol. titrant (mL)
Ascorbic acid standards	1			
	2			
	3			
				$\bar{X} =$
Blank	1			
	2			
	3			
				$\bar{X} =$
Sample	1			
	2			
	3			

15.3.2 Calculations

1. Using the data obtained in standardization of the dye, calculate the titer using the following formula:

$$\text{Titer} = F = \frac{\begin{array}{c}\text{mg ascorbic acid in volume}\\\text{of standard solution titrated}**\end{array}}{\left(\begin{array}{c}\text{average mL}\\\text{dye used to}\\\text{titrate standards}\end{array}\right) - \left(\begin{array}{c}\text{average mL}\\\text{dye used to}\\\text{titrate blank}\end{array}\right)}$$

**mg ascorbic acid in volume of standard solution titrated:

$$= (\text{mg of ascorbic acid} / 50\,\text{mL}) \times 2\,\text{mL}$$

2. Calculate the ascorbic acid content of the juice sample in mg/mL, using the equation that follows and the volume of titrant for each of your replicates. Calculate the mean and standard deviation of the ascorbic acid content for your juice (in mg/mL). Obtain from other lab members the mean ascorbic acid content (in mg/mL) for other types of juice. Use these mean values for each type of juice to express the vitamin C content of the juice samples as milligrams ascorbic acid/100 mL and as milligrams ascorbic acid/8 fl. oz. (29.56 mL/fl. oz.).

$$\text{mg ascorbic acid}/\text{mL} = (X - B) \times (F / E) \times (V / Y)$$

where:

X = mL for sample titration
B = average mL for sample blank titration
F = titer of dye
 (= mg ascorbic acid equivalent to 1.0 mL indophenol standard solution)
E = mL assayed (= 2 mL)
V = volume of initial assay solution (= 7 mL)
Y = volume of sample aliquot titrated (= 7 mL)

Ascorbic acid (AA) content for replicates of orange juice sample:

Replicate	mg AA/mL
1	
2	
3	
	$\bar{X}$ =
	SD =

Summary of ascorbic acid (AA) content of orange juice samples:

Sample identity	mg AA/mL	mg AA/100 mL	mg AA/8 fl. oz.
1			
2			
3			
4			

Example calculation:

Weight of ascorbic acid used = 50.2 mg

Average volume of titrant used:

Ascorbic acid standards = 15.5 mL
Blanks = 0.10 mL

Volume of titrant used for orange juice sample = 7.1 mL

$$\text{Titer} = F = \frac{\left[(50.2\,\text{mg} / 50\,\text{mL}) \times 2\,\text{mL}\right]}{(15.5\,\text{mL} - 0.10\,\text{mL})}$$

$$= 0.130\,\text{mg} / \text{mL}$$

$$\text{mg ascorbic acid} / \text{mL} = (7.1\,\text{mL} - 0.10\,\text{mL})$$
$$\times (0.130\,\text{mg} / 2\,\text{mL}) \times (7\,\text{mL} / 7\,\text{mL})$$
$$= 0.455\,\text{mg} / \text{mL}$$

$$0.455\,\text{mg} / \text{mL} = 45.4\,\text{mg} / 100\,\text{mL}$$

$$0.455\,\text{mg ascorbic acid} / \text{mL juice} \times 29.56\,\text{mL} / \text{fl.oz.} \times 8\,\text{fl.oz.}$$

$$= 107.6\,\text{mL ascorbic acid} / 8\,\text{fl.oz.}$$

15.4 QUESTIONS

1. By comparing results obtained for various orange juice products, did heat and/or oxygen exposure during processing and storage of the samples analyzed seem to affect the vitamin C content?
2. How do results you have available for the juice samples analyzed compare to (1) values listed on the nutrition label for the same juice product and (2) values in the US Department of Agriculture Nutrient Database for Standard Reference (web address: http://ndb.nal.usda.gov/)? For the nutrition label values, convert percent of Daily Value to mg/8 fl. oz., given that the Daily Value for vitamin C is 90 mg. Why might the vitamin C content determined for a specific orange juice product not match the value as calculated from percent of Daily Value on the nutrition label?

Ascorbic acid content of orange juices (mg AA/8 fl. oz):

Sample identity	Lab values	USDA database	Nutrition label
1			
2			
3			
4			

3. Why was it necessary to standardize the indophenol solution?
4. Why was it necessary to titrate blank samples?
5. Why might the vitamin C content as determined by this method be underestimated in the case of the heat-processed juice samples?

RESOURCE MATERIALS

AOAC International (2016) Official methods of analysis, 20th edn. (On-line). AOAC International, Rockville, MD
Pegg RB, Eitenmiller RR (2017) Vitamin analysis. Ch. 20. In: Nielsen SS (ed) Food analysis, 5th edn. Springer, New York

Water Hardness Testing by Complexometric Determination of Calcium

S. Suzanne Nielsen

Department of Food Science, Purdue University,
West Lafayette, IN, USA
e-mail: nielsens@purdue.edu

S.S. Nielsen, *Food Analysis Laboratory Manual*, Food Science Text Series,
DOI 10.1007/978-3-319-44127-6_16, © Springer International Publishing 2017

16.1 INTRODUCTION

16.1.1 Background

Ethylenediaminetetraacetate (EDTA) complexes with numerous mineral ions, including calcium and magnesium. This reaction can be used to determine the amount of these minerals in a sample by a complexometric titration. Endpoints in the titration are detected using indicators that change color when they complex with mineral ions. Calmagite and eriochrome black T (EBT) are such indicators that change from blue to pink when they complex with calcium and magnesium. In the titration of a mineral-containing solution with EDTA, the solution turns from pink to blue at the endpoint with either indicator. The pH affects a complexometric EDTA titration in several ways and must be carefully controlled. A major application of EDTA titration is testing the hardness of water, for which the method described is an official one (Standard Methods for the Examination of Water and Wastewater, Method 2340C; AOAC Method 920.196).

Hardness of water also can be tested by a more rapid test strip method. Such test strips are available from various companies. The strips contain EDTA and an indicator chemical to cause a color change when the calcium and magnesium in water react with the EDTA.

16.1.2 Reading Assignment

Ward, R.E., and Legako, J.F. 2017. Traditional methods for mineral analysis. Ch. 21, in *Food Analysis*, 5th ed. S.S. Nielsen (Ed.), Springer, New York.

16.1.3 Objective

Determine the hardness of water by EDTA titration and with Quantab® test strips.

16.2 EDTA TITRIMETRIC METHOD FOR TESTING HARDNESS OF WATER

16.2.1 Principle of Method

Ethylenediaminetetraacetic acid (EDTA) forms a Stable 1:1 complex with calcium or magnesium at pH 10. The metal ion indicators, calmagite and eriochrome black T (EBT), are pink when complexed to metal ions but blue when no metal ions are complexed to them. The indicators bind to metal ions less strongly than does EDTA. When the indicator is added to a solution containing metal ions, the solution becomes pink. When EDTA is added as titrant to the mineral-containing sample, metal ions preferentially complex with the EDTA, leaving the indicator without a metal ion to complex. When enough EDTA has been titrated to complex with all the metal ions present, the indicator appears blue. This blue color is the endpoint of the titration. The volume and concentration of the EDTA in the titration are used to calculate the concentration of calcium in the sample, which is expressed as mg calcium carbonate/l. Stoichiometry of the reaction is 1 mol of calcium complexing with 1 mol of EDTA.

16.2.2 Chemicals

	CAS no.	Hazards
Ammonium chloride (NH₄Cl)	12125-02-9	Harmful
Ammonium hydroxide (NH₄OH)	1336-21-6	Corrosive, dangerous for the environment
Calcium carbonate (CaCO₃)	471-34-1	
Calmagite [3-Hydroxy-4-(6-hydroxy-*m*-tolylazo) naphthalene-1-sulfonic acid]	3147-14-6	
Ethylenediaminetetraacetic acid, disodium salt (Na₂EDTA·2H₂O)	60-00-4	Irritant
Hydrochloric acid, concentrated (HCl)	7647-01-0	Corrosive
Magnesium chloride, hexahydrate (MgCl₂·6H₂O)	7791-18-6	
Magnesium sulfate, heptahydrate (MgSO₄·7H₂O)	10034-99-8	

16.2.3 Reagents

(**It is recommended that these solutions be prepared by the laboratory assistant before class.)

- Buffer solution**

  Dissolve 16.9 g NH₄Cl in 143 mL concentrated NH₄OH. In 50 mL deionized distilled (dd) water, dissolve 1.179 g Na₂EDTA·2H₂O (analytical reagent grade) and either 780 mg MgSO₄·7H₂O *or* 644 mg MgCl₂·6H₂O. Combine these two solutions with mixing and dilute to 250 mL with dd water. Store in tightly stoppered Pyrex or plastic bottle to prevent loss of ammonia (NH₃) or pickup of carbon dioxide (CO₂). Dispense this buffer solution with a repipette system. Discard buffer when 1–2 mL added to a sample fails to give pH 10.0±0.1 at the endpoint of the titration.

- Calcium standard solution, 1.00 mg CaCO₃/mL** (modified from official method; omit use of methyl red indicator)

  Use primary standard or special reagent that is low in heavy metals, alkalis, and magnesium. Dry CaCO₃ at 100 °C for 24 h. Accurately weigh ca. 1.0 g $CaCO_3$. Transfer to a 500-mL Erlenmeyer flask. Place a funnel in the neck of the flask and add HCl (1:1, conc. $HCl:H_2O$) a little at a time, until all the $CaCO_3$ has dissolved (make sure all the $CaCO_3$ in the neck of the flask has been washed down with HCl). Add 200 mL dd water and boil a few minutes to expel CO_2. Cool. Adjust to pH 3.8 with $3M$ NH_4OH or HCl (1:1, conc. $HCl : H_2O$), as required. Transfer quantitatively to a 1-L volumetric flask and dilute to volume with dd water (1 mL = 1.00 mg $CaCO_3$).

- EDTA standard solution, $0.01M$
  Weigh 3.723 g $Na_2EDTA \cdot 2H_2O$. Dilute to 1 L with dd water. Store in polyethylene (preferable) or borosilicate glass bottles. Standardize this solution using the calcium standard solution as described in the Procedure.
- Hydrochloric acid, 1:1 with water**
  To 10 mL of dd water, add 10 mL concentrated HCl. Mix carefully.
- Calmagite**
  Dissolve 0.10 g calmagite in 100 mL dd water. Use 1 mL per 30 mL solution to be titrated. Put in bottle with eye dropper.

16.2.4 Notes

In this experiment, calmagite will be used as the indicator dye rather than EBT. Unlike EBT, calmagite is stable in aqueous solution. Calmagite gives the same color change as EBT, but with a sharper endpoint.

To give a satisfactory endpoint, magnesium ions must be present. To ensure this, a small amount of neutral magnesium salt is added to the buffer.

The specified pH of 10.0 + 0.1 is a compromise situation. With increasing pH, the sharpness of the endpoint increases. However, at high pH, the indicator dye changes color and there is risk of precipitating calcium carbonate ($CaCO_3$) or magnesium hydroxide. The tendency toward $CaCO_3$ precipitation is the reason for the titration duration time limit of 5 min.

Fading or indistinct endpoints can be caused by interference from some metal ions. Certain inhibitors can be added before titration to reduce this interference, but the inhibitors specified are toxic (i.e., sodium cyanide) or malodorous. Magnesium salt of 1,2-cyclohexanediaminetetraacetic acid (MgCDTA), which selectively complexes heavy metals, may be substituted for these inhibitors. However, for samples with high concentrations of heavy metals, a non-EDTA method is recommended. In this experiment, inhibitors or MgCDTA will not be used.

16.2.5 Hazards, Precautions, and Waste Disposal

Adhere to normal laboratory safety procedures. Wear gloves and safety glasses at all times. The buffer solution, which contains ammonium hydroxide, should be disposed of as hazardous waste. Other wastes likely may be put down the drain using a water rinse, but follow good laboratory practices outlined by environmental health and safety protocols at your institution.

16.2.6 Supplies

(Used by students)

- Buret, 25 or 50 mL
- 9 Erlenmeyer flasks, 125 mL
- Funnel (to fill buret)
- Graduated cylinder, 50 mL
- 3 Graduated cylinders, 25 mL
- (Graduated cylinder of larger volumes may be necessary, for example, 100 mL or larger; size to be determined by trial in Sect. 16.2.8.2)
- Mechanical pipettor, 1000 μL, with plastic tips
- Pasteur pipette and bulb
- Spatula
- Volumetric flask, 1000 mL
- Volumetric pipette, 10 mL
- Weighing paper/boat

16.2.7 Equipment

- Analytical balance
- Drying oven, 100 °C
- Hot plate
- pH meter

16.2.8 Procedure

(Modified from Method 2340 Hardness, *Standard Methods for the Examination of Water and Wastewater*, 22nd ed.) (Instructions are given for analysis in triplicate.)

Standardization of EDTA Solution

1. Pipette 10 mL of calcium standard solution into each of three 125-mL Erlenmeyer flasks.
2. Adjust to pH 10.0 ± 0.05 with buffer solution. (If possible, do this pH adjustment with the buffer in an operating hood, due to its odor.) As necessary, use the HCl solution (1:1) in pH adjustment.
3. Add 1 mL of calmagite to each flask, and then titrate each flask with EDTA solution slowly, with continuous stirring, until last reddish tinge disappears, adding last few drops at 3–5 s intervals. Color at endpoint is blue in daylight and under daylight fluorescent lamp. Color may first appear lavender or purple, but will then turn to blue. Complete titration within 5 min from time of buffer addition.
4. Record the volume of EDTA solution used for each titration.

Titration of Water Sample

1. Dilute 25 mL tap water sample (or such volume as to require <15 mL titrant) to ca. 50 mL with dd water in 125-mL Erlenmeyer flask. For tap distilled water, test 50 mL, without dilution. Prepare samples in triplicate [Official method recommends the following: For water of low hardness (<5 mg/L), use 100–1000 mL specimen, proportionately larger amounts of reagents, micro-buret, and blank of distilled water equal to specimen volume.]
2. Adjust pH to 10±0.05 as described in Sect. 16.2.8.1, Step 2.
3. Titrate each sample with EDTA standard solution slowly, as described in Sect. 16.2.8.1, Step 3, for standardization of EDTA solution.
4. Record the volume of EDTA solution used for each titration.

16.2.9 Data and Calculations

Calculate molarity of calcium standard solution:

Molarity of calcium solution =

$$\frac{g\,CaCO_3}{(100.09\,g\,/\,mol)(liter\,solution)} = mol\,calcium\,/\,L$$

Standardization of EDTA solution:

Rep	Buret start (mL)	Buret end (mL)	Volume titrant (mL)	Molarity
1				
2				
3				
				$\bar{X}=$
				SD =

Calculate molarity of EDTA solution:

mol calcium = mol EDTA

$$M_1 V_1 = M_2 V_2$$

$$\left(M_{Ca\,solution}\right)\left(V_{Ca\,solution,L}\right)$$

$$= \left(M_{EDTA\,solution}\right)\left(V_{EDTA\,solution,L}\right)$$

Solve for $M_{EDTA\,solution}$

Titration of water sample with EDTA solution:

Rep	Dilution	Buret start (mL)	Buret end (mL)	Volume titrant (mL)	g Ca/L	mg CaCO₃/L
1						
2						
3						
					$\bar{X}=$	$\bar{X}=$
					SD =	SD =

Calcium content of water sample (g Ca/L and g CaCO$_3$/L):

$$mol\,calcium = mol\,EDTA$$

$$M_1 V_1 = M_2 V_2$$

$$\left(M_{Ca\,in\,sample}\right)\left(V_{sample,\,liter}\right) =$$

$$\left(M_{EDTA\,solution}\right)\left(V_{EDTA\,solution\,used\,in\,titration,\,L}\right)$$

Solve for $M_{Ca\,in\,sample}$:

$$M_{Ca\,in\,sample} \times 40.085\,g\,Ca\,/\,mol = g\,Ca\,/\,L$$

$$\left(g\,Ca\,/\,L\right)\left(100.09\,g\,CaCO_3\,/\,40.085\,g\,Ca\right)$$

$$\times \left(1000\,mg\,/\,g\right) = mg\,CaCO_3\,/\,L$$

16.2.10 Questions

1. If a sample of water is thought to have a hardness of approximately 250 mg/L CaCO$_3$, what size sample (i.e., how many mL) would you use so that you would use approximately 10 mL of your EDTA solution?
2. Why were you asked to prepare the CaCl$_2$ solution by using CaCO$_3$ and HCl rather than just weighing out CaCl$_2$?
3. In this EDTA titration method, would overshooting the endpoint in the titration cause an over- or underestimation of calcium in the sample? Explain your answer.

16.3   TEST STRIPS FOR WATER HARDNESS

16.3.1 Note

All information given is for AquaChek test strips, from Environmental Test Systems, Inc., a HACH Company, Elkhart, IN. Other similar test strips could be used. Any anion (e.g., magnesium, iron, copper) that will bind the EDTA may interfere with the AquaChek test. Very strong bases and acids also may interfere.

16.3.2 Principle of Method

The test strips have a paper, impregnated with chemicals, that is adhered to polystyrene for ease of handling. The major chemicals in the paper matrix are calmagite and EDTA, and minor chemicals are added to minimize reaction time, give long-term stability, and maximize color distinction between levels of water hardness. The strips are dipped into the water to test for total hardness caused by calcium and magnesium. The calcium displaces the magnesium bound to EDTA, and the released magnesium binds to calmagite, causing the test strip to change color.

16.3.3 Chemicals

	CAS no.	Hazards
Calcium carbonate (CaCO$_3$)	471-34-1	Harmful
Calmagite	3147-14-6	
Ethylenediaminetetraacetic acid, disodium salt (Na$_2$EDTA·2H$_2$0)	60-00-4	Irritant
Hydrochloric acid, concentrated (HCl)	7647-01-0	Corrosive
Other proprietary chemicals in test strip		

16.3.4 Reagents

(**It is recommended that this solution be prepared by the laboratory assistant before class.)

- Calcium standard solution, 1.000 mg CaCO$_3$/mL**
  Prepare as described in Sect.16.2.3, using CaCO$_3$ and concentrated HCl.

16.3.5 Hazards, Precautions, and Waste Disposal

No precautions are needed in use of the test strip. Adhere to normal laboratory safety procedures. Wastes likely may be put down the drain using a water rinse, but follow good laboratory practices outlined by environmental health and safety protocols at your institution.

16.3.6 Supplies

- AquaChek® Test Strips (Environmental Test Systems, Inc., a HACH Company, Elkhart, IN)
- 2 Beakers, 100 mL

16.3.7 Procedure

(*Note:* Test the same standard calcium solution as used in Sect. 16.2.8.1 and the same tap water and tap distilled water as used in Sect.16.2.8.2.)

1. Dip the test strip into a beaker filled with water or the standard calcium solution. Follow instructions on strip about how to read it, relating color to ppm CaCO$_3$.
2. Convert ppm CaCO$_3$ as determined with the test strips to mg CaCO$_3$/L and g Ca/L.

16.3.8 Data and Calculations

Sample	Rep (ppm CaCO$_3$) 1	2	3	Rep (mg CaCO$_3$/L) 1	2	3	Rep (g Ca/L) 1	2	3
Tap water									
Tap distilled water									
Standard Ca solution									

16.3.9 Question

1. Compare and discuss the accuracy and precision of the EDTA titration and test strip methods to measure calcium carbonate contents of the water samples and the calcium standard solution.

RESOURCE MATERIALS

Rice, EW, Baird RB, Eaton AD, Clesceri LS (eds) (2012) Standard methods for the examination of water and wastewater, 22st edn, Method 2340. American Public Health Association, American Water Works Association, Water Environment Federation, Washington, DC, pp. 2–37 to 2–39

Ward RE, Legako JF (2017) Traditional methods for mineral analysis. Ch. 21. In: Nielsen SS (ed) Food analysis, 5th edn. Springer, New York

Phosphorus Determination by Murphy-Riley Method

Young-Hee Cho (✉) • *S.Suzanne Nielsen*

*Department of Food Science, Purdue University,
West Lafayette, IN, USA
e-mail: cho173@purdue.edu; nielsens@purdue.edu*

S.S. Nielsen, *Food Analysis Laboratory Manual*, Food Science Text Series,
DOI 10.1007/978-3-319-44127-6_17, © Springer International Publishing 2017

17.1 INTRODUCTION

17.1.1 Background

Phosphorus is one of the important minerals found in foods. Murphy-Riley method, a dry ashing colorimetric method, has been widely used to measure the phosphorus content in natural waters as well as in foods. In this procedure, it is necessary to ash the sample prior to analysis. The method described here is applicable to most foods following ashing. If the food sample has a low magnesium content, several milliliters of saturated $Mg(NO_3)_2 \cdot 6H_2O$ in ethanol should be added to the sample prior to ashing to prevent volatilization and loss of phosphorus at the high temperatures used in ashing.

17.1.2 Reading Assignment

Ward, R.E., and Legako, J.F. 2017. Traditional methods for mineral analysis. Ch. 21, in *Food Analysis*, 5th ed. S.S. Nielsen (Ed.), Springer, New York.

17.1.3 Objective

Determine the phosphorus content of milk using a colorimetric method, with the Murphy and Riley reagent.

17.1.4 Principle of Method

Ammonium molybdate reacts with phosphorus to form phosphomolybdate. This complex is then reduced by ascorbic acid with antimony serving as a catalyst. The reduced phosphomolybdate complex has an intense blue color. The maximum absorptivity of this complex is at approximately 880 nm. This wavelength is above the operating range of most spectrophotometers. The absorptivity of the molybdate complex is great enough in the region of 600–700 nm, however, to allow determination of the phosphorus content of most foods at any wavelength within this region. Most spectrophotometers have a maximum usable wavelength of 600–700 nm.

17.1.5 Chemicals

	CAS no.	Hazard(s)
Ammonium molybdate	13106-76-8	Irritant
Potassium antimonyl tartrate	28300-74-5	Harmful
Ascorbic acid	50-81-7	
Monopotassium phosphate	7778-77-0	
Sulfuric acid	7664-93-9	Corrosive

17.1.6 Reagents

(** It is recommended that these solutions be prepared by the laboratory assistant before class. Glassware cleaned by phosphate-free detergents and double-distilled or distilled deionized water must be used in all dilutions and preparation of all reagents.)

- Ammonium molybdate**
  Dissolve 48 g of ammonium molybdate to 1000 mL in a volumetric flask.
- Potassium antimonyl tartrate**
  Dissolve 1.10 g of potassium antimonyl tartrate to 1000 mL in a volumetric flask.
- Ascorbic acid**
  Prepare fresh daily. Dissolve 2.117 g to 50 mL in a volumetric flask.
- Phosphorus stock solution and working standard solution**
  Dissolved 0.6590 g of dried (105 °C, 2 h) monopotassium phosphate (KH_2PO_4) to 1000 mL in a volumetric flask. This solution contains 0.150 mg phosphorus/mL (i.e., 150 µg/mL). To prepare a working standard solution, dilute 10 mL of this solution to 100 mL in a volumetric flask. The solution now contains 15 µg phosphorus/mL.
- 2.88 N sulfuric acid (H_2SO_4)**
  Prepare 200 mL of 2.88 N sulfuric acid (H_2SO_4) using concentrated sulfuric acid (H_2SO_4) and double-distilled or distilled deionized water. Always add concentrated acid to water, not water to concentrated acid. Do not use a mechanical pipette to pipette concentrated sulfuric acid and 2.88 N sulfuric acid, since this would corrode the pipettor.
- Murphy and Riley reagent
  Combine 14 mL of 2.88 N H_2SO_4, 2 mL of ammonium molybdate solution, and 2 mL of potassium antimonyl tartrate solution in a 50 mL Erlenmeyer flask.

17.1.7 Hazards, Precautions, and Waste Disposal

Adhere to normal laboratory safety procedures. Wear gloves, lab coat, and safety glasses at all time. Care should be exercised in pipetting any solution that contains antimony since it is a toxic compound. Waste containing antimony, sulfuric acid, and molybdate must be discarded as hazardous waste. Other wastes likely may be put down the drain using a water rinse, but follow good laboratory practices outlined by environmental health and safety protocols at your attention.

17.1.8 Supplies

(Used by students)

- 1 Crucible (preheated at 550 °C for 24 h)
- Glass funnel
- Glass stirring rods
- Kimwipes

- Mechanical adjustable volume pipettes, 1000 μL with pipette tips
- Nonfat liquid milk, 5 g
- Repipettor (for fast delivery of 2 mL H_2SO_4)
- Test tubes (13 × 100 mm)
- 1 Volumetric flask, 250 mL
- Whatman No. 41 ashless filter paper

17.1.9 Equipment

- Analytical balance, 0.1 mg sensitivity
- Forced draft oven
- Hot plate
- Muffle furnace
- Spectrophotometer
- Vortex mixer

17.2 PROCEDURE

(Instructions are given for analysis in duplicate.)

17.2.1 Ashing

(Based on note in Chap. 10, Sect. 10.2.7.3.)

1. Predry crucible at 550 °C for 24 h and weigh accurately.
2. Accurately weigh ca. 5 g of sample in the crucible.
3. Heat on the hot place until a majority of water has been evaporated.
4. Dry in a forced draft oven at 100 °C for 3 h.
5. Ash in a muffle furnace at 550 °C for 18–24 h.

17.2.2 Phosphorus Measurement

1. Preparation of Murphy and Riley reagent: Add 2 mL of ascorbic acid solution to pre-prepared 18 mL of Murphy and Riley reagent. Swirl to mix.
2. Standards: Prepare phosphorus standards using working standard solution (15 μg phosphorus/mL) and water as indicated in the table below. Pipet aliquots of the phosphorus standard into clean test tubes (duplicated for each concentration) and add water so each test tube contains 4 mL. Add 1 mL of Murphy and Riley (M&R) reagent to each tube. The final volume of each tube should be 5 mL. Mix with a vortex mixer. Allow the color to develop at room temperature for 10–20 min. Read the absorbance at 700 nm.

μg P/5 mL	15 μg P/mL	Vol. water (mL)	Vol. M&R (mL)
0 (blank)	0	4.0	1.0
1.5	0.1	3.9	1.0
3.0	0.2	3.8	1.0
4.5	0.3	3.7	1.0
6.0	0.4	3.6	1.0

3. Sample analysis: (Note: *Do not* use an automatic pipettor to obtain 2 mL of the sulfuric acid. This would corrode the pipettor.) Analyze one ashed sample. Carefully moisten ash in crucible with H_2O and then add 2 mL of 2.88 N H_2SO_4. Filter through ashless filter paper into a 250 mL volumetric flask. Thoroughly rinse crucible, ash, and filter paper. Dilute to volume with distilled water and mix. Analyze in duplicate by combining 0.2 mL of sample, 3.8 mL of distilled water, and 1 mL of Murphy and Riley reagent. Mix and allow to react for 10–20 min. Absorbance is read at 700 nm as per the phosphorus standard solutions.

17.3 DATA AND CALCULATIONS

1. Report data obtained, giving A_{700} of the duplicate standards and sample and the average A_{700}.

	Absorbance (700 nm)		
μg P/tube (5 mL)	1	2	Average
1.5			
3.0			
4.5			
6.0			
Sample			

2. Construct a standard curve for your phosphorus determination, expressed in terms of phosphorus (A_{700} vs. μg phosphorus/5 mL). Determine the equation of the line for the phosphorus standard curve.
3. Calculate the concentration of phosphorus in your milk sample expressed in terms of mg phosphorus/100 g sample. Show all calculations.

Sample calculation for milk sample:

Weight of milk sample: 5.0150 g
$A_{700} = 0.394$
Equation of the line:

$$y = 0.12x + 0.0066$$
$$y = 0.394$$
$$x = 3.3 \text{ μg P/5 mL}$$

P content of milk sample

$$= \frac{3.2\,\mu g\,P}{5\,mL} \times \frac{5\,mL}{0.2\,mL} \times \frac{250\,mL}{5.0150\,g}$$
$$= \frac{820\,\mu g\,P}{g} = \frac{82\,mg\,P}{100\,g}$$

17.4 QUESTIONS

1. How does your value compare to literature value (US Department of Agriculture Nutrient Database for Standard Reference) for phosphorus content of milk?

2. If you had not been told to use a 250 mL volumetric flask to prepare your ashed milk sample, how could you have calculated the dilution scheme was appropriate if you wanted to use 0.2 mL ashed milk sample in the assay. US Department of Agriculture Nutrient Database for Standard Reference indicates ca. 101 mg phosphorus/100 g. Show all calculations.

3. What are the possible sources of error using this method to determine the phosphorus content of milk?

4. What was the function of ascorbic acid in the assay?

RESOURCE MATERIALS

Murphy J, Riley JP (1962) A modified single solution method for the determination of phosphate in natural waters. Anal. Chim. Acta 27:31–36

Ward RE, Legako JF (2017) Traditional methods for mineral analysis. Ch. 21, In: Nielsen SS (ed) Food Analysis, 5th edn. Springer, New York

Iron Determination by Ferrozine Method

Charles E. Carpenter (⊠) • *Robert E. Ward*
Department of Nutrition, Dietetics and Food Sciences, Utah State University,
Logan, UT, USA
e-mail: chuck.carpenter@usu.edu; robert.ward@usu.edu

S.S. Nielsen, *Food Analysis Laboratory Manual*, Food Science Text Series,
DOI 10.1007/978-3-319-44127-6_18, © Springer International Publishing 2017

18.1   INTRODUCTION

18.1.1   Background

Chromogens are chemicals that react with compounds of interest and form colored products that can be quantified using spectroscopy. Several chromogens that selectively react with minerals are available. In this lab ferrozine is used to measure ferrous iron in an ashed food sample. The relationship between the absorbance of the chromogen-mineral complex is described by Beer's Law. In this procedure, a standard curve is generated with a stock iron solution to quantify the mineral in beef samples.

In this experiment, meat samples are first ashed to dissociate the iron bound to proteins, and the ash residue is solubilized in dilute HCl. The acid is necessary to keep the mineral in solution. Ferrozine complexes only with ferrous iron and not with ferric iron. Prior to the reaction with ferrozine, the solubilized ash is first treated with ascorbic acid to reduce iron to the ferrous form. This step is necessary with ashed samples, as this procedure would be expected to oxidize all the iron present in the meat. However, when other treatments are used to liberate iron, for example, trichloroacetic acid precipitation, comparison of samples treated with ascorbic acid and untreated samples could be done to determine the ratio of ferrous to ferric iron in foods.

18.1.2   Reading Assignment

Ward, R.E., and Legako, J.F. 2017. Traditional methods for mineral analysis. Ch. 21, in *Food Analysis*, 5th ed. S.S. Nielsen (Ed.), Springer, New York.

18.1.3   Objective

Determine the iron content of food samples using the ferrozine method.

18.1.4   Principle of Method

Ferrous iron in extracts or ashed samples reacts with ferrozine reagent to form a stable colored product which is measured spectrophotometrically at 562 nm. Iron is quantified by converting absorbance to concentration using a standard curve.

18.1.5   Chemicals

	CAS no.	Hazard(s)
3-(2-Pyridyl)-5,6-bis(4-phenylsulfonic acid)-1,2,-triazine (ferrozine; Sigma P-9762)	69898-45-9	
Ascorbic acid (Sigma 255564)	50-81-7	
Ammonium acetate (Aldrich 372331)	631-61-8	
Iron stock solution (e.g., 200 ppm) (Aldrich 372331)	4200-4205	

18.1.6   Reagents

- Ferrozine reagent, 1 mM in water
- Dissolve 0.493 g ferrozine reagent in water and dilute to 1 l in a volumetric flask.
- Ascorbic acid; 0.02% in 0.2N HCl, made fresh daily
- Ammonium acetate, 30% w/v
- Iron stock solution (10 µg iron/mL/0.1 N HCl)
- Solutions of 1.0N, 0.1N, and 0.2N HCl

18.1.7   Hazards, Precautions, and Waste Disposal

Adhere to normal laboratory safety procedures. Wear safety glasses at all times! Waste may be put down the drain using water rinse.

18.1.8   Supplies

- 16 Test tubes, 18 × 150 mm
- Meat sample
- Pipettes
- Porcelain crucible
- Volumetric flask

18.1.9   Equipment

- Analytical balance
- Hot plate
- Muffle furnace
- Spectrophotometer

18.2   PROCEDURE

(Instructions are given for analysis in duplicate.)

18.2.1   Ashing

1. In duplicate, place a ~5 g sample into the crucible and weigh accurately.
2. Heat on the hot plate until the sample is well charred and has stopped smoking.
3. Ash in muffle furnace at ca 550 °C until the ash is white.

18.2.2   Iron Measurement

1. Prepare standards of 10, 8, 6, 4, 2, and 0 µg iron/mL from a stock solution of 10 µg iron/mL. Make dilutions using ca 0.1 N HCl.
2. Dissolve ash in small amount of 1N HCl, and dilute to 50 mL in volumetric flask with 0.1N HCl.
3. In duplicate, put 0.500 mL of appropriately diluted samples and standards into 10 mL test tubes.
4. Add 1.250 mL ascorbic acid (0.02% in 0.2N HCl, made fresh daily). Vortex and let set for 10 min.
5. Add 2.000 mL 30% ammonium acetate. Vortex (pH needs to be >3 for color development).
6. Add 1.250 mL ferrozine (1 mM in water). Vortex and let set in dark for 15 min.

7. Use water to zero the spectrophotometer at 562 nm (single-beam instrument) or place in the reference position (dual-beam instrument). Take two readings (repeated measures, msmt) for each tube at 562 nm.

18.3   DATA AND CALCULATIONS

Weight of original samples:

(1)_____ g   (2)_____ g

Absorbance of standards and samples:

Standards (µg iron/ml)	Absorbance		
	Rep 1	Rep 2	Mean
0			
2			
4			
6			
8			
10			
Meat sample			
1			
2			

Calculation of total iron in sample:

1. Plot absorbance of standards on the y-axis versus µg iron/mL on the x-axis.
2. Calculate the iron concentration in the ash solution from the standard curve: (abs − y intercept)/slope = µg iron/mL ash solution.
3. Calculate iron in the sample using measured iron value from standard curve and meat sample:

$$C_{sample} = (\mu g \ iron/mL \ ash \ solution) \times (50 \ mL \ ash \ solution/g \ meat) = \mu g \ iron/g \ meat$$

18.4   QUESTION

1. How else could iron be determined using the ash digest? What would be the advantages and disadvantages of the ferrozine method versus the other method you identified?

RESOURCE MATERIALS

Ward RE and Legako, JF (2017) Traditional methods for mineral analysis. Ch. 21. In: Nielsen SS (ed) Food analysis, 5th edn. Springer, New York

chapter 19

Sodium Determination Using Ion-Selective Electrodes, Mohr Titration, and Test Strips

S. Suzanne Nielsen

Department of Food Science, Purdue University,
West Lafayette, IN, USA
e-mail: nielsens@purdue.edu

S.S. Nielsen, *Food Analysis Laboratory Manual*, Food Science Text Series,
DOI 10.1007/978-3-319-44127-6_19, © Springer International Publishing 2017

19.1 INTRODUCTION

19.1.1 Background

Sodium content of foods can be determined by various methods, including an ion-selective electrode (ISE), the Mohr or Volhard titration procedure, or indicator test strips. These methods are official methods of analysis for numerous specific products. All these methods are faster and less expensive procedures than analysis by atomic absorption spectroscopy or inductively coupled plasma-optical emission spectroscopy. This experiment allows one to compare sodium analysis of several food products by ISE, Mohr titration, and Quantab® Chloride Titrators.

19.1.2 Reading Assignment

Ward, R.E., and Legako, J.F. 2017. Traditional methods for mineral analysis. Ch. 21, in *Food Analysis*, 5th ed. S.S. Nielsen (Ed.), Springer, New York.

19.2 ION-SELECTIVE ELECTRODES

19.2.1 Objective

Determine the sodium content of various foods with sodium and/or chloride ion-selective electrodes.

19.2.2 Principle of Method

The principle of ISE is the same as for measuring pH, but by varying the composition of the glass in the sensing electrode, the electrode can be made sensitive to sodium or chloride ions. Sensing and reference electrodes are immersed in a solution that contains the element of interest. The electrical potential that develops at the surface of the sensing electrode is measured by comparing the reference electrode with a fixed potential. The voltage between the sensing and reference electrodes relates to the activity of the reactive species. Activity (A) is related to concentration (C) by $A = \gamma C$, where γ is the activity coefficient, which is a function of ionic strength. By adjusting the ionic strength of all test samples and standards to a nearly constant (high) level, the Nernst equation can be used to relate electrode response to concentration of the species being measured.

19.2.3 Chemicals

	CAS No.	Hazards
Ammonium chloride (NH$_4$Cl)	12125-02-9	Harmful
Ammonium hydroxide (NH$_4$OH)	1336-21-6	Corrosive, dangerous for environment
Nitric acid (HNO$_3$)	7697-37-2	Corrosive
Potassium nitrate (KNO$_3$)	7757-79-1	
Sodium chloride (NaCl)	7647-14-5	Irritant
Sodium nitrate (NaNO$_3$)	7631-99-4	Harmful, oxidizing

19.2.4 Reagents

(**If these solutions are not purchased, it is recommended that these solutions be prepared by the laboratory assistant before class.)

Note: You can use a chloride and/or sodium ion-selective electrode, with the appropriate associated solutions (commercially available from companies that sell the electrodes): electrode rinse solution, ionic strength adjuster, reference electrode fill solution, standard solution, and electrode storage solution.

- Electrode rinse solution**
  For sodium electrode, dilute 20-mL ionic strength adjuster to 1 L with deionized distilled (dd) water. For chloride electrode, deionized distilled water.
- Ionic strength adjuster (ISA)
  For sodium electrode, 4M NH$_4$Cl and 4M NH$_4$OH. For chloride electrode, 5M NaNO$_3$.
- Nitric acid, 0.1N
  Dilute 6.3-mL conc. HNO$_3$ to 1 L with dd water.
- Reference electrode fill solution**
  For sodium electrode, 0.1M NH$_4$Cl. For chloride electrode, 10% KNO$_3$.
- Standard solutions**: 1000 ppm, sodium and/or chloride
  Use the 1000 ppm sodium or chloride solution to prepare 50 mL for each of the following concentrations: 10, 20, 100, 500, and 1000 ppm sodium or chloride.

19.2.5 Hazards, Precautions, and Waste Disposal

Adhere to normal laboratory safety procedures. Wear gloves and safety glasses at all times. Ammonium hydroxide waste should be discarded as hazardous waste. Other waste likely can be put down the drain using a water rinse, but follow good laboratory practices outlined by environmental health and safety protocols at your institution.

19.2.6 Supplies

- 16–18 Beakers, 250 mL (or sample cups to hold 100 mL)
- Food products: catsup, cottage cheese, potato chips, and sports drink (e.g., Gatorade, white or clear)
- Graduated cylinder, 100 mL
- Magnetic stir bars
- Pipette bulb or pump
- 3 Spatulas
- 16–18 Volumetric flasks, 100 mL
- 2 Volumetric flasks, 50 mL
- Volumetric pipette, 2 mL
- 9 Volumetric pipettes, 5 mL

- Watch glass
- Weighing paper

19.2.7 Equipment

- Analytical balance
- Direct concentration readout ISE meter (i.e., suitable meter with millivolt accuracy to 0.1 mV)
- Heating plate with stirrer
- Magnetic stirrer
- Chloride electrode (e.g., Van London-pHoenix Company, Houston, TX, Chloride Ion Electrode, Cat. # CL01502)
- Sodium electrode (e.g., Van London-pHoenix Company, Houston, TX, Sodium Ion Electrode, Cat. # NA71502)

19.2.8 Procedure

(Replicate the preparation and analysis of standards and samples as specified by an instructor.)

19.2.8.1 *Sample Preparation (General Instructions)*

1. Treat specific samples as described below (i.e., prehomogenized and/or diluted if necessary, as per Technical Services of ISE manufacturer), and then add 5 g or 5 mL of prepared sample to a 100-mL volumetric flask, then add dd 2-mL ISA, and dilute to volume with dd water. (See instructions specific for each type of food product below. Samples with high fat levels may require fat removal. Consult technical services of the company that manufactures the ISE.)

   Specific Samples:

      Sports drink: No dilution is required before use.

      Catsup: Accurately weigh ca. 1 g catsup into 50-mL volumetric flask, and dilute to volume with dd water. Mix well.

      Cottage cheese: Accurately weigh ca. 1 g of finely grated cheese into a 250-mL beaker containing a stir bar. Add 100-mL 0.1 N HNO_3. Cover beaker with a watch glass and boil gently for 20 min on stirrer/hot plate in a hood. Remove from hot plate and cool to room temperature in the hood.

      Potato chips: Accurately weigh ca. 5 g of potato chips into a 250-mL beaker. Crush chips with a glass stirring rod. Add 95-mL boiling dd water and stir. Filter water extract into a 100-mL volumetric flask, using a funnel with glass wool. Let cool to room temperature and dilute to volume.

2. Prepare standards by adding 5-mL standard of proper dilution (e.g., 10, 20, 100, 500, and

1000 ppm sodium or chloride) to a 100-mL volumetric flask. Add 2-mL ISA, then dilute to volume with dd water.

> *Note:* Sample/standard preparation calls for identical 1:20 dilution of each (i.e., 5 mL diluted to 100 mL). Therefore, since samples and standards are treated the same, no correction for this dilution needs to be made in calibration or calculation of results.

19.2.8.2 *Sample Analysis by ISE*

1. Condition sodium electrode as specified by the manufacturer.
2. Assemble, prepare, and check sodium and reference electrodes as described in electrode instruction manuals.
3. Connect electrodes to meter according to meter instruction manual.
4. For instruments with direct concentration readout capability, consult meter manual for correct direct measurement procedures.
5. Using the pH meter set on mV scale, determine the potential (mV) of each standard solution (10, 20, 100, 500, and 1000 ppm), starting with the most dilute standard. Use a uniform stirring rate, with a magnetic stir bar in each solution, placed on a magnetic stir plate.
6. Rinse electrodes with electrode rinse solution between standards.
7. Measure samples and record the mV reading. As you rinse electrodes with electrode rinse solution between measurements, be careful not to get rinse solutions into the hole for outer fill solution in the reference electrode (or ensure that the hole is covered).
8. After use, store sodium electrode and reference electrode as specified by manufacturer.

19.2.9 Data and Calculations

1. Prepare a standard curve, with electrode response plotted against concentration on a log scale. (Plot actual concentration values on the log scale, not log values.) Concentrations may be determined by reading directly off the standard curve or using a calculated equation of the line.
2. Use the standard curve and the mV readings for the samples to determine the sodium and/or chloride concentrations in ppm for the food samples as analyzed.
3. Convert the ppm sodium and/or chloride values for the food samples to mg/mL for the sports drink, catsup, cheese, and potato chips.
4. Taking into account the dilution of the samples, calculate the sodium and/or chloride content for catsup, cheese, and potato chips (in mg/g) (on a wet weight basis). Summarize the data

and calculated results in one table. Show all sample calculations below each table.

5. Calculate sodium chloride content of each food, based on the (a) chloride content and/or (b) sodium content.

6. Calculate the sodium content of each food, based on the sodium chloride content.

7. Compare the sodium/sodium chloride contents of the foods you analyzed to those reported in the US Department of Agriculture (USDA) Nutrient Database for Standard Reference (http://ndb.nal.usda.gov).

19.2.10 Question

1. If you used both a sodium and chloride ISE, which electrode worked better, concerning accuracy, precision, and time to response? Explain your answer, with appropriate justification.

19.3 MOHR TITRATION

19.3.1 Objective

Determine the sodium content of various foods using the Mohr titration method to measure chloride content.

19.3.2 Principle of Method

The Mohr titration is a direct titration method to quantitate chloride ions and then to calculate sodium ions. The chloride-containing sample solution is titrated with a standard solution of silver nitrate. After the silver from silver nitrate has complexed with all the available chloride in the sample, the silver reacts with chromate that has been added to the sample, to form an orange-colored solid, silver chromate. The volume of silver used to react with the chloride is used to calculate the sodium content of the sample.

19.3.3 Chemicals

	CAS No.	Hazards
Potassium chloride (KCl)	7447-40-7	Irritant
Potassium chromate (K$_2$CrO$_4$)	7789-00-6	Toxic, dangerous for environment
Silver nitrate (AgNO$_3$)	7761-88-8	Corrosive, dangerous for environment

19.3.4 Reagents

(**It is recommended that these solutions be prepared by laboratory assistant before class.)

- Potassium chloride
- Potassium chromate, 10 % solution**

- Silver nitrate solution, ca. 0.1 M **
  Prepare approximately 400 mL of the ca. 0.1 M AgNO$_3$ (molecular weight (MW) 169.89) for each student or lab group. Students should accurately standardize the solution, as described in the Sect. 19.3.8.1.

19.3.5 Hazards, Precautions, and Waste Disposal

Wear gloves and safety glasses at all times, and use good lab technique. Potassium chromate may cause serious skin sensitivity reactions. The use of crystalline AgNO$_3$ or solutions of the silver salt can result in dark brown stains caused by photodecomposition of the salt to metallic silver. These stains are the result of poor technique on the part of the analyst, with spilled AgNO$_3$ causing discoloration of the floor. If you do spill this solution, immediately sponge up the excess solution and thoroughly rinse out the sponge at a sink. Then come back with the clean, rinsed sponge and mop up the area at least 3–4 times to remove all of the silver nitrate. Also, be sure to rinse all pipettes, burets, beakers, flasks, etc. to remove residual AgNO$_3$ when you are finished with this experiment. Otherwise these items also will stain, and drip stains are likely to appear on the floor. Potassium chromate and silver nitrate must be disposed of as a hazardous waste. Other waste likely can be put down the drain using a water rinse, but follow good laboratory practices outlined by environmental health and safety protocols at your institution.

19.3.6 Supplies

- 6 Beakers, 250 mL
- Brown bottle, 500 mL
- Buret, 25 mL
- 3 Erlenmeyer flasks, 125 mL
- 4 Erlenmeyer flasks, 250 mL
- Food products: cottage cheese (30 g), potato chips (15 g), and sports drink (15 mL) (e.g., Gatorade, white or clear)
- Funnel
- Glass wool
- Graduated cylinder, 25 mL
- Magnetic stir bars (to fit 125 or 250-mL flasks)
- Pipette bulb or pump
- Spatulas
- Weighing paper and boats
- Volumetric pipette, 1 mL

19.3.7 Equipment

- Analytical balance
- Hot plate
- Magnetic stir plate

19.3.8 Procedure

(Instructions are given for analysis in triplicate.)

19.3.8.1 *Standardization of ca. 0.1 M AgNO₃*

1. Transfer 400 mL of the $0.1\,M$ AgNO₃ solution to a brown bottle. This solution will be standardized and then used to titrate the food samples. Fill a buret with this AgNO₃ solution.
2. Prepare the primary standard (KCl, MW = 74.55) solution in triplicate. Accurately weigh to four decimal places about 100 mg KCl into three 125-mL Erlenmeyer flasks. Dissolve in dd water (about 25 mL), and add 2–3 drops of K_2CrO_4 solution. (Caution: potassium chromate may cause serious skin sensitivity reactions!)
3. Put a magnetic stir bar in each flask with the KCl solution, and place the beaker on a magnetic stir plate below the buret for titration. Using the AgNO₃ solution in the buret, titrate the KCl solutions to the appearance of the first permanent, pale, pink-orange color. (Note: you will first get a white precipitate, then green color, and then the pink-orange color.) This endpoint is due to the formation of Ag_2CrO_4. The solution must be vigorously stirred during the addition of the AgNO₃ solution to avoid erroneous results.
4. Record volume of AgNO₃.
5. Calculate and record molarity of AgNO₃.

$$\frac{g\,KCl}{(mL\,AgNO_3)} \times \frac{1\,mol\,KCl}{74.555\,g} \times \frac{1000\,mL}{1\,L}$$
$$= M\text{ of }AgNO_3/L = M\,AgNO_3$$

6. Label bottle of AgNO₃ with your name and the molarity of the solution.

19.3.8.2 *Sample Analysis by Mohr Titration*

Cottage Cheese

1. Accurately weigh 10 g of cottage cheese in triplicate into 250-mL beakers.
2. Add about 15 mL of warm dd water (50–55 °C) to each beaker. Mix to a thin paste using a glass stirring rod or spatula. Add another ca. 25-mL dd water to each beaker until the sample is dispersed.
3. Quantitatively transfer each solution to a 100-mL volumetric flask, rinsing beaker, and magnetic stir bar with dd water several times. Dilute to volume with dd water.
4. Filter each solution through glass wool. Transfer 50 mL of each solution to 250-mL Erlenmeyer flasks.

5. Add 1 mL of potassium chromate indicator to each 50 mL of filtrate.
6. Titrate each solution with standardized ca. $0.1\,M$ AgNO₃, to the first visible pale red-brown color that persists for 30 s. Record the volume of titrant used.

Potato Chips

1. Weigh accurately approximately 5 g of potato chips in duplicate into 250-mL beakers, then add 95-mL boiling dd water to each beaker.
2. Stir the mixture vigorously for 30 s, wait for 1 min, stir again for 30 s, then let cool to room temperature.
3. Filter each solution through glass wool. Transfer 50 mL of each solution to 250-mL Erlenmeyer flasks.
4. Add 1 mL of potassium chromate indicator to each 50 mL of filtrate.
5. Titrate each solution with standardized ca. $0.1\,M$ AgNO₃, to the first visible pale red-brown color that persists for 30 s. Record the volume of titrant used.

Sports Drink (Clear or White)

1. Pipette accurately 5 mL of sports drink in duplicate into 250-mL beakers, then add 95-mL boiling dd water to each beaker.
2. Stir the mixture vigorously for 30 s, wait 1 min, and stir again for 30 s.
3. Transfer 50 mL of each solution to 250-mL Erlenmeyer flasks.
4. Add 1 mL of potassium chromate indicator to each 50 mL of prepared sample.
5. Titrate each solution with standardized ca. $0.1\,M$ AgNO₃, to the first visible pale red-brown color that persists for 30 s. Record the volume of titrant used.

19.3.9 Data and Calculations

1. Calculate the chloride content and the sodium chloride content of each replicated sample, then calculate the mean and standard deviation for each type of sample. Express the values in terms of percent, wt/vol, for the cottage cheese and potato chips, and percent, vol/vol, for the sports drink. Note that answers must be multiplied by the dilution factor.

$$\%\,chloride = \frac{mL\text{ of }AgNO_3}{g\,(or\,mL)\,sample} \times \frac{mol\,AgNO_3}{L} \times \frac{35.5\,g\,Cl}{mol\,NaCl} \times \frac{1\,L}{1000\,mL} \times 100 \times dilution\,factor$$

$$\%\,sodium\,chloride\,(salt) = \frac{mL\,of\,AgNO_3}{g\,(or\,mL)\,sample} \times \frac{mol\,AgNO_3}{liter} \times \frac{58.5\,g}{mol\,NaCl} \times \frac{1\,L}{1000\,mL} \times 100 \times dilution\,factor$$

Sample	Rep	Buret start (mL)	Buret end (mL)	Vol. AgNO₃ (mL)	% Cl	% NaCl
Cottage cheese	1					
	2					
	3					
						$\bar{X} =$
						SD =
Potato chips	1					
	2					
	3					
						$\bar{X} =$
						SD =
Sports drink	1					
	2					
	3					
						$\bar{X} =$
						SD =

19.3.10 Questions

1. Show the calculations of how to prepare 400 mL of an approximately $0.1\,M$ solution of $AgNO_3$ (MW = 169.89).
2. Would this Mohr titration procedure as described above work well to determine the salt content of grape juice or catsup? Why or why not?
3. How did this method differ from what would be done using a Volhard titration procedure? Include in your answer what additional reagents would be needed.
4. Would overshooting the endpoint result in an over- or underestimation of the salt content using the (a) Mohr titration or (b) Volhard titration?

19.4   QUANTAB® TEST STRIPS

19.4.1   Objective

To measure the chloride content of foods using Quantab® Chloride Titrators, then calculate the sodium chloride content.

19.4.2   Principle of Method

Quantab® Chloride Titrators are thin, chemically inert plastic strips. These strips are laminated with an absorbent paper impregnated with silver nitrate and potassium dichromate, which together form brown-silver dichromate. When the strip is placed in an aqueous solution that contains chlorine, the liquid rises up the strip by capillary action. The reaction of silver dichromate with chloride ions produces a white column of silver chloride in the strip. When the strip is completely saturated with the liquid, a moisture-sensitive signal across the top of the titrator turns dark blue to indicate the completion of the titration. The length of

the white color change is proportional to the chloride concentration of the liquid being tested. The value on the numbered scale is read at the tip of the color change and then is converted to percent salt using a calibration table.

19.4.3   Chemicals

	CAS No.	Hazards
Sodium chloride (NaCl)	7647-14-5	Irritant

19.4.4   Reagents

- Sodium chloride stock solution
  Accurately weigh 5.00 g of dried sodium chloride and quantitatively transfer to a 100-mL volumetric flask. Dilute to volume with dd water and mix thoroughly.
- Sodium chloride standard solutions
  Dilute 2 mL of the stock solution to 1000 mL with dd water in a volumetric flask to create a 0.010 % sodium chloride solution to use as a standard solution with the low range Quantab® Chloride Titrators.
  Dilute 5 mL of the stock solution to 100 mL with dd water in a volumetric flask to create a 0.25 % sodium chloride solution to use as a standard solution with the high range Quantab® Chloride Titrators.

19.4.5   Supplies

- 5 Beakers, 200 mL
- Filter paper (when folded as a cone, should fit into a 200-mL beaker)
- Funnels
- Glass wool
- Glass stirring rod
- Graduated cylinder, 100 mL

- Quantab® Chloride Titrators, range: 0.05–1.0% NaCl; 300–6000 ppm Cl (High Range, HR) and 0.005–0.1% NaCl; 30–600 ppm Cl (Low Range, LR) (Environmental Test Systems/Hach Company, Elkhart, IN, 1-800-548-4381).
- Spatulas
- Sports drink, 10 mL (i.e., same one used in Sects. 19.2 and 19.3)
- 2 Volumetric flasks, 100 mL

19.4.6 Equipment

- Hot plate
- Top loading balance

19.4.7 Procedure

(Instructions are given for analysis in triplicate.)

19.4.7.1 *Standard Solutions of Sodium Chloride*

1. Transfer 50 mL of the 0.25% standard sodium chloride solution to a 200-mL beaker.
2. Fold a piece of filter paper into a cone-shaped cup and place it point end down into the beaker. This will allow liquid from the beaker to seep through the filter paper at the pointed end.
3. Using the 0.25% sodium chloride standard solution, place the lower end of the High Range Quantab® Strip (0.05–1.0%) into the filtrate within the pointed end of the filter paper cone, being sure not to submerge the titrator more than 1.0 in.
4. Thirty seconds after the moisture-sensitive signal string at the top of the titrator turns dark blue or a light brown, record the Quantab® reading at the tip of the yellow-white peak, to the nearest 0.1 units on the titrator scale.
5. Using the calibration chart included with the Quantab® package, convert the Quantab® reading to percent sodium chloride (NaCl) and to ppm chloride (Cl⁻). Note that each lot of Quantab® has been individually calibrated. Be sure to use the correct calibration chart (i.e., the control number on the product being used must match the control number on the bottle).
6. Repeat Steps 1–5 given above (Sect. 19.4.7.1) using the 0.01% sodium chloride standard solution with the Low Range Quantab® Strip.

19.4.7.2 *Sample Analysis with Quantab® Test Strips*

Cottage Cheese

1. Weigh accurately approximately 5 g of cottage cheese into a 200-mL beaker, then add 95-mL boiling dd water.

2. Stir mixture vigorously for 30 s, wait for 1 min, stir again for 30 s, and then let cool to room temperature.
3. Fold a piece of filter paper into a cone-shaped cup and place it point end down the beaker. This will allow liquid from the beaker to seep through the filter paper at the pointed end.
4. Testing with both the Low Range and the High Range Quantab® Test Strips, place the lower end of the Quantab® into the filtrate within the pointed end of the filter paper cone, being sure not to submerge the titrator more than 2.5 cm.
5. Thirty seconds after the moisture-sensitive signal string at the top of the titrator turns dark blue or a light brown, record the Quantab® reading at the tip of the yellow-white peak, to the nearest 0.1 units on the titrator scale.
6. Using the calibration chart included with the Quantab® package, convert the Quantab® reading to percent sodium chloride (NaCl) and to ppm chloride (Cl⁻). Note that each lot of Quantab® has been individually calibrated. Be sure to use the correct calibration chart (i.e., the control number on the product being used must match the control number on the bottle).
7. Multiply the result by the dilution factor 20 to obtain the actual salt concentration in the sample.

Potato Chips

1. Weigh accurately approximately 5 g of potato chips into a 200-mL beaker. Crush chips with a glass stirring rod. Add 95-mL boiling dd water and stir.
2. Filter water extract into a 100-mL volumetric flask, using a funnel with glass wool. Let cool to room temperature and dilute to volume. Transfer to a 200-mL beaker.
3. Follow Steps 3–7 from the procedure for cottage cheese, Sect. 19.4.7.2.

Catsup

1. Weigh accurately approximately 5 g of catsup into a 200-mL beaker. Add 95-mL boiling dd water and stir.
2. Filter water extract into a 100-mL volumetric flask. Let cool to room temperature and dilute to volume. Transfer to a 200-mL beaker.
3. Follow Steps 3–7 from the procedure for cottage cheese, Sect. 19.4.7.2.

Sports Drink

1. Weigh accurately approximately 5 mL of sports drink into a 200-mL beaker. Add 95-mL boiling dd water and stir.
2. Follow Steps 3–7 from procedure for cottage cheese, Sect. 19.4.7.2.

19.4.8 Data and Calculations

Rep	From calibration chart				Corrected for dilution factor			
	% NaCl		ppm Cl		% NaCl		ppm Cl	
	LR	HR	LR	HR	LR	HR	LR	HR
Catsup								
1								
2								
3								
					$\bar{X} =$	$\bar{X} =$	$\bar{X} =$	$\bar{X} =$
					SD =	SD =	SD =	SD =
Cottage cheese								
1								
2								
3								
					$\bar{X} =$	$\bar{X} =$	$X =$	$\bar{X} =$
					SD =	SD =	SD =	SD =
Potato chips								
1								
2								
3								
					$\bar{X} =$	$\bar{X} =$	$\bar{X} =$	$\bar{X} =$
					SD =	SD =	SD =	SD =
Sports drink								
1								
2								
3								
					$\bar{X} =$	$\bar{X} =$	$\bar{X} =$	$\bar{X} =$
					SD =	SD =	SD =	SD =

19.5   SUMMARY OF RESULTS

Summarize in a table the sodium chloride content (mean and standard deviation) of the various food products as determined by the three methods described in this experiment. Include in the table the sodium chloride contents of the foods from the nutrition label and those published in the USDA Nutrient Database for Standard Reference (web address: http://ndb.nal.usda.gov/).

Sodium chloride content (%) of foods by various methods:

Food Product		Ion-selective electrode	Mohr titration	Quantab® titrator	Nutrition label	USDA Database
Catsup	$\bar{X} =$					
	SD =					
Cottage cheese	$\bar{X} =$					
	SD =					
Potato chips	$\bar{X} =$					
	SD =					
Sports drink	$\bar{X} =$					
	SD =					

19.6   QUESTIONS

1. Based on the results and characteristics of the methods, discuss the relative advantages and disadvantages of each method of analysis for these applications.
2. Comparing your results to data from the nutrition label and USDA Nutrient Database, what factors might explain any differences observed?

Acknowledgments Van London-pHoenix Company, Houston, TX, is acknowledged for its contribution of the sodium and chloride ion-selective electrodes, and related supplies, for use in developing a section of this laboratory exercise. Environmental Test Systems/HACH Company, Elkhart, IN, is acknowledged for contributing the Quantab® Chloride Titrators for use in developing a section of this laboratory exercise.

RESOURCE MATERIALS

AOAC International (2016) Official methods of analysis, 20th edn. (On-line). Method 941.18, Standard solution of silver nitrate; Method 983.14, Chloride (total) in cheese. AOAC International, Rockville, MD

AOAC International (2016) Official methods of analysis, 20th edn. (On-line). Method 976.25, Sodium in foods for special dietary use, ion selective electrode method. AOAC International, Rockville, MD

AOAC International (2016) Official methods of analysis, 20th edn. (On-line). Method 971.19, Salt (chlorine as sodium chloride) in meat, fish, and cheese; Indicating strip method. AOAC International, Rockville, MD

Ward RE, Legako JF (2017) Traditional methods for mineral analysis. Ch. 21. In: Nielsen SS (ed) Food analysis, 5th edn. Springer, New York

Environmental Test Systems (2016) Quantab® Technical Bulletin. Chloride analysis for cottage cheese. Environmental Test Systems, Elkhart, IN

Van London-pHoenix Company, Houston, TX. Product literature.

Wehr HM, Frank JF (eds) (2004) Standard methods for the examination of dairy products, 17th edn., Part 15.053 Chloride (Salt). American Public Health Association, Washington, DC

20
Chapter

Sodium and Potassium Determinations by Atomic Absorption Spectroscopy and Inductively Coupled Plasma- Optical Emission Spectroscopy

S.Suzanne Nielsen (✉)
Department of Food Science, Purdue University,
West Lafayette, IN, USA
e-mail: nielsens@purdue.edu

S.S. Nielsen, *Food Analysis Laboratory Manual,* Food Science Text Series,
DOI 10.1007/978-3-319-44127-6_20, © Springer International Publishing 2017

20.1 INTRODUCTION

20.1.1 Background

The concentration of specific minerals in foods can be determined by a variety of methods. The aim of this lab is to acquaint you with the use of atomic absorption spectroscopy (AAS) and atomic emission spectroscopy (AES) for mineral analysis. As described in Yeung et al., 2017, AES is also commonly called optical emission spectroscopy (OES). The OES term is usually used in place of AES when it is combined with inductively coupled plasma (ICP). Hence, this chapter will refer to inductively coupled plasma-optical emission spectroscopy as ICP-OES.

This experiment specifies the preparation of standards and samples for determining the sodium (Na) and potassium (K) contents by AAS and ICP-OES. The samples suggested for analysis include two solid food products that require wet and/or dry ashing prior to analysis (catsup and potato chips) and one liquid food product that does not (a clear sports drink or a clear fruit juice).

Procedures for both wet ashing and dry ashing of the solid samples are described. Experience can be gained with both types of ashing and results can be compared between the two methods of ashing. Sodium results from this experiment can be compared with sodium results from analysis of the same products in the experiment that uses the more rapid methods of analysis of ion selective electrodes, the Mohr titration, and Quantab® test strips.

The limit of detection for sodium is 0.3 parts per billion (ppb) for flame AAS, 3 ppb by radial ICP-OES, and 0.5 ppb by axial ICP-OES. The limit of detection for potassium is 3 ppb for flame AAS, 0.2–20 ppb (depending on the model) by radial ICP-OES, and 1 ppb by axial ICP-OES. Other comparative characteristics of AAS and ICP-OES are described in Yeung et al., 2017.

20.1.2 Reading Assignment

Harris, G.K., and Marshall, M.R. 2017. Ash analysis. Ch. 16, in *Food Analysis*, 5th ed. S.S. Nielsen (Ed.), Springer, New York.

Yeung, C.K., Miller, D.D., and M.A. Rutzke. 2017. Atomic absorption spectroscopy, atomic emission spectroscopy, and inductively coupled plasma mass spectrometry. Ch. 9, in *Food Analysis*, 5th ed. S.S. Nielsen (Ed.), Springer, New York.

20.1.3 Note

If there is no access to an ICP-OES, a simple AES unit can be used, likely with the same standard solutions and samples prepared as described below.

20.1.4 Objective

The objective of this experiment is to determine the sodium and potassium contents of food products using atomic absorption spectroscopy and inductively coupled plasma-atomic emission spectroscopy.

20.1.5 Principle of Method

Atomic absorption is based on atoms *absorbing* energy, once heat energy from a flame has converted molecules to atoms. By absorbing the energy, atoms go from ground state to an excited state. The energy absorbed is of a specific wavelength from a hollow cathode lamp. One measures absorption as the difference between the amount of energy emitted from the hollow cathode lamp and that reaching the detector. Absorption is linearly related to concentration.

Atomic emission is based on atoms *emitting* energy, after heat energy from a flame has converted molecules to atoms and then raised the atoms from ground state to an excited state. The atoms emit energy of a specific wavelength as they drop from an excited state back to ground state. One measures the amount of emitted energy of a wavelength specific for the element of interest. Emission is linearly related to concentration.

20.1.6 Chemicals

	CAS no.	Hazards
Hydrochloric acid (HCl)	7647-01-1	Corrosive
Hydrogen peroxide, 30% (H_2O_2)	7722-84-1	Corrosive
Lanthanum chloride ($LaCl_3$)	10025-84-0	Irritant
Nitric acid (HNO_3)	7697-37-2	Corrosive
Potassium chloride (KCl) (for K std. solution)	7447-40-7	Irritant
Sodium chloride (NaCl) (for Na std. solution)	7647-14-5	Irritant

20.1.7 Reagents

(**It is recommended that these solutions be prepared by the laboratory assistant before class.)

- Potassium and sodium standard solutions, 1000 ppm**

    Used to prepare 100-mL solutions of each of the concentrations listed in Table 20.1. Each standard solution must contain 10 mL conc. HCl/100 mL final volume.

20.1.8 Hazards, Precautions, and Waste Disposal

Adhere to normal laboratory safety procedures. Wear safety glasses and gloves during sample preparation. Use acids in a hood.

20.1 table

Concentrations (ppm) of Na and K standard solutions for AAS and ICP-OES

AAS		ICP-OES	
Na	K	Na	K
0.20	0.10	50	50
0.40	0.50	100	100
0.60	1.00	200	200
0.80	1.50	300	300
1.00	2.00	400	400

20.1.9 Supplies

(Used by students)

- Two crucibles, previously cleaned and heated at 550 °C in a muffle furnace for 18 h (for dry ashing)
- Desiccator, with dry desiccant
- Digestion tubes (for wet ashing; size to fit digestion block)
- Filter paper, ashless
- Funnels, small (to filter samples)
- Plastic bottles, with lids, to hold 50 mL (or plastic sample tubes with lids, to hold 50 mL, to fit autosampler, if one is available)
- Eight volumetric flasks, 25 mL
- Four volumetric flasks, 50 mL
- Volumetric flask, 100 mL
- Volumetric pipettes, 2 mL, 4 mL, 5 mL, and 10 mL (2)
- Weigh boats/paper

20.1.10 Equipment

- Analytical balance
- Atomic absorption spectroscopy unit
- Digestion block (for wet ashing; set to 175 °C)
- Inductively coupled plasma-atomic absorption spectroscopy unit (or simple atomic absorption spectroscopy unit)
- Muffle furnace (for dry ashing; set to 550 °C)
- Water bath, heated to boil water (for dry ashing)

20.2   PROCEDURE

20.2.1 Sample Preparation: Liquid Samples

1. Put an appropriate volume of liquid sample in a 100-mL volumetric flask. For a sports drink, use 0.2 mL for both Na and K analysis by AAS. Use 50 mL for Na analysis and 80 mL for K analysis by ICP-OES.
2. Add 10 mL conc. HCl.
3. Add deionized distilled (dd) water to volume.

4. Shake well. (If there is any particulate matter present, the sample will need to be filtered through ashless filter paper.)
5. Make appropriate dilution and analyze (to sample for AAS, add $LaCl_3$ to final conc. of 0.1 %).

Liquid Blank:

Prepare a liquid blank sample to be assayed, following the sample preparation procedure but excluding the sample.

20.2.2 Sample Preparation: Solid Samples

Wet Ashing

Note: Digestion procedure described is a wet digestion with nitric acid and hydrogen peroxide. Other types of digestion can be used instead.

1. Label one digestion tube per sample plus one tube for the reagent blank (control).
2. Accurately weigh out 300–400 mg of each sample and place in a digestion tube. Prepare samples in duplicate or triplicate.
3. Pipette 5-mL concentrated nitric acid into each tube, washing the sides of the tube as you add the acid.
4. Set tubes with samples and reagent blank in digestion block. Turn on the digestion block and set to 175 °C to start the predigestion.
5. Swirl the samples gently once or twice during the nitric acid predigestion, using tongs and protective gloves.
6. Remove tubes from digestion block when brown gas starts to elute (or when solution begins to steam, if there is no brown gas) and set in the cooling rack. Turn off the digestion block.
7. Let the samples cool for at least 30 min. (Samples can be stored at this point for up to 24 h.)
8. Add 4 mL of 30 % hydrogen peroxide to each tube, doing only a few tubes at one time. Gently swirl the tubes. Put the tubes back in the digestion block. Turn on the digestion block still set to 175 °C.
9. Watch tubes closely for the start of the reaction, indicated by the appearance of rapidly rolling bubbles. As soon as the reaction starts, remove the tubes from the block and let the reaction continue in the cooling rack. (Caution: Some sample types will have a vigorous reaction, and for some the sample is lifted to the top of the tube, with the risk of boiling over.)
10. Repeat Steps 8 and 9 for all the samples and the reagent blank.

20.2 table

Dilution of samples for Na and K analysis by AAS and ICP-OES, using wet[a] or dry ashing[b]

Sample	Na		K	
	AAS	ICP-OES	AAS	ICP-OES
Catsup				
Wet ashing	Ashed sample diluted to 25 mL, then 0.2 mL diluted to 100 mL	Ashed sample diluted to 25 mL	Ashed sample diluted to 25 mL, then 0.4 mL diluted to 100 mL	Ashed sample diluted to 10 mL
Dry ashing	Ashed sample diluted to 25 mL, then 0.2 mL diluted to 100 mL	Ashed sample diluted to 50 mL	Ashed sample diluted to 25 mL, then 0.2 mL diluted to 100 mL	Ashed sample diluted to 25 mL
Cottage cheese				
Wet ashing	Ashed sample diluted to 25 mL, then 0.5 mL diluted to 100 mL	Ashed sample diluted to 10 mL	Ashed sample diluted to 25 mL, then 0.7 mL diluted to 100 mL	Ashed sample diluted to 5 mL
Dry ashing	Ashed sample diluted to 25 mL, then 0.2 mL diluted to 100 mL	Ashed sample diluted to 25 mL	Ashed sample diluted to 25 mL, then 0.5 mL diluted to 100 mL	Ashed sample diluted to 25 mL
Potato chips				
Wet ashing	Ashed sample diluted to 25 mL, then 0.2 mL diluted to 100 mL	Ashed sample diluted to 10 mL	Ashed sample diluted to 25 mL, then 0.2 mL diluted to 100 mL	Ashed sample diluted to 25 mL
Dry ashing	Ashed sample diluted to 25 mL, then 0.2 mL diluted to 100 mL	Ashed sample diluted to 25 mL	Ashed sample diluted to 50 mL, then 0.1 mL diluted to 100 mL	Ashed sample diluted to 50 mL

[a]For wet ashing, use ca. 300–400 mg sample
[b]For dry ashing, use ca. 1-g sample, dry matter (calculate based on moisture content)

11. Put all tubes in the digestion block, and leave until ca. 1–1.5 mL remains, and then remove each tube from the digestion block. Check the tubes every 10–15 min during this digestion. (If the tubes are left on the digestion block too long and they become dry, remove, cool, and *carefully* add ca. 2-mL concentrated nitric acid and continue heating.) Turn off the digestion block when all the tubes have been digested and removed.

12. Make appropriate dilution of samples with dd water in a volumetric flask as indicated in Table 20.2. (To sample for AAS, add $LaCl_3$ to final conc. of 0.1%.)

13. If necessary, filter samples using Whatman hardened ashless #540 filter paper into container appropriate for analysis by AAS or ICP-OES.

Dry Ashing

1. Accurately weigh out blended or ground ca. 1-g sample dry matter into crucible (i.e., take moisture content into account, so you have ca. 1-g dry product).

2. Pre-dry sample over boiling water bath.

3. Complete drying of sample in vacuum oven at 100 °C, 26-in. Hg, for 16 h.

4. Dry ash sample for 18 h. at 550 °C and then let cool in desiccator.

5. Dissolve ash in 10-mL HCl solution (1:1, $HCl:H_2O$).

6. Make appropriate dilution of samples with dd water in a volumetric flask as indicated in Table 20.2. (To sample for AAS, add $LaCl_3$ to final conc. of 0.1%.)

7. If necessary, filter samples using Whatman hardened ashless #540 filter paper into container appropriate for analysis by AAS or ICP-OES.

20.2.3 Analysis

Follow manufacturer's instructions for start-up, use, and shutdown of the AAS and ICP-OES. Take appropriate caution with the acetylene and flame in using AAS and the liquid or gas argon and the plasma in using the ICP-OES. Analyze standards, reagent blanks, and samples.

20.3   DATA AND CALCULATIONS

Note: Because of the nature of the differences between printouts for various ICP-OES manufacturers, the ICP operator should assist with interpretation of ICP-OES results. As specified under data handling instructions below, if ICP-OES emission data are available for standards, they should be recorded and plotted, for comparison to AAS standard curves. If ICP-OES emission data are available for samples, they should be converted to concentration data in ppm using the appropriate standard curve. If ICP-OES emission data are not available, report concentration in ppm.

Do all calculations for each duplicate sample individually, before determining a mean value on the final answer.

20.3.1   Standard Curve Data

Potassium standard curves				Sodium standard curves			
AAS		ICP-OES		AAS		ICP-OES	
ppm	Absorption	ppm	Emission	ppm	Absorption	ppm	Emission
50		1		50		1	
100		5		100		5	
200		10		200		10	
300		20		300		20	

20.3.2   Sample Data

Atomic absorption spectroscopy:

Sample	Rep	Sample size (g or mL)	Absorption	ppm	Dilution	Diluted (mg/mL or g/g)	Original (mg/mL or mg/g)
Liquid blank	1				–	–	–
	2				–	–	–
Sports drink	1						
	2						
Solid blank	1				–	–	–
	2				–	–	–
Catsup	1						
	2						
Cottage cheese	1						
	2						
Potato chips	1						
	2						

Inductively coupled plasma – optical emission spectroscopy:

Sample	Rep	Sample size (g or mL)	Emission	ppm	Dilution	Diluted (mg/mL or g/g)	Original (mg/mL or mg/g)
Liquid blank	1				–	–	–
	2				–	–	–
Sports drink	1						
	2						
Solid blank	1				–	–	–
	2				–	–	–
Catsup	1						
	2						
Cottage cheese	1						
	2						
Potato chips	1						
	2						

20.3.3 Data Handling

1. Prepare standard curves for sodium and potassium as measured by AAS.
2. Use the standard curves from AAS and the absorption readings of the samples to determine the concentrations in ppm of sodium and potassium for the food samples as analyzed (i.e., ashed and/or diluted).

   Note: For the AAS samples, you need to subtract the liquid blank absorbance from the sports drink sample values, and the solid blank absorbance from the catsup and potato chip sample values.
3. Prepare standard curves for sodium and potassium as measured by ICP-OES (if emission data are available).
4. Use the standard curves from ICP-OES and the emission readings of the samples to determine the concentrations in ppm of sodium and potassium for the food samples as analyzed (i.e., ashed and/or diluted). If emission data are not available for the samples, record the concentrations in ppm of sodium and potassium for the food samples as analyzed (i.e., ashed and/or diluted).
5. Convert the AAS and ICP-OES values for samples in ppm to mg/mL for the sports drink and to mg/g for the catsup and potato chips.
6. Calculate the sodium and potassium contents by AAS and ICP-OES for the original samples of sports drink (in mg/mL), catsup (in mg/g), and potato chips (mg/g) (on a wet weight basis). Summarize the data and calculated results in one table for AAS and ICP-OES. Show examples of all calculations below the table.

20.4 QUESTIONS

1. Compare the sodium and potassium values for catsup and potato chips to those reported in the US Department of Agriculture Nutrient Database for Standard Reference (http://ndb.nal.usda.gov/). Which method of analysis gives a value closer to that reported in the database for sodium and for potassium?
2. Describe how you would prepare the Na and K standard solutions for AES, using the 1000-ppm solutions of each, which are commercially available. If possible, all solutions for points in the standard curve should be made using different volumes of the same stock solution. Do not use volumes of less than 0.2 mL. Make all standards to the same volume of 100 mL. Note that each standard solution must contain 10 mL conc. HCl/100 mL final volume, as described under Reagents.
3. Describe how you would prepare a 1000-ppm Na solution, starting with commercially available solid NaCl.

RESOURCE MATERIALS

Harris, GK, Marshall MR (2017) Ash analysis. Ch. 16. In: Nielsen SS (ed) Food analysis, 5th edn. Springer, New York

Yeung CK, Miller DD, Rutzke MA (2017) Atomic absorption spectroscopy, atomic emission spectroscopy, and inductively coupled plasma mass spectrometry. Ch. 9. In: Nielsen (SS) Food analysis, 5th edn. Springer, New York

Standard Solutions and Titratable Acidity

S.Suzanne Nielsen

Department of Food Science, Purdue University,
West Lafayette, IN, USA
e-mail: nielsens@purdue.edu

S.S. Nielsen, *Food Analysis Laboratory Manual*, Food Science Text Series,
DOI 10.1007/978-3-319-44127-6_21, © Springer International Publishing 2017

21.1 INTRODUCTION

21.1.1 Background

Many types of chemical analyses are made using a method in which a constituent is titrated with a solution of known strength to an indicator endpoint. Such a solution is referred to as a standard solution. From the volume and concentration of standard solution used in the titration, and the sample size, the concentration of the constituent in the sample can be calculated.

The assay for titratable acidity is a volumetric method that uses a standard solution and, most commonly, the indicator phenolphthalein. In the titration, a standard solution of sodium hydroxide reacts with the organic acids present in the sample. The normality of the sodium hydroxide solution, the volume used, and the volume of the test sample are used to calculate titratable acidity, expressing it in terms of the predominant acid present in the sample. A standard acid such as potassium acid phthalate can be used to determine the exact normality of the standard sodium hydroxide used in the titration.

The phenolphthalein endpoint in the assay for titratable acidity is pH 8.2, where there is a significant color change from clear to pink. When colored solutions obscure the pink endpoint, a potentiometric method is commonly used. A pH meter is used to titrate such a sample to pH 8.2.

21.1.2 Reading Assignment

Tyl, C., and Sadler, G. D. 2017. pH and titratable acidity. Ch. 22, in *Food Analysis*, 5th ed. S.S. Nielsen (Ed.), Springer, New York.

21.1.3 Notes

1. Carbon dioxide (CO_2) acts as an interfering substance in determining titratable acidity, by the following reactions:

$$H_2O + CO_2 \leftrightarrow H_2CO_3 \text{ (carbonic acid)}$$
$$H_2CO_3 \leftrightarrow H^+ + HCO_3^- \text{ (bicarbonate)}$$
$$HCO_3^- \leftrightarrow H^+ + CO_3^{-2} \text{ (carbonate)}$$

In these reactions, buffering compounds and hydrogen ions are generated. Therefore, CO_2-free water is prepared and used for standardizing acids and base and for determining titratable acidity. An Ascarite® trap is attached to bottles of CO_2-free water, so that as air enters the bottle when water is siphoned out, the CO_2 is removed from the air.

2. Ascarite® is a silica base coated with NaOH, and it removes CO_2 from the air by the following reaction:

$$2NaOH + CO_2 \rightarrow Na_2CO_3 + H_2O$$

21.2 PREPARATION AND STANDARDIZATION OF BASE AND ACID SOLUTIONS

21.2.1 Objective

Prepare and standardize solutions of sodium hydroxide and hydrochloric acid.

21.2.2 Principle of Method

A standard acid can be used to determine the exact normality of a standard base and vice versa.

21.2.3 Chemicals

	CAS No.	Hazards
Ascarite	81133-20-2	Corrosive
Ethanol (CH_3CH_2OH)	64-17-5	Highly flammable
Hydrochloric acid (HCl)	7647-01-0	Corrosive
Phenolphthalein	77-09-8	Irritant
Potassium acid phthalate ($HOOCC_6H_4COOK$)	877-24-7	Irritant
Sodium hydroxide (NaOH)	1310-73-2	Corrosive

21.2.4 Reagents

(** It is recommended that these solutions be prepared by the laboratory assistant before class.)

(*Note:* Preparation of NaOH and HCl solutions is described under Procedure.)

- Ascarite trap**
  Put the Ascarite® in a syringe that is attached to the flask of CO_2-free water (see note about CO_2-free water).
- Carbon dioxide-free water**
  Prepare 1.5 L of CO_2-free water (per person or group) by boiling deionized, distilled (dd) water for 15 min in a 2-L Erlenmeyer flask. After boiling, stopper the flask with a rubber stopper through which is inserted in a tube attached to an Ascarite® trap. Allow the water to cool with Ascarite® protection.
- Ethanol, 100 mL
- Hydrochloric acid, concentrated
- Phenolphthalein indicator solution, 1 % **
  Dissolve 1.0 g in 100 mL ethanol. Put in bottle with eyedropper.
- Potassium acid phthalate (KHP)**
  3–4 g, dried in an oven at 120 °C for 2 h cooled and stored in a closed bottle inside a desiccator until use
- Sodium hydroxide, pellets

21.2.5 Hazards, Precautions, and Waste Disposal

Use appropriate precautions in handling concentrated acid and base. Otherwise, adhere to normal laboratory safety procedures. Wear gloves and safety glasses at all times. Waste likely may be put down the drain using a water rinse, but follow good laboratory practices outlined by environmental health and safety protocols at your institution.

21.2.6 Supplies

(Used by students)

- Beaker, 50 mL (for waste NaOH from buret)
- Beaker, 100 mL
- Buret, 25 or 50 mL
- 5 Erlenmeyer flasks, 250 mL
- Erlenmeyer flask, 1 L
- Funnel, small, to fit top of 25 or 50 mL buret
- Glass stirring rod
- Glass storage bottle, 100 mL
- Graduated cylinder, 50 mL
- Graduated cylinder, 1 L
- Graduated pipette, 1 mL
- Graduated pipette, 10 mL
- Parafilm®
- Pipette bulb or pump
- Plastic bottle, with lid, 50 or 100 mL
- Plastic bottle, with lid, 1 L
- Spatula
- Squirt bottle, with dd water
- Volumetric flask, 50 mL
- Volumetric flask, 100 mL
- Weighing paper/boat
- White piece of paper

21.2.7 Equipment

- Analytical balance
- Forced draft oven (heated to 120 °C)
- Hot plate

21.2.8 Calculations Required Before Lab

1. Calculate how much NaOH to use to prepare 50 mL of 25 % NaOH (wt/vol) in water (see Table 2.1, in Chap. 2 for definition of wt%).
2. Calculate how much concentrated HCl to use to prepare 100 mL of ca. 0.1 N HCl in water (concentrated HCl = 12.1 N).

21.2.9 Procedure

1. Prepare 25 % (wt/vol) NaOH solution: Prepare 50 mL of 25 % NaOH (wt/vol) in dd water. To do this, weigh out the appropriate amount of NaOH and place it in a 100-mL beaker. While adding about 40 mL of dd water, stir the NaOH pellets with a glass stirring rod. Continue stirring until all pellets are dissolved. Quantitatively transfer the NaOH solution into a 50-mL volumetric flask. Dilute to volume with dd water. The solution must be cooled to room temperature before final preparation. Store this solution in a plastic bottle and label appropriately.

2. Prepare ca. 0.1 N HCl solution: Prepare 100 mL of ca. 0.1 N HCl using concentrated HCl (12.1 N) and dd water. (Note: Do not use a mechanical pipettor to prepare this, since the acid can easily get into the shaft of the pipettor and cause damage.) To prepare this solution, place a small amount of dd water in a 100-mL volumetric flask, pipette in the appropriate amount of concentrated HCl, then dilute to volume with dd water. Mix well, and transfer into a glass bottle, seal bottle, and label appropriately.

3. Prepare ca. 0.1 N NaOH solution: Transfer 750 mL CO₂-free water to a 1-L plastic storage bottle. Add ca. 12.0 mL of well-mixed 25 % (wt/vol) NaOH solution prepared in Step 1. Mix thoroughly. This will give an approximately 0.1 N solution. Fill the buret with this solution using a funnel. Discard the first volume of the buret and then refill the buret with the NaOH solution.

4. Standardize ca. 0.1 N NaOH solution: Accurately weigh about 0.8 g of dried potassium acid phthalate (KHP) into each of three 250-mL Erlenmeyer flasks. Record the exact weights. Add ca. 50 mL of cool CO₂-free water to each flask. Seal the flasks with Parafilm® and swirl gently until the sample is dissolved. Add three drops of phenolphthalein indicator and titrate, against a white background, with the NaOH solution being standardized. Record the beginning and ending volume on the buret. Titration should proceed to the faintest tinge of pink that persists for 15 s. after swirling. The color will fade with time. Record the total volume of NaOH used to titrate each sample. Data from this part will be used to calculate the mean normality of the diluted NaOH solution.

5. Standardize ca. 0.1 N HCl solution: Devise a scheme to standardize (i.e., determine the exact N) the ca. 0.1 N HCl solution that you prepared in Step 2. Remember that you have your standardized NaOH to use. Do analyses in at least duplicate. Record the volumes used.

21.2.10 Data and Calculations

Using the weight of KHP and the volume of NaOH titrated in Sect. 21.2.9, Step 4, calculate the normality of the diluted NaOH solution as determined by each titration and then calculate the mean normality (molecular weight (MW) potassium acid phthalate = 204.228). The range of triplicate determinations for normality should be less than 0.2 % with good technique.

Rep	Weight of KHP (g)	Buret start (mL)	Buret end (mL)	Vol. NaOH titrated (mL)	N NaOH
1					
2					
3					
					$\bar{X} =$
					SD=

Sample calculation:

Weight of KHP = 0.8115 g
MW of KHP = 204.228 g/mol
Vol. of ca. 0.10 N NaOH used in titration = 39 mL
Mol KHP = 0.8115 g/204.228 g/mol
$\qquad$ = 0.003974 mol
Mol KHP = mol NaOH
$\qquad$ = 0.003974 mol = N NaOH × L NaOH
0.003978 mol NaOH/0.039 L NaOH = 0.1019 N

With the volumes of HCl and NaOH used in Sect. 21.2.9, Step 5, calculate the exact normality of the HCl solution as determined by each titration and then calculate the mean normality.

Rep	Vol. HCl (mL)	Vol. NaOH (mL)	N HCl
1			
2			
		$\bar{X} =$	

21.2.11 Questions

1. What does 25 % NaOH (wt/vol) mean? How would you prepare 500 mL of a 25 % NaOH (wt/vol) solution?
2. Describe how you prepared the 100 mL of ca. 0.1 N HCl. Show your calculations.
3. If you had not been told to use 12 mL of 25 % NaOH (wt/vol) to make 0.75 L of ca. 0.1 N NaOH, how could you have determined this was the appropriate amount? Show all calculations.
4. Describe in detail how you standardized your ca. 0.1 N HCl solution.

21.3   TITRATABLE ACIDITY AND pH

21.3.1   Objective

Determine the titratable acidity and pH of food samples.

21.3.2   Principle of Method

The volume of a standard base used to titrate the organic acids in foods to a phenolphthalein endpoint can be used to determine the titratable acidity.

21.3.3   Chemicals

	CAS No.	Hazards
Ascarite	81133-20-2	Corrosive
Ethanol (CH_3CH_2OH)	64-17-5	Highly flammable
Phenolphthalein	77-09-8	Irritant
Sodium hydroxide (NaOH)	1310-73-2	Corrosive

21.3.4   Reagents

(** It is recommended that these items/solutions be prepared by the laboratory assistant before class:)

- Ascarite trap**
  Put the Ascarite® in a syringe that is attached to the flask of CO_2-free water:
- Carbon dioxide-free water**
  Prepared and stored as described in Method A
- Phenolphthalein indicator solution, 1 %**
  Prepared as described in Method A
- Sodium hydroxide, ca. 0.1 N
  From Sect. 21.2.9, Step 4; exact N calculated
- Standard buffers, pH 4.0 and 7.0

21.3.5   Hazards, Precautions, and Waste Disposal

Adhere to normal laboratory safety procedures. Wear safety glasses at all times. Waste likely may be put down the drain using a water rinse, but follow good laboratory practices outlined by environmental health and safety protocols at your institution.

21.3.6   Supplies

- Apple juice, 60 mL
- 3 Beakers, 250 mL
- 2 Burets, 25 or 50 mL
- 4 Erlenmeyer flasks, 250 mL
- Funnel, small, to fit top of 25 or 50 mL buret
- Graduated cylinder, 50 mL
- Soda, clear, 80 mL
- 2 Volumetric pipettes, 10 or 20 mL

21.3.7   Equipment

- Hot plate
- pH meter

21.3.8 Procedure

21.3.8.1 Soda

Do at least duplicate determinations for unboiled soda and for boiled soda sample; open soda well before use to allow escape of carbon dioxide, so that the sample can be pipetted.

1. Unboiled soda: Pipette 20 mL of soda into a 250-mL Erlenmeyer flask. Add ca. 50 mL CO_2-free dd water. Add three drops of a 1% phenolphthalein solution and titrate with standardized NaOH (ca. 0.1N) to a faint pink color (NaOH in buret, from Method A). Record the beginning and ending volumes on the buret to determine the total volume of NaOH solution used in each titration. Observe the endpoint. Note whether the color fades.
2. Boiled soda: Pipette 20 mL of soda into a 250-mL Erlenmeyer flask. Bring the sample to boiling on a hot plate, swirling the flask often. Boil the sample only 30–60 s. Cool to room temperature. Add ca. 50 mL CO_2-free dd water. Add three drops of the phenolphthalein solution and titrate as described above. Record the beginning and ending volumes on the buret to determine the total volume of NaOH solution used in each titration. Observe the endpoint. Note whether the color fades.

21.3.8.2 Apple Juice

1. Standardize the pH meter with pH 7.0 and 4.0 buffers, using instructions for the pH meter available.
2. Prepare as described below three apple juice samples that will be compared:

    A – Apple juice (set aside, to recall original color)
    B – Apple juice; titrate with std. NaOH
    C – Apple juice; add phenolphthalein; titrate with std. NaOH; follow pH during titration

    Procedure: Into each of three (A, B, C) 250-mL beakers, pipette 20 mL of apple juice. Add ca. 50 mL CO_2-free water to each. To sample C, add three drops of a 1% phenolphthalein solution. Using two burets filled with the standardized NaOH solution (ca. 0.1N), titrate Samples B and C simultaneously. Follow the pH during titration of Sample C containing phenolphthalein. (*Note:* If only one buret is available, titrate Samples B and C sequentially, i.e., add 1 mL to B and then 1 mL to C.) Record the initial pH and the pH at ca. 1.0 mL intervals until a pH of 9.0 is reached. Also, observe any color changes that occur during the titration to determine when the phenolphthalein endpoint is reached. Sample A is intended to help you remember the original color of the apple juice. Sample B (without phenolphthalein) does not need to be followed with the pH meter but is to be titrated along with the other beakers to aid in observing color changes.

21.3.9 Data and Calculations

21.3.9.1 Soda

Using the volume of NaOH used, calculate the titratable acidity (TA) of each soda sample as percentage citric acid, and then calculate the mean TA of each type of sample (MW citric acid = 192.14; equivalent weight = 64.04) (Note: See Eq. 22.20 in Tyl and Sadler, 2017).

Rep	Buret start	Buret end	Vol. NaOH titrant	Color fades?	TA
Unboiled soda:					
1					
2					
					$\bar{X}=$
Boiled soda:					
1					
2					
					$\bar{X}=$

Sample calculation:

$$\% \text{Acid} = \frac{(\text{mL base titrant}) \times (N \text{ of base in mol / liter}) \times (\text{Eq. Wt. of acid})}{(\text{sample volume in mL}) \times 10}$$

N NaOH $= 0.1019\,N$
mL base $= 7$ mL
Eq. wt. citric acid $= 64.04$
Vol. sample $= 20$ mL

$$\% \text{ Acid} = (7\,\text{ml} \times 0.1019 \times 64.04)/(20\,\text{ml} \times 10)$$
$$= 0.276\% \text{ citric acid}$$

21.3.9.2 Apple Juice

Sample B
  Color change during titration:
  Color at end of titration:

Sample C (titrate to >pH 9.0)

mL NaOH	1	2	3	4	5	6	7	8	9	10
pH										

mL NaOH	11	12	13	14	15	16	17	18	19	20
pH										

Plot pH versus mL of $0.1N$ NaOH (but use the normality of your own NaOH solution) (pH on the y-axis) for the sample that contained phenolphthalein (Sample C). Interpolate to find the volume of titrant at pH 8.2 (the phenolphthalein endpoint).

Calculate the titratable acidity of the apple juice as percentage malic acid (MW malic acid = 134.09; equivalent weight = 67.04).

21.3.10 Questions

1. Soda samples. (a) Did any color changes occur in either the boiled or the unboiled sample within several minutes of the phenolphthalein endpoint being reached? (b) How did boiling the sample affect the determination of titratable acidity? (c) Explain the differences in color changes and titratable acidity between the two samples.
2. What caused the color changes in the apple juice titrated without any phenolphthalein present? (Hint: Consider the pigments in apples.) How would you recommend determining the endpoint in the titration of tomato juice?
3. You are determining the titratable acidity of a large number of samples. You ran out of freshly boiled dd H_2O with an Ascarite trap on the water container, so you switch to using tap distilled H_2O. Would this likely affect your results? Explain.
4. The electrode of your pH meter has a slow response time and seems to need cleaning, since it is heavily used for a variety of solutions high in proteins, lipids, and minerals. You would ideally check the electrode instructions for specific recommendations on cleaning, but the instructions were thrown away. (As the new lab supervisor, you have since started a policy of filing all instrument/equipment instructions.) What solutions would you use to try to clean the electrode?

RESOURCE MATERIALS

AOAC International (2016) Official methods of analysis, 20th edn. (On-line). AOAC International, Rockville, MD

Tyl C, Sadler GD (2017) pH and titratable acidity. Ch. 22. In: Nielsen SS (ed) Food analysis, 5th edn. Springer, New York

Fat Characterization

S.Suzanne Nielsen (✉)

Department of Food Science, Purdue University,
West Lafayette, IN, USA
e-mail: nielsens@purdue.edu

Michael C. Qian

Department of Food Science and Technology,
Oregon State University, Corvallis, OR, USA
e-mail: michael.qian@oregonstate.edu

Oscar A. Pike

Department of Nutrition, Dietetics, and Food Science,
Brigham Young University, Provo, UT, USA
e-mail: oscar_pike@byu.edu

S.S. Nielsen, *Food Analysis Laboratory Manual*, Food Science Text Series,
DOI 10.1007/978-3-319-44127-6_22, © Springer International Publishing 2017

22.1 INTRODUCTION

22.1.1 Background

Lipids in food are subjected to many chemical reactions during processing and storage. While some of these reactions are desirable, others are undesirable; so, efforts are made to minimize the reactions and their effects. The laboratory deals with characterization of fats and oils with respect to composition, structure, and reactivity.

22.1.2 Reading Assignment

Pike, O.A., and O'Keefe, S.F. 2017. Fat characterization, Ch. 23 in *Food Analysis*, 5th ed. S.S. Nielsen (Ed.), Springer, New York.

22.1.3 Overall Objective

The overall objective of this laboratory is to determine aspects of the composition, structure, and reactivity of fats and oils by various methods.

22.2 SAPONIFICATION VALUE

22.2.1 Objective

Determine the saponification number of fats and oils.

22.2.2 Principle of Method

Saponification is the process of treating a neutral fat with alkali, breaking it down to glycerol and fatty acids. The saponification value (or number) is defined as the amount of alkali needed to saponify a given quantity of fat or oil, expressed as mg potassium hydroxide to saponify 1 g sample. Excess alcoholic potassium hydroxide is added to the sample, the solution is heated to saponify the fat, the unreacted potassium hydroxide is back-titrated with standardized hydrochloric acid using a phenolphthalein indicator, and the calculated amount of reacted potassium hydroxide is used to determine the saponification value.

22.2.3 Chemicals

	CAS No.	Hazards
Ethanol	64-17-5	Highly flammable
Hydrochloric acid (HCl)	7647-01-0	Corrosive
Phenolphthalein	77-09-8	Irritant
Potassium hydroxide (KOH)	1310-58-3	Corrosive

22.2.4 Reagents

(**It is recommended that these solutions be prepared by the laboratory assistant before class.)

- Alcoholic potassium hydroxide, ca. $0.7N$ **
  Dissolve 40 g KOH, low in carbonate, in 1 L of distilled ethanol, keeping temperature below 15.5 °C while the alkali is being dissolved. The solution should be clear.
- Hydrochloric acid, ca. $0.5N$, accurately standardized **
  Prepare ca. $0.5N$ HCl. Determine the exact normality using solution of standard base.
- Phenolphthalein indicator solution **
  1%, in 95% ethanol

22.2.5 Hazards, Precautions, and Waste Disposal

Use hydrochloric acid in a fume hood. Otherwise, adhere to normal laboratory safety procedures. Wear safety glasses at all times. Wastes likely may be put down the drain using a water rinse, but follow good laboratory practices outlined by environmental health and safety protocols at your institution.

22.2.6 Supplies

(Used by students)

- Air (reflux) condenser (650 mm long, minimum)
- Beaker, 250 mL (to melt fat)
- Buchner funnel (to fit side-arm flask)
- Boiling beads
- 2 Burets, 50 mL
- Fat and/or oil samples
- Filter paper (to fit Buchner funnel, to filter oil and melted fat)
- 4 Flasks, 250–300 mL, to fit condenser
- Mechanical pipettor, 1000 µL, with plastic tips (or 1 mL volumetric pipette)
- Side-arm flask

22.2.7 Equipment

- Analytical balance
- Hot plate or water bath (with variable heat control)

22.2.8 Procedure

(Instructions are given for analysis in duplicate.)

1. Melt any solid samples. Filter melted fat sample and oil sample through filter paper to remove impurities.
2. Weigh accurately ca. 5 g melted fat or oil into each of two 250–300 ml flasks that will connect to a condenser. Record weight of sample. Prepare sample in duplicate.
3. Add accurately (from a buret) 50 mL of alcoholic KOH into the flask.

4. Prepare duplicate blank samples with just 50 mL of alcoholic KOH in a 250–300 mL flask.
5. Add several boiling beads to the flasks with fat or oil sample.
6. Connect the flasks with the sample to a condenser. Boil gently but steadily on a hot plate (or water bath) until the sample is clear and homogenous, indicating complete saponification (requires ca. 30–60 min). (Note: The fumes should condense as low as possible in the condenser; otherwise, a fire hazard will be created.)
7. Allow samples to cool somewhat. Wash down the inside of the condenser with a little deionized distilled (dd) water. Disconnect flask from condenser. Allow the samples to cool to room temperature.
8. Add 1 mL phenolphthalein to samples and titrate with 0.5N HCl (from a buret) until the pink color just disappears. Record the volume of titrant used.
9. Repeat Steps 5–8 with sample blanks. Reflux the blanks for the same time period as used for the sample.

22.2.9 Data and Calculations

Sample	Weight (g)	Titrant volume (mL)	Saponification value
1			
2			
			$\bar{X} =$

Oil/fat sample type tested:

Blank titration (mL):
   Sample 1 =
   Sample 2 =
   $\bar{X} =$

Calculate the saponification number (or value) of each sample as follows:

$$\text{Saponification value} = \frac{(B-S) \times N \times 56.1}{W}$$

where:

   Saponification value = mg KOH per g of sample
   B = volume of titrant (mL) for blank
   S = volume of titrant (mL) for sample
   N = normality of HCl (mmol/mL)
   56.1 = molecular weight (MW) of KOH (mg/mmol)
   W = sample mass (g)

22.2.10 Questions

1. What is meant by unsaponifiable matter in lipid samples? Give an example of such a type of compound.
2. What does a high versus a low saponification value tell you about the nature of a sample?

22.3 IODINE VALUE

22.3.1 Objective

Determine the iodine value of fats and oils.

22.3.2 Principle of Method

The iodine value (or number) is a measure of the degree of unsaturation, defined as the grams of iodine absorbed per 100 g sample. In the assay, a measured quantity of fat or oil dissolved in solvent is reacted with a measured excess amount of iodine or some other halogen, which reacts with the carbon-carbon double bonds. After a solution of potassium iodide is added to reduce excess ICl to free iodine, the liberated iodine is titrated with a standardized solution of sodium thiosulfate using a starch indicator. The calculated amount of iodine reacted with the double bonds is used to calculate the iodine value.

22.3.3 Chemicals

	CAS no.	Hazards
Acetic acid (glacial)	64-19-7	Corrosive
Carbon tetrachloride (CCl_4)	56-23-5	Toxic, dangerous for the environment
Chloroform	67-66-3	Harmful
Hydrochloric acid (HCl)	7647-01-0	Corrosive
Iodine	7553-56-2	Harmful, dangerous for the environment
Potassium dichromate ($K_2Cr_2O_7$)	7789-00-6	Toxic, dangerous for the environment
Potassium iodide (KI)	7681-11-0	
Sodium thiosulfate	7772-98-7	
Soluble starch	9005-25-8	

22.3.4 Reagents

(**It is recommended that these solutions be prepared by the laboratory assistant before class.)

- Potassium iodide solution, 15%
     Dissolve 150 g KI in dd water and dilute to 1 l.
- Sodium thiosulfate, 0.1N standardized solution (AOAC Method 942.27) **
     Dissolve ca. 25 g sodium thiosulfate in 1 L dd water. Boil gently for 5 min. Transfer while hot to

a storage bottle (make sure bottle has been well cleaned and is heat resistant). Store solution in a dark, cool place. Use the following procedure to standardize the sodium thiosulfate solution: Accurately weigh 0.20–0.23 g potassium dichromate ($K_2Cr_2O_7$) (previously dried for 2 h at 100 °C) into a glass-stoppered flask. Dissolve 2 g potassium iodide (KI) in 80 mL chlorine-free water. Add this water to the potassium dichromate. To this solution, add, with swirling, 20 mL ca. 1 M HCl and immediately place in the dark for 10 min. Titrate a known volume of this solution with the sodium thiosulfate solution, adding starch solution after most of the iodine has been consumed.

- Starch indicator solution, 1 % (prepare fresh daily)**
  Mix ca. 1 g soluble starch with enough cold dd water to make a thin paste. Add 100 mL boiling dd water. Boil ca. 1 min while stirring.
- Wijs iodine solution **
  Dissolve 10 g ICl_3 in 300 mL CCl_4 and 700 mL glacial acetic acid. Standardize this solution against 0.1 N sodium thiosulfate (25 mL of Wijs solution should consume 3.4–3.7 mEq of thiosulfate). Then, add enough iodine to the solution such that 25 mL of the solution will require at least 1.5 times the milliequivalency of the original titration. Place the solution in an amber bottle. Store in the dark at less than 30 °C.

22.3.5 Hazards, Precautions, and Waste Disposal

Carbon tetrachloride and potassium chromate are toxic and must be handled with caution. Use acetic acid and hydrochloric acid in a fume hood. Otherwise, adhere to normal laboratory safety procedures. Wear safety glasses at all times. Carbon tetrachloride, chloroform, iodine, and potassium chromate must be handled as hazardous wastes. Other wastes likely may be put down the drain using a water rinse, but follow good laboratory practices outlined by environmental health and safety protocols at your institution.

22.3.6 Supplies

(Used by students)

- 2 Beakers, 250 mL (one to melt fat; one to boil water)
- Buchner funnel (to fit side-arm flask)
- Buret, 10 or 25 mL
- Fat and/or oil samples
- Filter paper (to fit Buchner funnel, to filter melted fat and oil)
- 4 Flasks, 500 mL, glass stoppered
- Graduated cylinder, 25 mL
- Graduated cylinder, 100 mL

- Mechanical pipettor, 1000 µL, with plastic tips (or 1 mL volumetric pipette)
- Side-arm flask
- Volumetric pipette, 10 mL
- Volumetric pipette, 20 mL

22.3.7 Equipment

- Analytical balance
- Hot plate

22.3.8 Procedure

(Instructions are given for analysis in duplicate.)

1. Melt any samples that are solid at room temperature by heating to a maximum of 15 °C above the melting point. Filter melted fat sample and oil sample through filter paper to remove impurities.
2. Weigh accurately 0.1–0.5 g sample (amount used depends on expected iodine number) into each of two dry 500 mL glass-stoppered flasks. Add 10 mL chloroform to dissolve the fat or oil.
3. Prepare two blanks by adding only 10 mL chloroform to 500 mL glass-stoppered flasks.
4. Pipette 25 mL Wijs iodine solution into the flasks. (The amount of iodine must be 50–60 % in excess of that absorbed by the fat.)
5. Let flasks stand for 30 min in the dark with occasional shaking.
6. After incubation in the dark, add 20 mL potassium iodide solution to each flask. Shake thoroughly. Add 100 mL freshly boiled and cooled water, washing down any free iodine on the stopper.
7. Titrate the iodine in the flasks with standard sodium thiosulfate, adding it gradually with constant and vigorous shaking until the yellow color almost disappears. Then add 1–2 mL of starch indicator and continue the titration until the blue color entirely disappears. Toward the end of the titration, stopper the flask and shake violently so that any iodine remaining in the chloroform can be taken up by the potassium iodide solution. Record the volume of titrant used.

22.3.9 Data and Calculations

Sample	Weight (g)	Titrant volume (mL)	Iodine value
1			
2			
			$\bar{X}$ =

Oil/fat sample type tested:

Blank titration (mL):

Sample 1 =
Sample 2 =
$\bar{X} =$

Calculate the iodine value of each sample as follows:

$$\text{Iodine value} = \frac{(B-S) \times N \times 126.9}{W \times 1000} \times 100$$

where:

Iodine value = g iodine absorbed per 100 g of sample
B = volume of titrant (mL) for blank
S = volume of titrant (mL) for sample
N = normality of $Na_2S_2O_3$ (mol/1000 mL)
126.9 = MW of iodine (g/mol)
W = sample mass (g)

22.3.10 Questions

1. In the iodine value determination, why is the blank volume higher than that of the sample?
2. What does a high versus a low iodine value tell you about the nature of the sample?

22.4 FREE FATTY ACID VALUE

22.4.1 Objective

Determine the free fatty acid (FFA) value of fats and oils.

22.4.2 Principle of Method

Free fatty acid value, or acid value, reflects the amount of fatty acids hydrolyzed from triacylglycerols. Free fatty acid is the percentage by weight of a specific fatty acid. Acid value is defined as the milligrams of potassium hydroxide needed to neutralize the free acids present in 1 g of fat or oil. A liquid fat sample combined with neutralized 95% ethanol is titrated with standardized sodium hydroxide to a phenolphthalein endpoint. The volume and normality of the sodium hydroxide are used, along with the weight of the sample, to calculate the free fatty acid value.

22.4.3 Chemicals

	CAS no.	Hazards
Ethanol	64-17-5	Highly flammable
Phenolphthalein	77-09-8	Irritant
Sodium hydroxide (NaOH)	1310-73-2	Corrosive

22.4.4 Reagents

(**It is recommended that these solutions be prepared by the laboratory assistant before class.)

- Ethanol, neutralized
  Neutralize 95% ethanol to a permanent pink color with alkali and phenolphthalein.
- Phenolphthalein indicator**
  In a 100 mL volumetric flask, dissolve 1 g phenolphthalein in 50 mL 95% ethanol. Dilute to volume with dd water.
- Sodium hydroxide, 0.1 N, standardized **
  Use commercial product, or prepare as described in other laboratory manual chapters (Chaps. 2 and 21).

22.4.5 Hazards, Precautions, and Waste Disposal

Adhere to normal laboratory safety procedures. Wear safety glasses at all times. Wastes likely may be put down the drain using a water rinse, but follow good laboratory practices outlined by environmental health and safety protocols at your institution.

22.4.6 Supplies

(Used by students)

- Beaker, 250 mL (to melt fat)
- Buchner funnel (to fit side-arm flask)
- Buret, 10 mL
- 4 Erlenmeyer flasks, 250 mL
- Fat and/or oil samples
- Filter paper (to fit Buchner funnel, to filter melted fat and oil)
- Graduated cylinder, 100 mL
- Mechanical pipettor, 1000 µL, with plastic tips (or 1 mL volumetric pipette)
- Side-arm flask

22.4.7 Equipment

- Analytical balance
- Hot plate

22.4.8 Procedure

(Instructions are given for analysis in triplicate.)

1. Melt any samples that are solid at room temperature by heating to a maximum of 15 °C above the melting point. Filter melted fat sample and oil sample through filter paper to remove impurities.
2. As a preliminary test, accurately weigh ca. 5 g melted fat or oil into a 250 mL Erlenmeyer flask.
3. Add ca. 100 mL neutralized ethanol and 2 mL phenolphthalein indicator.

4. Shake to dissolve the mixture completely. Titrate with standard base (ca. 0.1 N NaOH), shaking vigorously until the endpoint is reached. This is indicated by a slight pink color that persists for 30 s. Record the volume of titrant used. Use the information below to determine if the sample weight you have used is correct for the range of acid values under which your sample falls. This will determine the sample weight to be used for Step 5.

The *Official Methods and Recommended Practices of the AOCS* (AOCS 2009) recommends the following sample weights for ranges of expected acid values:

FFA range (%)	Sample (g)	Alcohol (mL)	Strength of alkali
0.00–0.2	56.4±0.2	50	0.1 N
0.2–1.0	28.2±0.2	50	0.1 N
1.0–30.0	7.05±0.05	75	0.25 N

5. Repeat Steps 1–3 more carefully in triplicate, recording each weight of the sample and the volume of titrant.

22.4.9 Data and Calculations

Sample	Weight (g)	Titrant volume (mL)	FFA value
1			
2			
3			$\bar{X}$ =
			SD =

Oil/fat sample type tested:

Calculate the FFA value of each sample as follows:

$$\% \text{FFA (as oleic)} = \frac{V \times N \times 282}{W \times 1000} \times 100$$

where:

> % FFA = percent free fatty acid (g/100 g), expressed as oleic acid
> V = volume of NaOH titrant (mL)
> N = normality of NaOH titrant (mol/1000 mL)
> 282 = MW of oleic acid (g/mol)
> W = sample mass (g)

22.4.10 Questions

1. What is a high FFA value indicative of relative to product history?
2. Why is the FFA content of frying oil important?

3. In a crude fat extract, FFA are naturally present, but they are removed during processing to enhance the stability of the fat. State and describe the processing step that removes the FFA naturally present.

22.5 PEROXIDE VALUE

22.5.1 Objective

Determine the peroxide value of fats and oils, as an indicator of oxidative rancidity.

22.5.2 Principle of Method

Peroxide value is defined as the milliequivalents of peroxide per kilogram of fat, as determined in a titration procedure to measure the amount of peroxide or hydroperoxide groups. To a known amount of fat or oil, excess potassium iodide is added, which reacts with the peroxides in the sample. The iodine liberated is titrated with standardized sodium thiosulfate using a starch indicator. The calculated amount of potassium iodide required to react with the peroxide present is used to determine the peroxide value.

22.5.3 Chemicals

	CAS no.	Hazards
Acetic acid (glacial)	64-19-7	Corrosive
Chloroform	67-66-3	Harmful
Hydrochloric acid (HCl)	7647-01-0	Corrosive
Potassium chromate ($K_2Cr_2O_7$)	7789-00-6	Toxic, dangerous for environment
Potassium iodide (KI)	7681-11-0	
Sodium thiosulfate	7772-98-7	
Soluble starch	9005-25-8	

22.5.4 Reagents

(**It is recommended that these solutions be prepared by the laboratory assistant before class.)

- Acetic acid-chloroform solution
  Mix three volumes of concentrated acetic acid with two volumes of chloroform.
- Potassium iodide solution, saturated **
  Dissolve excess KI in freshly boiled dd water. Excess solid must remain. Store in the dark. Test before use by adding 0.5 mL acetic acid-chloroform solution, and then add 2 drops 1% starch indicator solution. If solution turns blue, requiring >1 drop 0.1 N thiosulfate solution to discharge color, prepare a fresh potassium iodide solution.
- Sodium thiosulfate, 0.1 N, standard solution (AOAC Method 942.27) **
  Dissolve ca. 25 g sodium thiosulfate pentahydrate in 1 L dd water. Boil gently for 5 min.

Transfer while hot to a storage bottle (make sure bottle has been well cleaned and is heat resistant). Store solution in a dark, cool place. Use the following procedure to standardize the sodium thiosulfate solution: Accurately weigh 0.20–0.23 g potassium chromate ($K_2Cr_2O_7$) (previously dried for 2 h at 100 °C) into a glass-stoppered flask. Dissolve 2 g potassium iodide (KI) in 80 mL chlorine-free water. Add this water to the potassium chromate. To this solution, add, with swirling, 20 mL ca. 1 M HCl, and immediately place in the dark for 10 min. Titrate a known volume of this solution with the sodium thiosulfate solution, adding starch solution after most of the iodine has been consumed.

- Starch indicator solution, 1 % (prepare fresh daily)**
Mix ca. 1 g soluble starch with enough cold dd water to make a thin paste. Add 100 mL boiling dd water. Boil ca. 1 min while stirring.

22.5.5 Hazards, Precautions, and Waste Disposal

Potassium chromate is toxic and must be handled with caution. Use hydrochloric acid in a fume hood. Otherwise, adhere to normal laboratory safety procedures. Wear gloves and safety glasses at all times. Chloroform and potassium chromate must be handled as hazardous wastes. Other wastes likely may be put down the drain using a water rinse, but follow good laboratory practices outlined by environmental health and safety protocols at your institution.

22.5.6 Supplies

(Used by students)

- Beaker, 250 mL (to melt fat)
- Buchner funnel (to fit side-arm flask)
- Buret, 25 mL or 50 mL
- 4 Erlenmeyer flasks, 250 mL, glass stoppered
- Fat and/or oil samples
- Filter paper (to fit Buchner funnel, to filter melted fat and oil)
- 2 Graduated cylinders, 50 mL
- Mechanical pipettor, 1000 μL, with plastic tips (or 1 mL volumetric pipette)
- Side-arm flask

22.5.7 Equipment

- Analytical balance
- Hot plate

22.5.8 Procedure

(Instructions are given for analysis in duplicate.)

  1. Melt any samples that are solid at room temperature by heating to a maximum of 15 °C

above the melting point. Filter melted fat sample and oil sample through filter paper to remove impurities.
2. Accurately weigh ca. 5 g fat or oil (to the nearest 0.001 g) into each of two 250 mL glass-stoppered Erlenmeyer flasks.
3. Add 30 mL acetic acid-chloroform solution and swirl to dissolve.
4. Add 0.5 mL saturated KI solution. Let stand with occasional shaking for 1 min. Add 30 mL dd water.
5. Slowly titrate samples with 0.1 N sodium thiosulfate solution, with vigorous shaking until yellow color is almost gone.
6. Add ca. 0.5 mL 1 % starch solution, and continue titration, shaking vigorously to release all iodine from chloroform layer, until blue color just disappears. Record the volume of titrant used. (If <0.5 mL of the sodium thiosulfate solution is used, repeat determination.)
7. Prepare (omitting only the oil) and titrate a blank sample. Record the volume of titrant used.

22.5.9 Data and Calculations

Sample	Weight (g)	Titrant volume (mL)	Peroxide value
1			
2			
			$\overline{X} =$

Oil/fat sample type tested:

Blank titration (mL):

    Sample 1 =
    Sample 2 =
    $\overline{X} =$

Calculate the peroxide value of each sample as follows:

$$\text{Peroxide value} = \frac{(S - B) \times N}{W} \times 1000$$

where:

    Peroxide value = mEq peroxide per kg of sample
    S = volume of titrant (mL) for sample
    B = volume of titrant (mL) for blank
    N = normality of $Na_2S_2O_3$ solution (mEq/mL)
    1000 = conversion of units (g/kg)
    W = sample mass (g)

22.5.10 Questions

  1. What are some cautions in using peroxide value to estimate the amount of autoxidation in foods?

2. The peroxide value method was developed for fat or oil samples. What must be done to a food sample before measuring its peroxide value using this method?

22.6 THIN-LAYER CHROMATOGRAPHY SEPARATION OF SIMPLE LIPIDS

22.6.1 Objective

Separate and identify the lipids in some common foods using thin-layer chromatography (TLC).

22.6.2 Principle of Method

Like all types of chromatography, TLC is a separation technique that allows for the distribution of compounds between a mobile phase and a stationary phase. Most classes of lipids can be separated from each other by adsorption chromatography on thin layers. In TLC, a thin layer of stationary phase is bound to an inert support (i.e., glass plate, plastic, or aluminum sheet). The sample and standards are applied as spots near one end of the plate. For ascending chromatography, the plate is placed in a developing chamber, with the end of the plate nearest the spots being placed in the mobile phase at the bottom of the chamber. The mobile phase migrates up the plate by capillary action, carrying and separating the sample components. The separated bands can be visualized or detected and compared to the separation of standard compounds.

22.6.3 Chemicals

	CAS no.	Hazards
Acetic acid	64-19-7	Corrosive
Diethyl ether	60-29-7	Harmful, extremely flammable
Hexane	110-54-3	Harmful, highly flammable, dangerous for the environment
Sulfuric acid	7664-93-9	Corrosive

22.6.4 Reagents

- Chloroform/methanol, 2:1, v/v
- Mobile phase
  Hexane/diethyl ether/acetic acid, 78:20:2
- Standards
  Triacylglycerol, fatty acid, cholesteryl ester, and cholesterol
- Sulfuric acid solution
  Concentrated H_2SO_4, in 50 % aqueous solution

22.6.5 Hazards, Precautions, and Waste Disposal

Use acetic acid and sulfuric acid in a fume hood. Diethyl ether is extremely flammable, is hygroscopic, and may form explosive peroxides. Otherwise, adhere to normal laboratory safety procedures. Wear safety glasses at all times. Diethyl ether and hexane must be handled as hazardous wastes. Other wastes likely may be put down the drain using a water rinse, but follow good laboratory practices outlined by environmental health and safety protocols at your institution.

22.6.6 Supplies

- Capillary tubes (or syringes) (to apply samples to plates)
- Developing tank, with lid
- Filter paper, Whatman No. 1 (to line developing tank)
- Oil/fat food samples (e.g., hamburger, safflower oil) (prepare at a concentration of 20 µg/mL in 2:1 v/v chloroform-methanol solution)
- Pencil
- Thin-layer chromatography plates: Silica Gel 60, 0.25 mm thick coating on glass backing, 20 × 20 cm (EM Science)

22.6.7 Equipment

- Air blower (e.g., blow hair dryer)
- Oven

22.6.8 Procedure

Preparation of Silica Gel Plates

1. Place plates in oven at 110 °C for 15 min, then cool to ambient temperature (5 min).
2. With a pencil, draw a line to mark the origin, 2.5 cm from the bottom of the plate.
3. Make marks with a pencil to divide the plate into 10 "lanes" of equal width.
4. Use capillary tubes or syringes to apply approximately 10 µl of each standard and sample to a separate lane (use the middle eight lanes). The application should be done as a streak across the center of the lane origin. This is best accomplished with four spots of 2.5 µl each.
5. Below the origin line, write the identity of the sample/standard in each lane.
6. Allow spots to dry. You may accelerate drying by using a low-temperature air blower.
7. Write your name in the top right corner of the plate.

Development of Plates

1. Line the developing tank with Whatman no. 1 or similar filter paper.
2. Pour the mobile phase gently over the filter paper until the depth of solvent in the tank is approximately 0.5 cm. About 200 mL is required.
3. Place the lid on the tank and allow 15 min for the atmosphere in the tank to become saturated with solvent vapor.
4. Place the spotted TLC plate in the developing tank and allow it to develop until the solvent front reaches a point about 2 cm from the top of the plate.
5. Remove the plate from the tank and *immediately* mark the position of the solvent front. Evaporate the solvent in the fume hood.

Visualization of Lipids

1. In a well-ventilated fume hood, spray lightly with 50 % aqueous H_2SO_4. Allow to dry.
2. Heat plate for 5–10 min at 100–120 °C. Remove from oven, cool, and inspect. Handle the plate with caution as the surface still contains sulfuric acid.
3. Mark all visible spots at their center, and note the color of the spots.

22.6.9 Data and Calculations

For the spots of each of the standards and the samples, report the distance from the origin for the spot. Also for each spot, calculate the R_f value, as the distance from the origin to the spot divided by the distance from the origin to the solvent front. Using the R_f value of the standards, identify as many of the spots (bands) in the samples as possible.

Standard	Distance from origin	R_f value
Triacylglycerol		
Fatty acid		
Cholesteryl ester		
Cholesterol		

Sample spot number	Distance from origin	R_f value	Identity

Oil/fat sample type tested:

22.6.10 Questions

1. Explain the chemical structure of an ester of cholesterol.
2. Besides the four fat constituents used as standards, what other fat constituents might be found using a TLC method such as this?

Acknowledgments This laboratory exercise was developed with input from Dr Arun Kilara with Arun Kilara Worldwide, Northbrook, IL.

RESOURCE MATERIALS

AOCS (2009) Official methods and recommended practices of the AOCS, 6th edn. American Oil Chemists' Society, Champaign, IL

Pike OA, O'Keefe SF (2017) Fat characterization. Ch. 23. In: Nielsen SS (ed) Food analysis, 5th edn. Springer, New York

Proteins: Extraction, Quantitation, and Electrophoresis

Denise M. Smith

School of Food Science, Washington State University,
Pullman, WA, USA
e-mail: denise.smith@wsu.edu

S.S. Nielsen, *Food Analysis Laboratory Manual*, Food Science Text Series,
DOI 10.1007/978-3-319-44127-6_23, © Springer International Publishing 2017

23.1 INTRODUCTION

23.1.1 Background

Electrophoresis can be used to separate and visualize protein banding patterns. In sodium dodecyl sulfate-polyacrylamide gel electrophoresis (SDS-PAGE), proteins are dissociated into subunits and then separated based on size within a gel matrix by applying an electric field. It is usually necessary to apply a sample volume to the gel that contains a known amount of total protein to allow comparisons between samples. While it is possible to use an official method (e.g., Kjeldahl, N combustion) to determine total protein, it is often convenient to use a rapid colorimetric method of protein analysis that requires only a small amount of sample. The bicinchoninic acid (BCA) assay method will be used for this purpose.

In this experiment, sarcoplasmic muscle proteins are extracted with a $0.15 M$ salt solution, the protein content of the extract is measured by the BCA colorimetric assay, and the proteins in the fish extracts are separated and visualized by SDS-PAGE. Visualization of the protein banding patterns makes it possible to distinguish among different types of fish since many fish have a characteristic protein subunit pattern. For example, one might use this technique as part of a scheme to detect economic adulteration when an inexpensive fish species is substituted for a more expensive fish in the wholesale or retail marketplace.

23.1.2 Reading Assignment

Chang, S.K.C., and Zhang Y. 2017. Protein analysis. Ch.18, in *Food Analysis*, 5th ed. S.S. Nielsen (Ed.), Springer, New York.

Smith, D.M. 2017. Protein separation and characterization. Ch. 24, in *Food Analysis*, 5th ed. S.S. Nielsen (Ed.), Springer, New York.

23.1.3 Objective

Extract proteins from the muscles of freshwater and saltwater fish, measure the protein content of the extracts, separate the proteins by electrophoresis, and then compare the different protein banding patterns that result based on subunit size and relative quantity.

23.1.4 Principle of Method

Sarcoplasmic proteins can be extracted from fish muscle with $0.15 M$ salt. Protein content of the extract can be determined by the colorimetric BCA assay method. In this assay, protein reduces cupric ions to cuprous ions under alkaline conditions. The amount of cuprous ions formed is proportional to the amount of protein present. The cuprous ions react with the BCA reagent to form a purple color that can be quantified spectrophotometrically and related to the protein content by comparison to a standard curve. Proteins in the extract can be separated by SDS-PAGE. Proteins become negatively charged when bound to SDS, so they move through the gel matrix toward the anode (pole with positive charge) at a rate based on size alone. The molecular mass of a given protein subunit can be estimated by comparing its electrophoretic mobility with proteins of known molecular weight using a standard curve. A linear relationship is obtained by plotting the logarithm of the molecular mass of standard proteins against their respective electrophoretic mobilities (R_f).

23.1.5 Notes

This experiment may be done over two laboratory sessions. The protein can be extracted, quantified, and prepared for electrophoresis in the first session. The prepared protein samples can be frozen until electrophoresis is completed in the second laboratory session. Alternatively, in a single laboratory session, one group of students could do the protein extraction and quantitation, while a second group of students prepares the electrophoresis gels. Also, different groups of students could be assigned different fish species. Multiple groups could run their samples on a single electrophoresis gel. The gels for electrophoresis can be purchased commercially (e.g., Bio-Rad, Mini-PROTEAN TGX precast gels, 12% resolving gel) or made as described below.

Some fish species work better than others for preparing the extracts and comparing differences in protein patterns. Catfish (freshwater) and tilapia (saltwater) work well as extracts and show some differences. Trout gives very thick extracts. Freshwater and saltwater salmon show few differences.

23.1.6 Chemicals

	CAS No.	Hazards
Sample Extraction:		
Sodium chloride (NaCl)	7647-14-5	Irritant
Sodium phosphate, monobasic (NaH$_2$PO$_4$ · H$_2$O)	7558-80-7	Irritant
Protein Determination (BCA method):		
Bicinchoninic acid		
Bovine serum albumin (BSA)	9048-46-8	
Copper sulfate (CuSO$_4$)	7758-98-7	Irritant
Sodium bicarbonate (NaHCO$_3$)	144-55-8	
Sodium carbonate (Na$_2$CO$_3$)	497-19-8	Irritant
Sodium hydroxide (NaOH)	1310-73-2	Corrosive
Sodium tartrate	868-18-8	

Electrophoresis:

Acetic acid (CH$_3$COOH)	64-19-7	Corrosive
Acrylamide	79-06-1	Toxic
Ammonium persulfate (APS)	7727-54-0	Harmful, oxidizing
Bis-acrylamide	110-26-9	Harmful
Bromophenol blue	115-39-9	
Butanol	71-36-3	Harmful
Coomassie Blue R-250	6104-59-2	
Ethylenediaminetetraacetic acid, disodium salt (Na$_2$EDTA · 2H$_2$O)	60-00-4	Irritant
Glycerol (C$_3$H$_8$O$_3$)	56-81-5	
Glycine	56-40-6	
Hydrochloric acid (HCl)	7647-01-0	Corrosive
β-Mercaptoethanol	60-24-2	Toxic
Methanol (CH$_3$OH)	67-56-1	Highly flammable
Protein molecular weight standards (e.g., Bio-Rad 161–0374, Precision Plus Protein Dual Color Standards, 10–250 KD)		
Sodium dodecyl sulfate (SDS, dodecyl sulfate, sodium salt)	151-21-3	Harmful
N, N, N′, N′-Tetramethylethylenediamine (TEMED)	110-18-9	Highly flammable, corrosive
Tris base	77-86-1	

23.2   REAGENTS

(** It is recommended that these solutions be prepared by laboratory assistant before class.)

23.2.1   Sample Extraction

- Extraction buffer, 300 mL/fish species **
    Buffer of 0.15*M* sodium chloride, 0.05*M* sodium phosphate, pH 7.0 (Students asked to show the calculations for this buffer later in Questions.)

23.2.2   Protein Determination (BCA Method)

Commercial BCA test kits can be purchased, such as the Pierce BCA Protein Assay Kit (cat. No. 23225, Rockford, IL) which includes procedures to complete either test tube (2 mL working reagent/tube) or microplate assays (200 uL working reagent/well).
    This kit includes:
- Bovine serum albumin (BSA) standard ampules, 2 mg/mL in 0.9% saline and 0.05% sodium azide as a preservative
- BCA Reagent A: Contains sodium carbonate, sodium bicarbonate, BCA detection reagent, and sodium tartrate in 0.1 M sodium hydroxide
- Reagent B: 4% cupric sulfate solution

23.2.3   Electrophoresis

(*Note*: Running buffers (Tris/glycine/SDS), sample buffers (Laemmli sample buffer), and precast gels can be purchased commercially or made in the laboratory. β-Mercaptoethanol may need to be added to a commercial sample preparation buffer.)

- Acrylamide: bis-acrylamide solution**
    29.2 g acrylamide and 2.4 g methylene bis-acrylamide, with dd water to 100 mL
- Ammonium persulfate (APS), 7.5%, in dd water, 1 mL, prepared fresh daily**
- Bromophenol blue, 0.05%
- Coomassie Brilliant Blue stain solution**
    Purchase premixed or prepare: 454 mL dd water, 454 mL methanol, 92 mL acetic acid, and 1.5 g Coomassie Brilliant Blue R-250. Note: Bio-Rad offers Coomassie G-250 stains that do not require traditional destaining with methanol or acetic acid. A water wash stops the staining process.
- Destain solution**
    850 mL dd water, 75 mL methanol, and 75 mL acetic acid
- EDTA, disodium salt, 0.2*M*, 50 mL **
- Glycerol, 37% (use directly)
- Electrophoresis sample preparation buffer **
    Purchase premixed or prepare: 1 mL of 0.5*M* Tris (pH 6.8), 0.8 mL glycerol, 1.6 mL 10% SDS, 0.4 mL β-mercaptoethanol, and 0.5 mL 0.05% (wt/vol) bromophenol blue, diluted to 8 mL with dd water.
- Sodium dodecyl sulfate, 10% solution in dd water, 10 mL**
- TEMED (use directly)
- Tray (running) buffer, 25 mM Tris, 192 mM glycine, 0.1% SDS, pH 8.3**
    Purchase premixed or prepare
- Tris buffer, 1.5*M*, pH 8.8, 50 mL (separating (resolving) gel buffer)**
    Purchase premixed or prepare
- Tris buffer, 0.5*M*, pH 6.8, 50 mL (stacking gel buffer)**
    Purchase premixed or prepare

    Gel preparation: Use the formula that follows and the instructions in the procedure to prepare two 8.4×5.0 cm SDS-PAGE slab gels, 15% acrylamide, 0.75 mm thick. Actual formulation will depend on size of slab gels.

Reagent	Separating gel 15% gel	Stacking gel 4.5% gel
Acrylamide: bis-acrylamide	2.4 mL	0.72 mL
10% SDS	80 µL	80 µL

Reagent	Separating gel 15% gel	Stacking gel 4.5% gel
1.5 M Tris, pH 8.8	2.0 mL	–
0.5 M Tris, pH 6.8	–	2.0 mL
dd water	3.6 mL	5.3 mL
37% glycerol	0.15 mL	–
10% APS[a]	40 µL	40 µL
TEMED	10 µL	10 µL

[a]APS added to separating and stacking gels after all other reagents are combined, solution is degassed, and each gel is ready to be poured

23.2.4 Hazards, Precautions, and Waste Disposal

Acrylamide monomers may cause cancer and are very toxic in contact with skin and if swallowed. β-Mercaptoethanol is harmful if swallowed, toxic in contact with the skin, and irritating to eyes. Adhere to normal laboratory safety procedures. Wear gloves and safety glasses at all times. Acrylamide and β-mercaptoethanol wastes must be disposed of as hazardous wastes. Gloves and pipette tips in contact with acrylamide and β-mercaptoethanol also should be handled as hazardous wastes. Other waste likely can be washed down the drain with a water rinse, but follow good laboratory practices outlined by environmental health and safety protocols at your institution.

23.3 SUPPLIES

(Used by students)

23.3.1 Sample Extraction

- Beaker, 250 mL
- Centrifuge tubes, 50 mL
- Cutting board
- Erlenmeyer flask, 125 mL
- Graduated cylinder, 50 mL
- Filter paper, Whatman No. 1
- Fish, freshwater (e.g., catfish) and saltwater species (e.g., tilapia)
- Funnel
- Knife
- Pasteur pipettes and bulbs
- Test tube with cap
- Weighing boat

23.3.2 Protein Determination (BCA Method)

- Beaker, 50 mL
- Graduated cylinder, 25 mL
- Mechanical, adjustable volume pipettor, 1000 µL, with plastic tips
- Test tubes

23.3.3 Electrophoresis

- Beaker, 250 mL (for boiling samples)
- Two Erlenmeyer flasks, 2 L (for stain and destain solutions)
- Glass boiling beads (for boiling samples)
- Graduated cylinder, 100 mL
- Graduated cylinder, 500 mL
- Hamilton syringe (to load samples on gels)
- Mechanical, adjustable volume pipettors, 1000 µL, 100 µL, and 20 µL, with plastic tips
- Pasteur pipettes, with bulbs
- Rubber stopper (to fit 25-mL sidearm flasks)
- Two sidearm flasks, 25 mL
- Test tubes or culture tubes, small size, with caps
- Tubing (to attach to vacuum system to degas gel solution)
- Weighing paper/boats

23.3.4 Equipment

- Analytical balance
- Aspirator system (for degassing solutions)
- Blender
- Centrifuge
- Electrophoresis unit
- pH meter
- Power supply
- Spectrophotometer
- Top loading balance
- Vortex mixer
- Water bath

23.4 PROCEDURE

(Single sample extracted)

23.4.1 Sample Preparation

1. Coarsely cut up about 100-g fish muscle (representative sample) with a knife. Accurately weigh out 90 g on a top loading balance.
2. Blend one part fish with three parts extraction buffer (90-g fish and 270 mL extraction buffer) for 1.0 min in a blender. (Note: Smaller amounts of fish and buffer, but in the same 1:3 ratio, can be used for a small blender.)
3. Pour 30 mL of the muscle homogenate into a 50-mL centrifuge tube. Label tube with tape. Balance your tube against a classmate's sample. Use a spatula or Pasteur pipette to adjust tubes to an equal weight.
4. Centrifuge the samples at 2000× g for 15 min at room temperature. Collect the supernatant.
5. Filter a portion of the supernatant by setting a small funnel in a test tube. Place a piece of Whatman No. 1 filter paper in the funnel and moisten it with the extraction buffer. Filter the

supernatant from the centrifuged sample. Collect about 10 mL of filtrate in a test tube. Cap the tube.

6. Determine protein content of filtrate using the BCA method and prepare sample for electrophoresis (see below).

23.4.2 BCA Protein Assay

(Instructions are given for duplicate analysis of each concentration of standard and sample.)

1. Prepare the Working Reagent for the BCA assay by combining Pierce Reagent A with Pierce Reagent B, 50:1 (v/v), A:B. For example, use 50-mL Reagent A and 1.0-mL Reagent B to prepare 51.0-mL Working Reagent, which is enough for the BSA standard curve and testing the extract from one type of fish. (Note: This volume is adequate for assaying duplicates of five standard samples and two dilutions of each of two types of fish.)
2. Prepare the following dilutions of the supernatant (filtrate from Procedure, Sample Preparation, Step 5): dilutions of 1:5, 1:10, and 1:20 in *extraction buffer*. Mix well.
3. In test tubes, prepare duplicates of each reaction mixture of diluted extracts and BSA standards (using 2 mg BSA/mL solution) as indicated in the table that follows:

Tube identity	dd water (μL)	BSA Std. (μL)	Fish extract (μL)	Working reagent (mL)
Blank	100	0	–	2.0
Std. 1	80	20	–	2.0
Std. 2	60	40	–	2.0
Std. 3	40	60	–	2.0
Std. 4	20	80	–	2.0
Std. 5	0	100	–	2.0
Sample 1:5	50	–	50	2.0
Sample 1:10	50	–	50	2.0
Sample 1:20	50	–	50	2.0

4. Mix each reaction mixture with a vortex mixer and then incubate in a water bath at 37°C for 30 min.
5. Read the absorbance of each tube at 562 nm using a spectrophotometer.
6. Use the data from the BSA samples to create a standard curve of absorbance at 562 nm versus μg protein/tube. Determine the equation of the line for the standard curve. Calculate the protein concentration (μg/mL) of the extract from each fish species using the equation of the line from the BSA standard curve and the absorbance value for a dilution of the fish extract that had an absorbance near the middle point on the standard curve. Remember to correct for dilu-

tion used. (Note: Do not use the diluted samples for electrophoresis. Use the original extract prepared as described below.)

23.4.3 Electrophoresis

1. Assemble the electrophoresis unit according to the manufacturer's instructions.
2. Use the table that follows the list of electrophoresis reagents to combine appropriate amounts of all reagents for the *separating* gel, except APS, in a sidearm flask. Degas the solution and then add APS. Proceed immediately to pour the solution between the plates to create the separating (resolving) gel. Pour the gel to a height approximately 1 cm below the bottom of the sample well comb. Immediately, add a layer of butanol across the top of the separating gel, adding it carefully so as not to disturb the upper surface of the separating gel. This butanol layer will prevent a film from forming and help obtain an even surface. Allow the separating gel to polymerize for 30 min and then remove the butanol layer just before the stacking gel is ready to be poured.
3. Use the table that follows the list of electrophoresis reagents to combine appropriate amounts of all reagents for the *stacking* gel, except APS, in a sidearm flask. Degas the solution for 15 min (as per the manufacturer's instructions) and then add APS. Proceed immediately to pour the solution between the plates to create the stacking gel. Immediately, place the well comb between the plates and into the stacking gel. Allow the stacking gel to polymerize for 30 min before removing the well comb. Before loading the samples into the wells, wash the wells twice with dd water.
4. Mix the fish extract samples well (filtrate from Sect. 23.4.1, Step 5). For each sample, combine 0.1-mL sample with 0.9-mL electrophoresis sample buffer in screw cap culture tube. Apply cap.
5. Heat capped tubes for 3 min in boiling water.
6. Apply 10 and 20 μg protein of each fish extract to wells of the stacking gel using a syringe. Calculate the volume to apply based on the protein content of the extract and the dilution used when preparing the extract in electrophoresis sample buffer.
7. Apply 10 μl of molecular weight standards to one sample well.
8. Follow the manufacturer's instructions to assemble and run the electrophoresis unit. When the line of Bromophenol Blue tracking dye has reached the bottom of the separating gel, shut off the power supply. Disassemble the electrophoresis unit, and carefully remove the separating gel from between the plates. Place the gel in a flat dish with the Coomassie Brilliant Blue stain solution. Allow the gel to stain for at least 30 min. (If possible, place the dish with the gel

on a gentle shaker during staining and destaining.) Pour off the stain solution and then destain the gel for at least 2 h using the destain solution with at least two changes of the solution.

9. Measure the migration distance (cm) from the top of the gel to the center of the protein band for the molecular weight standards and for each of the major protein bands in the fish extract samples. Also measure the migration distance of the bromophenol blue tracking dye from the top of the gel.

10. Observe and record the relative intensity of the major protein bands for each fish extract.

23.5   DATA AND CALCULATIONS

23.5.1   Protein Determination

Tube identity	Absorbance	µg protein/ tube	µg/mL sample
Std. 1, 20 µL BSA			
Std. 1, 20 µL BSA			
Std. 2, 40 µL BSA			
Std. 2, 40 µL BSA			
Std. 3, 60 µL BSA			
Std. 3, 60 µL BSA			
Std. 4, 80 µL BSA			
Std. 4, 80 µL BSA			
Std. 5, 100 µL BSA			
Std. 5, 100 µL BSA			
Sample 1:5			
Sample 1:5			
			$\bar{X} =$
Sample 1:10			
Sample 1:10			
			$\bar{X} =$
Sample 1:20			
Sample 1:20			
			$\bar{X} =$

Sample calculation for fish extract protein concentration:

For fish extract diluted 1:20 and 50 µL analyzed with absorbance of 0.677:

Equation of the line: $y = 0.0108x + 0.0022$
                                      If $y = 0.677$, $x = 62.48$

$$C_i = C_f (V_2 / V_1)(V_4 / V_3)$$

(See Chap. 3 in this laboratory manual; $C_i =$ initial concentration; $C_f =$ final concentration)

$C_i = (62.48\text{-ug protein/tube}) \times (20\ \text{ml}/1\ \text{mL}) \times$
     $(\text{tube}/50\ \text{uL})$
   $= 24.99 \mu g$ protein/uL fish extract

Sample calculation to determine the volume of prepared extract to apply 20 ug protein to each sample well:

How many µL are needed to get 20 µg protein? Remember the electrophoresis sample buffer dilution is 1:10:

$$24.99\ \mu g\ \text{protein} / \mu L \times Z\ \mu L \times (1\ mL / 10\ mL) = 20\ \mu g$$

$$Z = 8.00\ \text{uL}$$

23.5.2   Electrophoresis

1. Calculate the relative mobility of three major protein bands and all the molecular weight standards. To determine the relative mobility (R_f) of a protein band, divide its migration distance from the top of the gel to the center of the protein band by the migration distance of the bromophenol blue tracking dye from the top of the gel:

$$R_f = \frac{\text{distance of protein migration}}{\text{distance of tracking dye migration}}$$

Sample identity	Distance of protein migration	Distance of tracking dye migration	Relative mobility	Molecular weight
Molecular weight standards				
1				
2				
3				
4				
5				
Fish species				
Freshwater				
Saltwater				

2. Prepare a standard curve by plotting relative mobility (x-axis) versus log molecular weight of standards (y-axis).

3. Using the standard curve, estimate the molecular weight of the major protein subunits in the freshwater and saltwater fish extracts.

23.6   QUESTIONS

1. Describe how you would prepare 1 L of the buffer used to extract the fish muscle proteins (0.15 M sodium chloride, 0.05 M sodium phosphate, pH 7.0). Show all calculations.

2. Discuss the differences between the fish species, regarding the presence or absence of major protein bands identified by the molecular mass and the relative amounts of these proteins.

RESOURCE MATERIALS

Bio-Rad (2013) A Guide to Polyacrylamide Gel Electrophoresis and Detection. Bulletin 6040 Rev B. Bio-Rad Laboratories, http://www.bio-rad.com/webroot/web/pdf/lsr/literature/Bulletin_6040.pdf Accessed June 24, 2015

Chang SKC, Zhang Y (2017) Protein analysis. Ch. 18. In: Nielsen SS (ed) Food analysis, 5th edn. Springer, New York

Etienne M et al. (2000) Identification of fish species after cooking by SDS-PAGE and urea IEF: a collaborative study. J Agr Food Chem 48:2653–2658

Etienne M et al. (2001) Species identification of formed fishery products and high pressure-treated fish by electrophoresis: a collaborative study. Food Chem 72:105–112

Laemmli UK (1970) Cleavage of structural proteins during the assembly of the head of bacteriophage T4. Nature 227:680–685

Olsen BJ and Markwell J (2007) Assays for the determination of protein concentration. Unit 3.4 Basic Protocol 3. BCA Assay. Current Protocols in Protein Science John Wiley and Sons, New York

Pierce (2013) Instructions: Pierce BCA protein assay kit. Pierce Biotechnology, Rockford, IL https://tools.lifetechnologies.com/content/sfs/manuals/MAN0011430_Pierce_BCA_Protein_Asy_UG.pdf Accessed June 24, 2015

Piñeiro C et al (1999) Development of a sodium dodecyl sulfate-polyacrylamide gel electrophoresis reference method for the analysis and identification of fish species in raw and heat-processed samples: a collaborative study. Electrophoresis 20:1425–1432

Smith DM (2017) Protein separation and characterization. Ch. 24. In: Nielsen SS (ed) Food analysis, 5th edn. Springer, New York

Chapter

Glucose Determination by Enzyme Analysis

Charles E. Carpenter (✉) • Robert E. Ward

Department of Nutrition, Dietetics and Food Sciences, Utah State University,
Logan, UT, USA
e-mail: chuck.carpenter@usu.edu; robert.ward@usu.edu

S.S. Nielsen, *Food Analysis Laboratory Manual,* Food Science Text Series,
DOI 10.1007/978-3-319-44127-6_24, © Springer International Publishing 2017

24.1 INTRODUCTION

24.1.1 Background

Enzyme analysis is used for many purposes in food science and technology. Enzyme activity is used to indicate adequate processing, to assess enzyme preparations, and to measure constituents of foods that are enzyme substrates. In this experiment, the glucose content of corn syrup solids is determined using the enzymes, glucose oxidase and peroxidase. Glucose oxidase catalyzes the oxidation of glucose to form hydrogen peroxide (H_2O_2), which then reacts with a dye in the presence of peroxidase to give a stable-colored product.

As described, this experiment uses individual, commercially available reagents, but enzyme test kits that include all the reagents to quantitate glucose also are available as a package. Enzyme test kits also are available to quantitate various other components of foods. Companies that sell enzyme test kits usually provide detailed instructions for the use of these kits, including information about the following: (1) principle of the assay, (2) contents of the test kit, (3) preparation of solutions, (4) stability of solutions, (5) procedure to follow, (6) calculations, and (7) further instructions regarding dilutions and recommendations for specific food samples.

24.1.2 Reading Assignment

BeMiller, J.N. 2017. Carbohydrate analysis. Ch. 19, in *Food Analysis*, 5th ed. S.S. Nielsen (Ed.), Springer, New York.

Reyes-De-Coreuera, J.I., and Powers, J.R. 2017. Application of enzymes in food analysis. Ch. 25, in *Food Analysis*, 5th ed. S.S. Nielsen (Ed.), Springer, New York.

24.1.3 Objective

Determine the glucose content of food products using the enzymes, glucose oxidase and peroxidase.

24.1.4 Principle of Method

Glucose is oxidized by glucose oxidase to form hydrogen peroxide, which then reacts with a dye in the presence of peroxidase to give a stable-colored product that can be quantitated spectrophotometrically (coupled reaction).

24.1.5 Chemicals

	CAS No.	Hazards
Acetic acid (Sigma A6283)	64-19-7	Corrosive
o-Dianisidine·2HCl (Sigma D3252)	20325-40-0	Tumor causing, carcinogenic
D-Glucose (Sigma G3285)	50-99-7	
Glucose oxidase (Sigma G6641)	9001-37-0	
Horseradish peroxidase (Sigma P6782)	9003-99-0	
Sodium acetate (Sigma S2889)	127-09-3	
Sulfuric acid (Aldrich 320501)	7664-93-9	Corrosive

24.1.6 Reagents

(**It is recommended that these solutions be prepared by the laboratory assistant before class.)

- Acetate buffer**, 0.1 M, pH 5.5
    Dissolve 8 g sodium acetate in ca. 800 mL water in a 1-L beaker. Adjust pH to 5.5 using 1 M HCl. Dilute to 1 L in a volumetric flask.
- Glucose test solution**
    In a 100-mL volumetric flask, dissolve 20 mg glucose oxidase (~300–1000 units), 40 mg horseradish peroxidase, and 40 mg o-dianisidine·2HCl in the 0.1 M acetate buffer. Dilute to volume with the acetate buffer and filter as necessary.
- Glucose standard solution, 1 mg/mL
    Use commercial D-glucose solution (e.g., Sigma).
- Sulfuric acid, diluted** (1 part H_2SO_4 + 3 parts water)
    In a 500-mL beaker in the hood, add 150 mL water and then add 50 mL H_2SO_4. This will generate a lot of heat.

24.1.7 Hazards, Precautions, and Waste Disposal

Concentrated sulfuric acid is extremely corrosive; avoid contact with skin and clothes and breathing vapors. Acetic acid is corrosive and flammable. Wear safety glasses at all times and corrosive-resistant gloves. Otherwise, adhere to normal laboratory safety procedures. The o-dianisidine 2HCl must be disposed of as hazardous waste. Other waste likely may be put down the drain using a water rinse, but follow good laboratory practices outlined by environmental health and safety protocols at your institution.

24.1.8 Supplies

- Beaker, 1 L
- Corn syrup solids (or high fructose corn syrup), 0.5 g
- 5 Spatulas
- 14 Test tubes, 18 × 150 mm, heavy walled to keep from floating in water bath
- Test-tube rack

- 2 Volumetric flasks, 100 mL
- Volumetric flask, 250 mL
- Volumetric pipette, 10 mL
- 2 volumetric flasks, 1 L
- Weighing paper

24.1.9 Equipment

- Analytical balance
- Mechanical, adjustable volume pipettors, 200, 1000, and 5000 μL, with tips
- pH meter
- Spectrophotometer
- Water bath, 30 °C

24.2 PROCEDURE

(Instructions are given for analysis in duplicate.)

1. Prepare dilutions for standard curve. Use the adjustable pipettors to deliver aliquots of glucose standard solution (1 mg/mL) and deionized distilled (dd) water as indicated in the table below into clean test tubes. These dilutions will be used to create a standard curve of 0–0.2 mg glucose/mL.

	mg glucose/mL				
	0	0.05	0.10	0.15	0.20
mL glucose std. solution	0	0.150	0.300	0.450	0.600
mL dd water	3.000	2.850	2.700	2.550	2.400

2. Prepare sample solution and dilutions. Accurately weigh ca. 0.50 g corn syrup solids and dilute with water to volume in a 250-mL volumetric flask (Sample A). Using volumetric pipettes and flasks, dilute 10.00 mL of Sample A to 100 mL with water (Sample B). These sample dilutions will let you determine glucose concentrations in samples containing 1–100 % glucose.
3. Add 1.000 mL of water to each of 14 test tubes. In duplicate, add 1.000 mL of the individual standard and sample dilutions to the test tubes.
4. Put all tubes in the water bath at 30 °C for 5 min. Add 1.000 mL glucose test solution to each tube at 30 s intervals.
5. After exactly 30 min, stop the reactions by adding 10 mL of the diluted H_2SO_4. Cool to room temp.
6. Zero spectrophotometer with water in the reference position using a double beam spectrophotometer. Take two readings (repeated measures, msmt) using separate aliquots from each tube.

24.3 DATA AND CALCULATIONS

Weight of original sample: _____ g

Absorbance of standard solutions:

		(mg glucose/mL)				
Tube	Msmt	0	0.05	0.10	0.15	0.20
1	1					
	2					
2	1					
	2					
Average absorbance:						

Absorbance of samples:

Tube	Msmt	Sample A	Sample B
1	1		
	2		
2	1		
	2		
Average absorbance:			

Calculation of glucose concentration in sample:

1. Plot absorbance of standards on the y-axis versus mg glucose/mL on the x-axis.
2. Calculate the concentration of glucose for the sample dilution A or B that had an absorbance within the working range of the standard curve: $(Abs - y\text{-intercept})/\text{slope} = $ mg glucose/mL.
3. Calculate the glucose concentration in the original sample, as a percentage.

Example calculations:

Original sample of 0.512 g

Average measured absorbance sample dilution B: 0.200

Calculation from standard curve: 0.200–0.003/(2.98 mL/mg glucose) = 0.066 mg glucose/mL B)

$$C_{sample} = (0.066 \text{ mg glucose/mL B})$$
$$\times (100 \text{ mL B}/10 \text{ mL A})$$
$$\times (250 \text{ mL A}/512 \text{ mg sample})$$
$$= 0.323 \text{ mg glucose/mg sample}$$
$$= 32.3\% \text{ glucose}$$

24.4 QUESTIONS

1. Explain why this experiment is said to involve a coupled reaction. Write in words the equations for the reactions. What conditions must be in place to ensure accurate results for such a coupled reaction?
2. How do the results obtained compare to specifications for the commercial product analyzed?

RESOURCE MATERIALS

BeMiller JN (2017) Carbohydrate analysis. Ch. 19. In: Nielsen SS (ed) Food analysis, 5th edn. Springer, New York

Reyes-De-Coreuera JI, and Powers JR (2017) Application of enzymes in food analysis. Ch. 26. In: Nielsen SS (ed) Food analysis, 5rd edn. Springer, New York

Gliadin Detection by Immunoassay

Y.H. Peggy Hsieh (✉) • Qinchun Rao

Department of Nutrition, Food and Exercise Sciences, Florida State University,
Tallahassee, FL, USA
e-mail: yhseih@fsu.edu; qrao@fsu.edu

S.S. Nielsen, *Food Analysis Laboratory Manual*, Food Science Text Series,
DOI 10.1007/978-3-319-44127-6_25, © Springer International Publishing 2017

25.1   INTRODUCTION

25.1.1   Background

Immunoassays are very sensitive and efficient tests that are commonly used to identify a specific protein. Examples of applications in the food industry include identification of proteins expressed in genetically modified foods, allergens, or proteins associated with a disease, including celiac disease. This genetic disease is associated with up to 1% of the world's population and more than two million Americans. These individuals react immunologically to wheat proteins, and consequently their own immune systems attack and damage their intestines. This disease can be managed if gluten is avoided in foods. Gluten consists of alcohol-insoluble glutenins and alcohol-soluble prolamins, which are mainly found in wheat and also in oat, barley, rye, and other grain flours and related starch derivatives. Rice and corn are two common grains that do not contain significant gluten and are well tolerated by those with celiac disease. Wheat protein makes up 7–15% of a wheat grain. Prolamins in wheat are called gliadin. About 40% of the wheat proteins are various forms of gliadin protein.

The immune system of animals can respond to many foreign substances by the development of specific antibodies. Antibodies bind strongly to and assist in the removal of a foreign substance in the body. Animals make antibodies against many different "antigens," defined as foreign substance that will elicit a specific immune response in the host. These include foreign proteins, peptides, carbohydrates, nucleic acids, lipids, and many other naturally occurring or synthetic compounds.

Immunoassays are tests that take advantage of the remarkably specific and strong binding of antibodies to antigens. Immunoassays can be used to determine the presence and quantity of either antibody or antigen. Antibodies that identify a specific protein (antigen) can be developed by injection of a laboratory animal with this protein, much as humans are vaccinated against a disease. These antigen-specific antibodies can be used to identify the antigen in a food (e.g., detection of gliadin in food products) through the appropriate use of a label, such as an enzyme or fluorescent molecules linked covalently to either the antibody or a reference antigen. This type of immunoassay concept also can be used to determine the presence of specific antibodies in blood. For example, by analyzing for the presence of gliadin-specific antibodies in an individual's blood, one can determine if the individual has celiac disease.

25.1.2   Reading Assignment

Hsieh, Y-H.P. and Rao, Q. 2017. Immunoassays. Ch.27, in *Food Analysis*, 5th ed. S.S. Nielsen (Ed.), Springer, New York.

25.1.3   Objective

Determine the presence of gliadin in various food products using a rabbit anti-gliadin antibody horseradish peroxidase conjugate in a dot blot immunoassay.

25.1.4   Principle of Method

A simple dot blot immunoassay will be used in this lab to detect gliadin in food samples. Dot blot assays use nitrocellulose (NC) paper for a solid phase. Initially, gliadin proteins are isolated by differential centrifugation, in which most of the non-gliadin proteins are washed away with water and sodium chloride (NaCl) solutions, and then the gliadin is extracted with a detergent solution. A drop of the food sample extract or standard antigen (gliadin) is applied to the NC paper, where it adheres nonspecifically. The remaining binding sites on the NC paper are then "blocked" using a protein unrelated to gliadin, such as bovine serum albumin (BSA) to minimize the nonspecific binding. The bound gliadin antigen in the food spot then can be reacted with an antigen-specific antibody-enzyme conjugate. Theoretically, this antibody probe then will bind only to gliadin antigen bound already to the NC paper. Next, the strip is washed free of unbound antibody-enzyme conjugate and then placed in a substrate solution in which an enzymatically catalyzed precipitation reaction can occur. Brown-colored "dots" indicate the presence of gliadin-specific antibody and hence gliadin antigen. The color intensity indicates the amount of antigen present in the food sample. The stronger the color, the higher amount of the antigen (gliadin) in the food sample extract.

25.1.5   Chemicals

	CAS no.	Hazards
Bovine serum albumin (BSA)	9048-46-8	
Chicken egg albumin (CEA)	9006-59-1	
3,3'-Diaminobenzidine tetrahydrochloride (DAB)	7411-49-6	
Gliadin standard protein	9007-90-3	
Hydrogen peroxide, 30% (H_2O_2)	7722-84-1	Oxidizing, corrosive
Rabbit anti-gliadin immunoglobulin conjugated to horseradish peroxidase (RAGlg-HRP, Sigma, A1052)		
Sodium chloride (NaCl)	7647-14-5	Irritant
Sodium dodecyl sulfate (SDS)	151-21-3	Harmful
Sodium phosphate, monobasic ($NaH_2PO_4 \cdot H_2O$)	7558-80-7	Irritant

	CAS no.	Hazards
Tris(hydroxymethyl) aminomethane (TRIS)	77-86-1	Irritant
Tween-20 detergent	9005-64-5	

25.1.6 Reagents

(**It is recommended that these solutions be prepared by the laboratory assistant before class.)

* Blocking solution **
    3% (g/mL) BSA in PBST; 5–10 mL per student
* DAB substrate**
    60 mg DAB dissolved in 100 mL 50 mM TRIS (pH 7.6) and then filtered through filter paper Whatman #1. (*Note*: DAB may not completely dissolve in this buffer if it is the free base form instead of the acid form. Just filter out the undissolved DAB and it will still work well.) Just 5 min prior to use, add 100 µL 30% H_2O_2, 5–10 mL per student.
* Gliadin antibody probe **
    1:500 (mL/mL) diluted RAGIg-HRP using 0.5% (g/mL) BSA in PBST; 5–10 mL per student
* Gliadin extraction detergent**
    1% (g/mL) SDS in water; 10 mL per student
* Gliadin standard protein, 4000 µg/mL, in 1% SDS**
    One vial of 150 µL per student
* Negative control sample**
    3% CEA (or other non-gliadin protein) in PBST; 100 µL per student
* Phosphate-buffered saline (PBS)**
    0.05M sodium phosphate, 0.9% (g/mL) NaCl, pH 7.2
* Phosphate-buffered saline + Tween 20 (PBST) **
    0.05M sodium phosphate, 0.9% (g/mL) NaCl, 0.05% (mL/mL) Tween 20, pH 7.2, 250 mL per student

25.1.7 Hazards, Precautions, and Waste Disposal

Adhere to normal laboratory safety procedures. Wear gloves and safety glasses at all times. Handle the DAB substrate with care. Wipe up spills and wash hands thoroughly. The DAB, SDS, and hydrogen peroxide wastes should be disposed of as hazardous wastes. Other wastes likely may be put down the drain using a water rinse, but follow good laboratory practices outlined by environmental health and safety protocols at your institution.

25.1.8 Supplies

(Used by students)

* 1–10 µL, 10–100 µL, and 200–1000 µL positive displacement pipettors

* Disposable tips, for pipettors
* Filter paper, Whatman #1
* Food samples (e.g., flour, crackers, cookies, starch, pharmaceuticals, etc.)
* Funnels, tapered glass
* Mechanical, adjustable volume pipettors, for 2 µL, 100 µL, and 1000 µL ranges, with plastic tips (glass capillary pipettes can be substituted for 2 µL pipettors)
* Microcentrifuge tubes, 1.5 mL, two per sample processed
* Nitrocellulose paper (Bio-Rad 162–0145) cut into 1.7×2.3 cm rectangular strips (NC strips)
* Petri dishes, 3.5 cm
* Test tubes, 13×100 mm, six per student
* Test tube rack, one per student
* Tissue paper
* Tweezers, one set per student
* Wash bottles, one per two students for PBST
* Wash bottles, one per two students for distilled water

25.1.9 Equipment

* Mechanical platform shaker
* Microcentrifuge
* pH meter
* Vortex mixer

25.2 PROCEDURE

25.2.1 Sample Preparation

(*Note*: Sample preparation by the students may take place on a separate day prior to the immunoassay. In this initial sample preparation lab, the principles of differential centrifugation, with respect to the Osborne protein classification system, may be studied. Sample preparation and immunoassay may not be reasonable to achieve in one day. If only one day can be allocated for this lab, the samples can be prepared ahead of time for the students by the technical assistant, and this lab will demonstrate concept and techniques of a simple immunoassay.)

1. Weigh accurately (record the mass) about 0.1 g of flour, starch, or a ground processed food and add to a 1.5 mL microcentrifuge tube. Add 1.0 mL distilled water and vortex for 2 min. Place in the microcentrifuge with other samples and centrifuge at 800×g for 5 min. Discard the supernatant (albumins). Repeat.
2. Add 1.0 mL of 1.5M NaCl to the pellet from Step 1 and resuspend it by vortexing for 2 min. If the pellet is not resuspending, dislodge it with a spatula. Centrifuge at 800×g for 5 min. Discard the supernatant (globulins). Repeat.
3. Add 1.0 mL of 1% SDS detergent to the pellet from Step 2 and resuspend it to extract the gliadins. Vortex for 2 min. Centrifuge at 800×g for

5 min. Carefully pipette off most of the supernatant and transfer to a clean microcentrifuge tube. Discard the pellet.

25.2.2 Standard Gliadin

The standard pure gliadin is dissolved at a concentration of 4000 µg/mL in 1% SDS. To provide for a series of standards to compare unknown samples, dilute the standard serially by a factor of 10 in 13×100 mm test tubes to make 400 µg/mL, 40 µg/mL, and 4 µg/mL standards in 1% SDS. Use 100 µL of the highest standard transferred to 900 µL of 1% SDS detergent for the first tenfold dilution. Repeat this procedure serially to produce the last two standards. As each standard is made, mix it well on a vortex mixer.

25.2.3 Nitrocellulose Dot Blot Immunoassay

[*Note*: The nitrocellulose (NC) strips should only be handled with tweezers to prevent binding of proteins and other compounds from your fingers. Hold the strips with the tips of the tweezers on the corners of the nitrocellulose to avoid damaging or interfering with the spotted surface.]

1. Mark two NC strips with a pencil into six equal boxes each (see drawing below).
2. Pipette 2 µL each of sample, standard or negative controls onto the NC paper. Lay the NC strips flat onto some tissue.

   (a) On NC strip A, pipette 2 µL of four different SDS food sample extracts.
   (b) On NC strip B, pipette 2 µL of four different gliadin standards.
   (c) On the remaining two squares (5 and 6) on strips A and B, add 2 µL of 1% SDS and 2 µL of the 3% CEA protein negative control, respectively.
   (d) Let the spots air-dry on the tissue.

   Below are diagrams of the NC strips marked off in boxes and numbered by pencil. The circles are *not* penciled in, but rather they represent where the 2 µL sample or standard spots will be applied.

3. Place both NC strips into a petri dish containing 5 mL of blocking solution (3% BSA in PBST) and let incubate for 20 min on the mechanical shaker so that the NC strips are moving around slightly in the solution.
4. Rinse the strips well with PBST using a wash bottle over a sink, holding the strips by the tips of their corners with tweezers.
5. Place the NC strips into a petri dish containing about 5 mL of 1:500 diluted RAGIg-HRP conjugate, and incubate for 60 min on the mechanical shaker.

A series (food sample):	B series (gliadin standards):
1 = food sample 1 @ 1× dilution	1 = gliadin standard @ 4000 µg/mL
2 = food sample 2 @ 1× dilution	2 = gliadin standard @ 400 µg/mL
3 = food sample 3 @ 1× dilution	3 = gliadin standard @ 40 µg/mL
4 = food sample 4 @ 1× dilution	4 = gliadin standard @ 4 µg/mL
5 = 1% SDS control	5 = 1% SDS control
6 = 3% protein negative control (CEA)	6 = 3% protein negative control (CEA)

6. Wash the NC strips with PBST using a wash bottle and then incubate them for 5 min in a clean petri dish half full with PBST. Rinse again well with PBST, and rinse one last time with distilled water.
7. Add NC strips to a petri dish containing the 5 mL of DAB/H_2O_2 substrate, and watch for the development of a brown stain. (*Note*: Handle the substrate with care. Wipe up spills, wash hands thoroughly, and wear gloves. Although there is no specific evidence that DAB is a carcinogenic compound, it should be treated as if it were.) Stop the reaction in 10–15 min, or when the background nitrocellulose color is becoming noticeably brown, by rinsing each strip in distilled water.
8. Let the NC strips air-dry on tissue paper.

25.3 DATA AND CALCULATIONS

Make any observations you feel are pertinent to this laboratory. Attach the developed NC strips to your lab report with a transparent tape.

Describe the results based on observations of the degree of brown-colored stain in standards and samples relative to negative controls. You can use a crude quantitative rating system like +++, ++, +, ±, and − to describe and report the relative intensities of the dot reactions (*note*: the brown dot images will fade in several days).

Make very crude approximations of the quantity of gliadin in each substance relative to (more or less than) the standard gliadin dots. Comment on this crude estimate relative to the food product's gluten status [i.e., gluten free or not; *Codex Alimentarius* (http://www.codexalimentarius.net/) defines less than 20 mg of gluten/kg in total, based on the food as sold or distributed to the consumer to be classified as "gluten-free"].

Tabulate your results in a manner that is easy to interpret.

To make the gluten status estimation, you must know the values of both the concentration of the food sample (g food/mL extraction solution) extracted and the concentration of the gliadin standards (mg gliadin/mL extraction solution) to which you are making a comparison.

Example calculation:

If the food sample has a concentration of 100 mg/mL and reacts equivalently to a 4 µg/mL gliadin standard, it can be estimated that 4 µg gliadin is in 100 mg food sample, because both are applied at equal volumes so the two concentrations can be related fractionally (i.e., 4 µg gliadin/100 mg or 40 mg gliadin/kg food sample). The gliadin content of gluten is generally taken as 50%. Since this sample has a gliadin concentration higher than the limit set by *Codex Alimentarius* (10 mg gliadin/kg or 20 mg gluten/kg food), the food cannot be considered as "gluten-free."

25.4   QUESTIONS

1. Draw a set of symbolic pictures representing the stages of the dot blot assay used in this laboratory, including the major active molecular substances being employed (i.e., nitrocellulose solid matrix, antigen, BSA blocking reagent, antibody-enzyme conjugate, substrate, product).

2. Why should you block unbound sites on nitrocellulose with 3% BSA in a special blocking step after applying samples to the membrane?
3. Why is a spot of 1% SDS (gliadin extraction detergent) used in the dot blot?
4. Why is a protein negative control spot (3% CEA) used in the dot blot?
5. Describe the basic role of horseradish peroxidase enzyme (i.e., why is it attached to the rabbit antibody), and what roles do DAB and H_2O_2 play in the development of the colored dot reaction in this immunoassay? Do not describe the actual chemical reaction mechanisms, but rather explain why a color reaction can ultimately infer a gliadin antigen that is present on the nitrocellulose paper.

Acknowledgment This laboratory exercise was initially developed for the first edition of the laboratory manual by Mr Gordon Grant and Dr Peter Sporns, Department of Agricultural, Food, and Nutritional Science, University of Alberta, Edmonton, Alberta, Canada. Dr. Hsieh, who has now updated the laboratory for two editions, gratefully acknowledges the original contribution.

RESOURCE MATERIALS

Hsieh Y-HP (2017) Immunoassays. Ch.27. In: Nielsen SS (ed) Food analysis, 5th edn. Springer, New York

Miletic ID, Miletic VD, Sattely-Miller EA, Schiffman SS (1994) Identification of gliadin presence in pharmaceutical products. J Pediatr Gastroenterol Nutr 19:27–33

Rubio-Tapia, A.; Ludvigsson, J. F.; Brantner, T. L.; Murray, J. A.; Everhart, J. (2012) The prevalence of celiac disease in the United States. Gastroenterology 142: S181-S182

Sdepanian VL, Scaletsky ICA, Fagundes-Neto U, deMorais MB (2001) Assessment of gliadin in supposedly gluten-free foods prepared and purchased by celiac patients. J Pediatr Gastroenterol Nutr 32:65–70

Skerritt JH, Hill AS (1991) Enzyme immunoassay for determination of gluten in foods: collaborative study. J Assoc Off Anal Chem 74:257–264

WHO and FAO (2008) CODEX STAN 118-1981-Standard for Foods for Special Dietary Use for Persons Intolerant to Gluten

Viscosity Measurements of Fluid Food Products

Helen S. Joyner

School of Food Science, University of Idaho,
Moscow, ID, USA
e-mail: hjoyner@uidaho.edu

S.S. Nielsen, *Food Analysis Laboratory Manual*, Food Science Text Series,
DOI 10.1007/978-3-319-44127-6_26, © Springer International Publishing 2017

26.1 INTRODUCTION

26.1.1 Background

Whether working in product development, quality control, or process design and scale-up, rheology plays an integral role in the manufacturing of high-quality food products. Rheology is a science based on fundamental physical relationships concerned with how all materials respond to applied forces or deformations. Flow behavior is one such response to force or deformation.

Determination and control of the flow properties of fluid foods are critical for optimizing processing conditions and obtaining the desired sensory qualities for the consumer. Transportation of fluids (pumping) from one location to another requires pumps, piping, and fittings such as valves, elbows, and tees. Proper sizing of this equipment depends on a number of elements but primarily on the flow properties of the product. For example, the equipment used to pump a dough mixture would be very different from that used for milk. Additionally, rheological properties are fundamental to many aspects of food safety. During continuous thermal processing of fluid foods, the amount of time the food is in the system (known as the residence time or RT), and therefore the amount of heating or "thermal dose" received, is directly related to its flow properties.

The rheological properties of a fluid are a function of composition, temperature, and other processing conditions, such as the speed the fluid travels through a pipe. Identifying how these parameters influence flow properties may be accomplished by measuring viscosity with a viscometer or rheometer. In this laboratory, we will measure the viscosity of two fluid foods using common rheological instruments widely used throughout the food industry. These instruments include a Brookfield rotational viscometer, a Zahn cup, and a Bostwick consistometer.

26.1.2 Reading Assignment

Joyner, H.S., and Daubert, C.R. 2017. Rheological principles for food analysis. Ch. 29, in *Food Analysis*, 5th ed. S.S. Nielsen (Ed.), Springer, New York.

Singh, R.P., and Heldman, D.R. 2001. *Introduction to Food Engineering*, 3rd ed., pp. 69–78, 144–157. Academic Press, San Diego, CA.

26.1.3 Objectives

1. Explain the basic principles of fluid rheology.
2. Gain experience measuring fluid viscosity using different equipment.
3. Describe the effects of temperature and (shear) speed on viscosity.

26.1.4 Supplies

- 6 beakers, 250 mL
- 3 beakers, 600 mL
- French salad dressing
- Honey (make sure there are no additional ingredients aside from honey)
- Thermometer or thermocouple with digital reader
- Stopwatch

26.1.5 Equipment

- Brookfield rotational viscometer model LV and spindle #3
- Brookfield Zahn cup #5 (hole diameter of at least 0.2 in)
- Bostwick consistometer
- Refrigerator

26.2 PROCEDURE

26.2.1 Brookfield Viscometer Measurements

1. Fill a 250 mL beaker with 200 mL honey and the two remaining 250 mL beakers with 200 mL salad dressing each. Label the beakers appropriately. Place one of the beakers of salad dressing in a refrigerator at least 1 h prior to analysis. The remaining beakers shall be allowed to equilibrate to room temperature.
2. Prior to evaluating the samples, make sure the viscometer is level. Use the leveling ball and circle on the viscometer to check.
3. On the data sheet provided, record the viscometer model number and spindle size, product information (type, brand, etc.), and the sample temperature. Because rheological properties are strongly dependent on temperature, sample temperatures must be measured and recorded prior to performing measurements.
4. Immerse the spindle into the test fluid (i.e., honey, salad dressing) up to the notch cut in the shaft. The viscometer motor should be off.
5. Zero the digital viscometers before turning on the motor.
6. Set the motor at the lowest speed revolutions per minute (rpm) setting. Once the digital display shows a stable value, record the percentage of full-scale torque reading. Increase the rpm setting to the next speed and again record the percentage of full-scale torque reading. Repeat this procedure until the maximum rpm setting has been reached or 100% (but not higher) of the full-scale torque reading is obtained. *Do not increase the speed once 100% of the full-scale torque has been reached.*

7. Stop the motor, then slowly raise the spindle from the sample. Remove the spindle and clean with soap and water, then dry. Be sure not to use any abrasive scrubbers or soaps on the spindle, as these can scratch the spindle and result in measurement errors.

8. Calculate the viscosity using the spindle factors. A factor exists for each spindle-speed combination (Table 26.1):
For every dial reading (percentage full-scale torque), multiply the display value by the corresponding factor to calculate the viscosity with units of mPa-s.

Example:
A Ranch salad dressing was tested with a Brookfield LV viscometer equipped with spindle #3. At a speed of 6 rpm, the display read 40.6%. For these conditions, the viscosity is:

$$\eta = (40.6)(200) = 8120 \, mPa-s = 8012 \, Pa-s$$

9. Repeat Steps 2–8 to test all samples.
10. Once all the data have been collected for the salad dressing and honey at room temperature, remove the salad dressing sample from the refrigerator, and repeat Steps 2–8. Be sure to record the sample temperature before performing any measurements.
11. You may choose to run the samples in duplicate or triplicate. Data from samples collected under identical conditions may be pooled to generate an average reading.

26.2.2 Zahn Cup Measurements

1. Fill a 600 mL beaker with 450 mL honey and the two remaining 600 mL beakers with 450 mL salad dressing each. The beakers must be tall enough so that the Zahn cup can be completely immersed in the sample. Label the beakers appropriately. Place one of the beakers of salad dressing in a refrigerator at least 2 h prior to

26.1
table   Factors for Brookfield model LV (spindle #3)

Speed (rpm)	Factor
0.3	4000
0.6	2000
1.5	800
3	400
6	200
12	100
30	40
60	20

analysis. The remaining beakers shall be allowed to equilibrate to room temperature.

2. On the data sheet provided, record the Zahn cup hole diameter and volume, product information (type, brand, etc.), and the sample temperature. Because rheological properties are strongly dependent on temperature, sample temperatures must be measured and recorded prior to performing measurements.

3. Completely immerse the Zahn cup into the test fluid (i.e., honey, salad dressing) so that the top of the cup is below the surface of the fluid. Allow the cup to fill with fluid. The cup must remain in the fluid for at least 5 min before measurements begin to equilibrate it to the sample temperature.

4. Put a finger through the ring at the top of the handle on the cup and lift it straight up out of the beaker. Start the stopwatch as soon as the top of the cup breaks the surface of the fluid.

5. Allow the cup to drain into the beaker with the bottom of the cup no more than six inches above the surface of the fluid.

6. Stop the stopwatch as soon as there is a clear break in the stream of fluid flowing out of the cup. Record the time at break, also called the Zahn time.

7. Clean the cup with soap and water, then dry. Be sure not to use any abrasive scrubbers or soaps on the cup, as these can scratch the cup and result in measurement errors.

8. Calculate the viscosity using the time conversion factors and the density of the test fluid. There are two conversion factors, K and C, to convert the Zahn time to viscosity in centistokes (cS) using the following equation:

$$\nu = K(t-C)$$

where:

ν = kinematic viscosity (cS)
t = Zahn time (s)

For a Brookfield Zahn Cup #5:

$K = 23$
$C = $ zero

The kinematic viscosity can be converted to viscosity with units on mPa-s by multiplying by the specific gravity of the fluid. The specific gravities of honey and French dressing are approximately 1.42 and 1.10, respectively.

Example:
A Ranch salad dressing was tested with a #5 Brookfield Zahn cup. The stream broke at 80 s. Given that the specific gravity of the dressing is about 1.2, the viscosity is:

$$\eta = \nu(SG) = (K(t-C))(SG)$$
$$\eta = (23(80-0))(1.2) = 2208\, mPa - s = 2.208\, Pa - s$$

9. Repeat Steps 2–8 to test all samples.
10. Once all the data have been collected for the salad dressing and honey at room temperature, remove the salad dressing sample from the refrigerator, and repeat Steps 2–8. Be sure to record the sample temperature before performing any measurements.
11. You may choose to run the samples in duplicate or triplicate. Data from samples collected under identical conditions may be pooled to generate an average reading.

26.2.3 Bostwick Consistometer Measurements

1. Fill a 250 mL beaker with 200 mL honey and the two remaining 250 mL beakers with 200 mL salad dressing each. Label the beakers appropriately. Place one of the beakers of salad dressing in a refrigerator at least 1 h prior to analysis. The remaining beakers shall be allowed to equilibrate to room temperature.
2. Prior to evaluating the samples, make sure the consistometer is at the correct angle. Level the consistometer by adjusting the leveling screws under the consistometer until the leveling bubble is in the center of the black circle.
3. On the data sheet provided, record the consistometer make and model, product information (type, brand, etc.), and the sample temperature. Because rheological properties are strongly dependent on temperature, sample temperatures must be measured and recorded prior to performing measurements.
4. Close the product gate and hold it down while pulling the lever arm as far up as it can go.
5. Pour the test fluid (i.e., honey, salad dressing) in the product reservoir until it reaches the top of the product gate.
6. Push the lever arm down to raise the product gate; start the stopwatch as you push the lever arm down. Record the consistency or the distance the fluid traveled 30 s after the gate is raised. The numbers on the slope are distances in cm.
7. Clean the consistometer with soap and water, then dry. Be sure the consistometer is completely dry before using it again.
8. Repeat Steps 2–7 to test all samples.
9. Once all the data have been collected for the salad dressing and honey at room temperature, remove the salad dressing sample from the refrigerator, and repeat Steps 2–7. Be sure to

record the sample temperature before performing any measurements.
10. You may choose to run the samples in duplicate or triplicate. Data from samples collected under identical conditions may be pooled to generate an average reading.

26.3 DATA

26.3.1 Brookfield Viscometer

Date:
Product information:
Viscometer make and model:
Spindle size:

Sample	Sample temp (°C)	Spindle speed (rpm)	% Reading	Factor	Vicosity (mPa-s)

26.3.2 Zahn Cup

Date:
Product information:
Zahn cup hole diameter:
Zahn cup volume:

Sample	Sample temp (°C)	Zahn time (s)	Vicosity (cS)	Fluid density (kg/m³)	Vicosity (mPa-s)

26.3.3 Bostwick Consistometer

Date:
Product information:
Consistometer make and model:

Sample	Sample temp (°C)	Consistency (cm)

26.4 CALCULATIONS

26.4.1 Brookfield Viscometer

1. Sketch the experimental apparatus and label the major parts.
2. Calculate the viscosity of the test fluids at each rpm.
3. Plot viscosity versus rpm for each fluid on a single graph.
4. Label the plots with the type of fluid (e.g., Newtonian, pseudoplastic, Herschel-Bulkley)

based on the response of viscosity to speed (rpm). *Keep in mind that speed is proportional to shear rate. In other words, as the speed is doubled, the shear rate is doubled.*

26.4.2 Zahn Cup

1. Sketch the experimental apparatus and label the major parts.
2. Calculate the viscosity of the test fluids.
3. Compare the viscosity of the test fluids with the Zahn cup to the viscosities measured with the Brookfield viscometer. What speed in rpm would the Brookfield need to be run at for each fluid to match the viscosity measured using the Zahn cup?

26.4.3 Bostwick Consistometer

1. Sketch the experimental apparatus and label the major parts.
2. Compare the consistency of the test fluids with the consistometer to the viscosities measured with the Brookfield viscometer. Do the data match? What rpm gives you the best match?

26.5 QUESTIONS

1. What is viscosity?
2. What is a Newtonian fluid? What is a non-Newtonian fluid? Were your materials Newtonian or non-Newtonian? Explain your choice.
3. Describe the importance of viscosity and flow properties in food processing, quality control, and consumer satisfaction. How might food composition impact its viscosity or flow properties? What ingredient(s) in the salad dressing may have caused deviations from Newtonian behavior? What effect does temperature have on the viscosity of fluid foods?
4. For samples at similar temperatures and identical speeds, was the viscosity of honey ever less than the viscosity of salad dressing? Is this behavior representative of the sample rheology at all speeds?
5. Why is it important to test samples at more than 1 speed?

6. Compare the viscosities determined from the three different viscosity measurement methods. Are they similar? Do they give you the same information about the fluids? If there are differences in the viscosity results, what might be responsible for the differences?
7. Which viscosity measurement method gives you the most information about the test fluids? How could you adjust the other two measurement methods to give you more information about the test fluid?
8. Consistency is not the same as viscosity. Considering that different fluids have different flow behaviors, do you think it would be possible to develop an equation to convert consistency to viscosity? Why or why not?
9. A fluid food product was designed to have a certain viscosity profile. When the product was being developed, the viscosity was measured at several different speeds on a Brookfield viscometer. Now the product is in full-scale production in two locations. The quality control team in Location 1 uses a Zahn cup to check the viscosity of each batch of product. The quality control team at Location 2 uses a Bostwick consistometer to check the viscosity of each batch of product. What issues might occur if there is a problem with product viscosity at one of the production facilities or if the two production facilities want to compare product viscosity data?

Acknowledgment This laboratory was adapted from a laboratory created by:

Dr. Christopher R. Daubert, Department of Food Bioprocessing & Nutritional Sciences, North Carolina State University, Raleigh, NC, USA

Dr. Brian E. Farkas, Department of Food Science, Purdue University, West Lafayette, IN, USA

RESOURCE MATERIALS

Joyner, HS, Daubert CR (2017) Rheological principles for food analysis. Ch 29. In: Nielsen SS (ed) Food analysis, 5th edn. Springer, New York

Singh RP, Heldman DR (2001) Introduction to food engineering, 3rd edn. Academic Press, San Diego, CA, pp 69–78, 144–157

CIE Color Specifications Calculated from Reflectance or Transmittance Spectra

M. Monica Giusti (✉)

Department of Food Science and Technology, The Ohio State University,
Columbus, OH, USA
e-mail: giusti.6@osu.edu

Ronald E. Wrolstad • Daniel E. Smith

Department of Food Science and Technology, Oregon State University,
Corvallis, OR, USA
e-mail: ron.wrolstad@oregonstate.edu; dan.smith@oregonstate.edu

S.S. Nielsen, *Food Analysis Laboratory Manual*, Food Science Text Series,
DOI 10.1007/978-3-319-44127-6_27, © Springer International Publishing 2017

27.1 INTRODUCTION

27.1.1 Background

Food color is arguably one of the most important determinants of acceptability and is, therefore, an important specification for many food products. The development of compact and easy to use colorimeters and spectrometers has made the quantitative measurement of color a routine part of product development and quality assurance.

There are several widely employed systems of color specification: notably Munsell, Commission Internationale de l'Eclairage (CIE) tristimulus, and the more recent CIEL*a*b* system. The Munsell system relies on matching with standard color chips. Value, hue, and chroma are employed to express lightness, "color," and saturation, respectively. The CIE tristimulus system uses mathematical coordinates (X, Y, and Z) to represent the amount of red, green, and blue primaries required by a "standard observer" to give a color match. These coordinates can be combined to yield a two-dimensional representation (chromaticity coordinates x and y) of color. The CIEL*a*b* system employs $L*$ (lightness), $a*$ (red-green axis), and $b*$ (yellow-blue axis) to provide a visually linear color specification.

Available software, often incorporated into modern instruments, enables the investigator to report data in any of the above notations. Understanding the different color specification systems, and the means of interconversion, aids the food scientist in selecting an appropriate means of reporting and comparing color measurements.

27.1.2 Reading Assignment

Wrolstad, R.E., and Smith, D.E. 2017. Color analysis. Ch. 31, in *Food Analysis*, 5th ed. S.S. Nielsen (Ed.), Springer, New York.

27.1.3 Objectives

1. Learn how to calculate the following CIE color specifications from reflectance and transmission spectra:
   (a) Tristimulus values X, Y, and Z
   (b) Chromaticity coordinates x and y and luminosity, Y
   (c) Dominant wavelength (λ_d) and % purity (using the chromaticity diagram)
2. Using readily available software, interconvert between the CIE Y and chromaticity coordinates and other color specification systems including Munsell and CIEL*a*b*.

27.1.4 Materials

1. % transmittance (%T) spectrum (A spectrum of syrup from Maraschino cherries colored with radish extract is provided, Table 27.1.)

table **27.1**	% Transmittance[a] and reflectance[b] data for Maraschino cherry	
λ_{nm}	%T Maraschino cherry syrup	%R Maraschino cherries
400	1.00	0.34
410	2.00	0.34
420	2.70	1.08
430	3.40	0.89
440	3.80	1.14
450	3.50	1.06
460	2.40	0.85
470	1.30	0.83
480	0.60	0.7
490	0.30	0.77
500	0.30	0.75
510	0.30	0.8
520	0.30	0.85
530	0.30	0.77
540	0.40	0.86
550	1.30	0.82
560	6.60	0.99
570	7.60	1.42
580	13.6	2.19
590	22.4	4.29
600	33.9	7.47
610	46.8	11.2
620	59.0	15.0
630	68.6	17.8
640	74.9	20.2
650	78.8	21.8
660	81.1	23.2
670	82.7	25.1
680	84.2	26.3
690	84.8	27.8
700	85.7	28.4

[a]1 cm pathlength; Shimadzu Model UV160A Spectrophotometer
[b]Hunter ColorQuest 45/0 Colorimeter, illuminant D_{65}, reflectance mode, specular included, 10° observer angle

2. % reflectance spectrum (%R) (A spectrum of Maraschino cherries colored with radish extract is provided, Table 27.1.)
3. CIE chromaticity diagram (Fig. 27.1), Munsell conversion charts, or appropriate interconversion software

Examples:

(a) An online applet http://www.colorpro.com/info/tools/labcalc.htm is a graphical tool that permits the user to adjust tristimulus values by means of slider bars. Corresponding values of CIE $L*a*b*$ and Lch equivalents and a visual representation of the associated color are displayed.
(b) A second application, http://www.colorpro.com/info/tools/rgbcalc.htm, provides the same

1964
Chromaticity
diagram (10
supplemental
standard
observer)

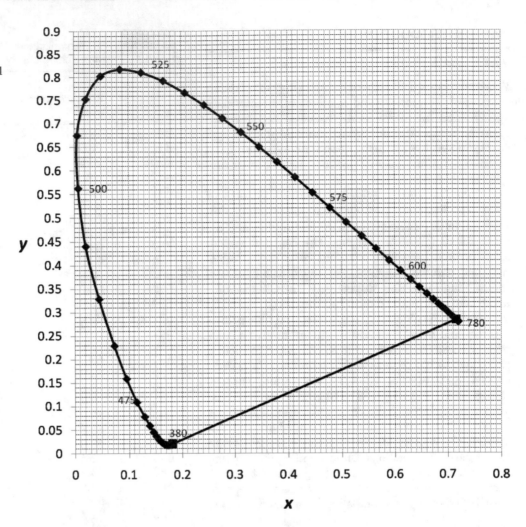

slider adjustment of RGB values with conversion to equivalent values in other systems.

(c) Convert L^*, a^*, and b^* values to other notations: http://www.colorpro.com/info/tools/convert.htm#TOP.

(d) A free evaluation copy of software that permits entry of numeric values for any of tristimulus, Munsell, CIE $L^*a^*b^*$, and chromaticity (x, y) coordinates, with conversion to the other systems can be obtained from http://www.xrite.com/. An annual license for this program (CMC) is available for purchase from http://wallkillcolor.com.

27.1.5 Optional

Spectra of other samples can be acquired using the following instruments:

1. Visible spectrophotometer (transmittance spectrum)
2. Spectrophotometers for color analyses (commonly called colorimeter) operated in transmittance or reflectance mode

27.2 PROCEDURE

27.2.1 Weighted Ordinate Method

1. Determine percent transmittance (%T) or percent reflectance (%R) at the specified wavelengths (e.g., every 10 nm between 400 and 700 nm). [*Note*: Example data for transmittance and reflectance (Table 27.1) are provided and can be used for these calculations.]

2. Multiply %T (or %R) by $E\bar{x}$, $E\bar{y}$, and $E\bar{z}$ (see Tables 27.2 or 27.3, respectively, for %T or %R). These factors incorporate both the CIE spectral distribution for illuminant D_{65} and the 1964 CIE standard supplemental observer curves for x, y, and z.

3. Sum the values $\%T\left(or\%R\right)E\bar{x}$, $\%T\left(or\%R\right)E\bar{y}$, and $\%T\left(or\%R\right)E\bar{z}$ to give X, Y, and Z, respectively (Table 27.4). The sums of each are divided by the sum of $E\bar{y}$ (760.7). (By doing this, the three values are normalized to $Y=100$, which is "perfect" white; objects are specified relative to

27.2 table

Calculation of CIE specifications by the weighted ordinate method: % transmittance

λ_{nm}	%T	$E\overline{x}$	$E\overline{x}\bullet$%T	$E\overline{y}$	$E\overline{y}\bullet$%T	$E\overline{z}$	$E\overline{z}\bullet$%T
400	0.60		0.10		2.50		
410	3.20		0.30		14.90		
420	8.80		0.90		41.80		
430	13.00		1.60		64.20		
440	19.10		3.10		98.00		
450	20.40		4.90		109.70		
460	16.50		7.00		95.00		
470	10.20		9.70		68.90		
480	4.20		13.30		40.60		
490	0.80		16.90		20.70		
500	0.20		24.10		11.40		
510	2.10		33.80		6.20		
520	7.00		45.00		3.60		
530	15.70		58.10		2.00		
540	26.10		66.60		0.90		
550	38.10		71.40		0.30		
560	48.70		68.90		0.00		
570	57.50		62.60		0.00		
580	67.30		57.70		0.00		
590	73.50		51.10		0.00		
600	79.90		46.80		0.00		
610	76.30		39.10		0.00		
620	63.50		29.50		0.00		
630	46.00		20.10		0.00		
640	30.20		12.60		0.00		
650	18.30		7.30		0.00		
660	10.70		4.20		0.00		
670	5.70		2.20		0.00		
680	2.70		1.10		0.00		
690	1.20		0.50		0.00		
700	0.6		0.2		0.00		
SUM				760.7			

1964 CIE color matching functions for 10° standard supplemental observer, illuminant D_{65}:

$X = E\overline{x}\bullet$%T $/ E\overline{y} =$

$Y = E\overline{y}\bullet$%T $/ E\overline{y} =$

$Z = E\overline{z}\bullet$%T $/ E\overline{y} =$

27.3 table

Calculation of CIE specifications by the weighted ordinate method: % reflectance

λ_{nm}	%R	$E\overline{x}$	$E\overline{x}\bullet$%R	$E\overline{y}$	$E\overline{y}\bullet$%R	$E\overline{z}$	$E\overline{z}\bullet$%R
400	0.60		0.10		2.50		
410	3.20		0.30		14.90		
420	8.80		0.90		41.80		
430	13.00		1.60		64.20		
440	19.10		3.10		98.00		
450	20.40		4.90		109.70		
460	16.50		7.00		95.00		
470	10.20		9.70		68.90		
480	4.20		13.30		40.60		
490	0.80		16.90		20.70		
500	0.20		24.10		11.40		
510	2.10		33.80		6.20		
520	7.00		45.00		3.60		
530	15.70		58.10		2.00		
540	26.10		66.60		0.90		
550	38.10		71.40		0.30		
560	48.70		68.90		0.00		
570	57.50		62.60		0.00		
580	67.30		57.70		0.00		
590	73.50		51.10		0.00		
600	79.90		46.80		0.00		
610	76.30		39.10		0.00		
620	63.50		29.50		0.00		
630	46.00		20.10		0.00		
640	30.20		12.60		0.00		
650	18.30		7.30		0.00		
660	10.70		4.20		0.00		
670	5.70		2.20		0.00		
680	2.70		1.10		0.00		
690	1.20		0.50		0.00		
700	0.6		0.2		0.00		
SUM				760.7			

1964 CIE color matching functions for 10° standard supplemental observer, illuminant D_{65}:

$X = E\overline{x}\bullet$%R $/ E\overline{y} =$

$Y = E\overline{y}\bullet$%R $/ E\overline{y} =$

$Z = E\overline{z}\bullet$%R $/ E\overline{y} =$

27.4 table	CIE color specifications worksheet for Maraschino cherry sample	
	%T	%R
X		
Y (luminosity)		
Z		
X + Y + Z		
x		
y		
λ_d		
% Purity		
Munsell notation		
CIE L*		
a*		
b*		
Hue angle, arctanb/a		
Chroma, $(a^{*2} + b^{*2})^{1/2}$		

luminosity of perfect white rather than the absolute level of light.)

4. Determine chromaticity coordinates x and y as follows:

$$x = (X)/(X + Y + Z) \qquad y = (Y)/(X + Y + Z)$$

5. *Luminosity* is the value of Y following the normalization described above.

27.2.2 Expression in Other Color Specification Systems

Plot the x and y coordinates on the CIE chromaticity diagram (Fig. 27.1) and determine dominant wavelength and % purity:

Dominant wavelength = λ_d = wavelength of spectrally pure light that if mixed with white light will match a color; analogous to hue.

On the CIE chromaticity diagram (Fig. 27.1), draw a straight line from illuminant D_{65}, extending through the sample point to the perimeter of the diagram. The point on the perimeter will be the dominant wavelength.

Coordinates for illuminant D_{65}: $x = 0.314$
$y = 0.331$

% purity = ratio of distance (a) from the illuminant to the sample over the distance ($a + b$) from the illuminant to the spectrum locus. Analogous to *chroma*.

Determine Munsell *value*, *hue*, and *chroma* with chromaticity coordinates x and y. Also convert these data to their *L*a*b** equivalents. Calculate chroma and hue as indicated.

$$Chroma = (a^{*2} + b^{*2})^{1/2}$$
$$Hue\ angle = arctan b^*/a^*$$

27.3 QUESTIONS

Assume D_{65} illuminant and 10° supplemental standard observer for all measurements in the questions below.

1. What is the analogous term in the Munsell system to luminosity in the CIE system?
2. What are the dominant wavelength and % purity of a food with chromaticity coordinates $x = 0.450$ and $y = 0.350$?
3. A lemon is found to have values of $L^* = 75.34$, $a^* = 4.11$, and $b^* = 68.54$. Convert to corresponding chromaticity coordinates x and y and plot on the 1964 chromaticity diagram.
4. Which has the greater hue angle, an apple with coordinates $L^* = 44.31$, $a^* = 47.63$, and $b^* = 14.12$ or $L^* = 47.34$, $a^* = 44.5$, and $b^* = 15.16$? Which apple has the greater value of chroma?

RESOURCE MATERIALS

Berns RS (2000) Billmeyer and Saltzman's principles of color technology, 3rd edn. Wiley, New York

Judd DB, Wyszecki G (1975) Color in business, science and industry, 3rd edn. Wiley, New York

Wrolstad RE, Smith DE (2017) Color analysis. Ch. 31. In: Nielsen SS (ed) Food analysis, 5th edn. Springer, New York

Extraneous Matter Examination

S. Suzanne Nielsen

Department of Food Science, Purdue University,
West Lafayette, IN, USA
e-mail: nielsens@purdue.edu

S.S. Nielsen, *Food Analysis Laboratory Manual*, Food Science Text Series,
DOI 10.1007/978-3-319-44127-6_28, © Springer International Publishing 2017

28.1 INTRODUCTION

28.1.1 Background

Extraneous materials are any foreign substances in foods that are associated with objectionable conditions or practices in production, storage, or distribution of foods. Extraneous materials include (a) filth or objectionable matter contributed by animal contamination (rodent, insect, or bird matter) or unsanitary conditions, (b) decomposed material or decayed tissues due to parasitic or nonparasitic causes, and (c) miscellaneous matter (sand, soil, glass, rust, or other foreign substances). Bacterial contamination is excluded from these substances.

Filth is classified according to its extractability. Light filth is oleophilic and lighter than water (separated from product by floating it in an oil-aqueous mixture). Insect fragments, rodent hairs, and feather barbules are examples of light filth. Heavy filth is heavier than water and separated from the product by sedimentation based on different densities of filth, food particles, and immersion liquids ($CHCl_3$, CCl_4, etc.). Examples of heavy filth are sand, soil, and nutshell fragments. Sieved filth involves particles separated from the product by the use of selected mesh sizes. Whole insects, stones, sticks, and bolts are examples of sieved filth.

Various methods of isolation of extraneous matter from various food commodities can be found in the *Official Methods of Analysis* of the AOAC International and in the *Approved Methods of Analysis* of the AACC International. Presented here are a few procedures for some food commodities, with descriptions based on AOAC methods, but the quantities reduced to half.

28.1.2 Reading Assignment

Dogan, H., and Subramanyam, B. 2017. Extraneous matter. Ch 34, in *Food Analysis*, 5th ed. S.S. Nielsen (Ed.), Springer, New York.

28.1.3 Notes

Regulatory examination of samples by the Food and Drug Administration (FDA) would be based on replicate samples using official methods, including the specified sample size. However, for instructional purposes, the costs associated with adequate commercial 1-L Wildman trap flasks, reagents, and food samples specified in official methods may be prohibitive. Procedures given below are based on AOAC methods, but all quantities are reduced to half, and a 500-mL Wildman trap flask (vs. 1-L trap flask) is specified in most procedures. Commercially available 1-L trap flasks with the standard stopper rod would ideally be used (with all quantities in the procedures doubled). However, 500-mL trap flasks can be made for use in

this experiment. To do this, drill a hole through a rubber stopper of a size just too large for a 500-mL Erlenmeyer flask. Thread a heavy string through the hole in the stopper, and knot both ends of the string. Coat the sides of the rubber stopper with glycerin and *carefully* force it (with larger end of stopper pointed up) through the top of the flask. Note that the string could be a trap for contaminants such as rodent hair and insect fragments.

For the parts of this laboratory exercise that require filter paper, S&S #8 (Schleicher & Schuell, Inc., Keene, NH) is recommended. It meets the specifications set forth in the AOAC Method 945.75 Extraneous Materials (Foreign Matter) in Products, Isolation Techniques Part B(i), which suggests using "smooth, high wet strength, rapid acting filter paper ruled with oil-, alcohol-, and water-proof lines 5 mm apart." The S&S #8 ruled filter paper is 9 cm in diameter and fits well into the *top* of standard 9-cm *plastic* petri dishes. The *bottom* of the plastic petri dish can be used as a protective cover over the sample filter paper in the top of the petri dish. The top of the plastic petri dish provides a 9 cm flat surface (as opposed to glass petri dishes) for examining the filter paper, making it easier to view the plate without having to continuously refocus the microscope. The 5-mm ruled lines provide a guide for systematically examining and enumerating contaminants on the filter paper at 30× magnification. To obtain a moist surface on which contaminants can be manipulated and observed, apply a small amount of glycerin: 60% alcohol (1:1) solution to the top of the petri dish before transferring the filter paper from the Buchner funnel. Using both overhead and substage lighting with the microscope will assist in identifying contaminants.

28.1.4 Objective

The objective of this laboratory is to utilize techniques to isolate the extraneous matter from various foods: cottage cheese, jam, infant food, potato chips, and citrus juice.

28.1.5 Principle of Methods

Extraneous materials can be separated from food products by particle size, sedimentation, and affinity for oleophilic solutions. Once isolated, extraneous materials can be examined microscopically.

28.2 EXTRANEOUS MATTER IN SOFT CHEESE

28.2.1 Chemicals

	CAS no.	Hazards
Phosphoric acid (H_3PO_4)	7664-38-2	Corrosive

28.2.2 Reagents

- Phosphoric acid solution, 400–500 mL
  Combine 1 part phosphoric acid with 40 parts deionized distilled (dd) water (vol/vol).

28.2.3 Hazards, Precautions, and Waste Disposal

Adhere to normal laboratory safety procedures. Wear safety glasses at all times. Waste likely may be put down the drain using a water rinse, but follow good laboratory practices outlined by environmental health and safety protocols at your institution.

28.2.4 Supplies

- Beaker, 1 L (for phosphoric acid solution)
- Beaker, 600 mL (to boil water)
- Buchner funnel
- Cottage cheese, 115 g
- Filter paper
- Heavy gloves
- Pipette, 10 mL (to prepare phosphoric acid solution)
- Pipette bulb or pump
- Spoon
- Sidearm flask, 500 mL or 1 L
- Stirring rod
- Tap water, ca. 500 mL (boiling)
- Tweezers
- Volumetric flask, 500 mL (to prepare phosphoric acid solution)
- Weighing boat

28.2.5 Equipment

- Hot plate
- Microscope
- Top loading balance
- Water aspirator system

28.2.6 Procedure

(Based on AOAC Method 960.49, Filth in Dairy Products)

1. Weigh out 115-g cottage cheese and add it to 400–500-mL boiling phosphoric acid solution (1+40 mixture) in a 1-L beaker, stirring with a glass stirring rod continuously to disperse the cottage cheese.
2. Filter the mixture through filter paper in a Buchner funnel, using a vacuum created by a water aspirator. Do not let the mixture accumulate on the paper, and continually wash filter with a stream of hot water to prevent clogging. Make sure the cheese mixture is hot as it is filtered. When filtration is impeded, add hot water or phosphoric acid solution (1+40 mixture) until

the paper clears. [May also use dilute (1–5%) alkali or hot alcohol to aid in filtration.] Resume addition of sample and water until sample is filtered.
3. Examine filter paper microscopically.

28.3 EXTRANEOUS MATTER IN JAM

28.3.1 Chemicals

	CAS no.	Hazards
Heptane (12.5 mL)	142-82-5	Harmful, highly flammable, dangerous for the environment
Hydrochloric acid, concentrated (HCl) (5 mL)	7647-01-0	Corrosive

28.3.2 Hazards, Precautions, and Waste Disposal

Heptane is an extremely flammable liquid; avoid open flames, breathing vapors, and contact with skin. Otherwise, adhere to normal laboratory safety procedures. Wear safety glasses at all times. Dispose of heptane waste as hazardous waste. Other waste may be put down the drain using a water rinse.

28.3.3 Supplies

- 2 Beakers, 250 mL (for weighing jam and heating water)
- Buchner funnel
- Filter paper
- Glass stirring rod
- Graduated cylinder, 100 mL
- Ice water bath (to cool mixture to room temperature)
- Jam, 50 g
- Graduated pipette, 10 mL (for heptane)
- Pipette bulb or pump
- Sidearm flask, 500 mL or 1 L
- Spoon
- Thermometer
- Tweezers
- Volumetric pipette, 5 mL (for conc. HCl)
- Waste jar (for heptane)
- Water, dd, 100 mL (heated to 50 °C)
- Wildman trap flask, 500 mL

28.3.4 Equipment

- Hot plate
- Microscope
- Top loading balance
- Water aspirator system

28.3.5 Procedure

(Based on AOAC Method 950.89, Filth in Jam and Jelly)

1. Empty contents of jam jar into beaker and mix thoroughly with glass stirring rod.
2. Weigh 50 g of jam into a beaker, add ca. 80 mL dd water at 50 °C, transfer to a 500-mL trap flask (use the other ca. 20 mL dd water to help make transfer), add 5 mL conc. HCl, and boil for 5 min.
3. Cool to room temperature (with an ice water bath).
4. Add 12.5 mL heptane and stir thoroughly.
5. Add dd water to a level so heptane is just above rubber stopper when in the "trap" position.
6. Trap off the heptane, and filter the heptane through filter paper in a Buchner funnel using vacuum created by a water aspirator.
7. Examine filter paper microscopically.

28.4 EXTRANEOUS MATTER IN INFANT FOOD

28.4.1 Chemicals

	CAS no.	Hazards
Light mineral oil (10 mL)	8012-95-1	

28.4.2 Hazards, Precautions, and Waste Disposal

Adhere to normal laboratory safety procedures. Wear safety glasses at all times. Waste may be put down the drain using water rinse.

28.4.3 Supplies

- Baby food, ~113 g (one jar)
- Buchner funnel
- Filter paper
- Glass stirring rod
- Graduated cylinder, 10 or 25 mL
- Pipette bulb or pump
- Sidearm flask, 500 mL or 1 L
- Spoon
- Tweezers
- Volumetric pipette, 10 mL
- Water, deaerated, 500 mL
- Wildman trap flask, 500 mL

28.4.4 Equipment

- Microscope
- Water aspirator system

28.4.5 Procedure

(Based on AOAC Method 970.73, Filth in Pureed Infant Food, A. Light Filth)

1. Transfer 113 g (one jar) of baby food to a 500-mL trap flask.
2. Add 10 mL of light mineral oil and mix thoroughly.
3. Fill the trap flask with deaerated water (can use dd water) at room temperature.
4. Let stand 30 min, stirring four to six times during this period.
5. Trap off mineral oil in a layer above the rubber stopper and then filter the mineral oil through filter paper in a Buchner funnel using vacuum created by a water aspirator.
6. Examine filter paper microscopically.

28.5 EXTRANEOUS MATTER IN POTATO CHIPS

28.5.1 Chemicals

	CAS no.	Hazards
Ethanol, 95%	64-17-5	Highly flammable
Heptane (9 mL)	142-82-5	Harmful, highly flammable, dangerous to environment
Petroleum ether (200 mL)	8032-32-4	Harmful, highly flammable, dangerous to environment

28.5.2 Reagents

- Ethanol, 60%, 1 L
  Use 95% ethanol to prepare 1 L of 60% ethanol; dilute 632 mL of 95% ethanol with water to 1 L.

28.5.3 Hazards, Precautions, and Waste Disposal

Petroleum ether, heptane, and ethanol are fire hazards; avoid open flames, breathing vapors, and contact with skin. Otherwise, adhere to normal laboratory safety procedures. Wear safety glasses at all times. Heptane and petroleum ether wastes must be disposed of as hazardous wastes. Other waste may be put down the drain using a water rinse.

28.5.4 Supplies

- Beaker, 400 mL
- Buchner funnel
- Filter paper
- Glass stirring rod
- Graduated cylinder, 1 L (to measure 95% ethanol)
- Ice water bath

- Potato chips, 25 g
- Sidearm flask, 500 mL or 1 L
- Spatula
- Wildman trap flask, 500 mL
- Tweezers
- Volumetric flask, 1 L (to prepare 60 % ethanol)
- Waste jars (for heptane and petroleum ether)

28.5.5 Equipment

- Hot plate
- Microscope
- Top loading balance
- Water aspirator system

28.5.6 Procedure

(Based on AOAC Method 955.44, Filth in Potato Chips)

1. Weigh 25 g of potato chips into a 400-mL beaker.
2. With a spatula or glass stirring rod, crush chips into small pieces.
3. In a hood, add petroleum ether to cover the chips. Let stand 5 min. Decant petroleum ether from the chips through filter paper. Again add petroleum ether to the chips, let stand 5 min, and decant through filter paper. Let petroleum ether evaporate from chips in hood.
4. Transfer chips to a 500-mL trap flask, add 125 mL 60 % ethanol, and boil for 30 min. Mark initial level of ethanol on flask. During boiling and at the end of boiling, replace ethanol lost by evaporation as a result of boiling.
5. Cool in ice water bath.
6. Add 9 mL heptane, mix, and let stand for 5 min.
7. Add enough 60 % ethanol to the flask so that only the heptane layer is above the rubber stopper. Let stand to allow heptane layer to form at the top, trap off the heptane layer, and filter it through filter paper in a Buchner funnel.
8. Add 9 mL more heptane to solution. Mix and then let stand until heptane layer rises to the top. Trap off the heptane layer, and filter it through filter paper (i.e., new piece of filter paper, not piece used in Parts 3 and 4) in a Buchner funnel.
9. Examine the filter paper microscopically.

28.6 EXTRANEOUS MATTER IN CITRUS JUICE

28.6.1 Supplies

- Beaker, 250 mL
- Buchner funnel
- Cheesecloth
- Citrus juice, 125 mL
- Graduated cylinder, 500 mL or 1 L
- Sidearm flask, 250 mL
- Tweezers

28.6.2 Equipment

- Microscope
- Water aspirator system

28.6.3 Procedure

[Based on AOAC Method 970.72, Filth in Citrus and Pineapple Juice (Canned), Method A. Fly Eggs and Maggots]

1. Filter 125 mL of juice through a Buchner funnel fitted with a double layer of cheesecloth. Filter with a vacuum created by a water aspirator. Pour the juice slowly to avoid accumulation of excess pulp on the cheesecloth.
2. Examine material on cheesecloth microscopically for fly eggs and maggots.

28.7 QUESTIONS

1. Summarize the results for each type of food analyzed for extraneous materials.
2. Why are contaminants such as insect fragments found in food, when the Pure Food and Drug Act prohibits adulteration?

RESOURCE MATERIALS

AOAC International (2016) Official methods of analysis, 20th edn. (On-line). AOAC International, Rockville, MD
Dogan H, Subramanyam B. (2017) Extraneous matter. Ch 34. In: Nielsen SS (ed) Food analysis, 5th edn. Springer, New York

3
part

Answers to Practice Problems

Answers to Practice Problems in Chap. 2, Preparation of Reagents and Buffers

Catrin Tyl (✉) • Baraem P. Ismail
Department of Food Science and Nutrition, University of Minnesota,
St. Paul, MN, USA
e-mail: tylxx001@umn.edu; bismailm@umn.edu

S.S. Nielsen, *Food Analysis Laboratory Manual*, Food Science Text Series,
DOI 10.1007/978-3-319-44127-6_29, © Springer International Publishing 2017

1. (a) This problem can be solved by using Eq. (2.5): The molecular weight of NaH_2PO_4 is 120 g/mol. Make sure to use the same units throughout. Molecular weights are stated in mol/g; the unit of concentration is in mol/L; thus, the 500 mL should also be converted into L:

$$m[g] = M\left[\frac{mol}{L}\right] \times v[L] \times MW\left[\frac{g}{mol}\right] \quad (2.5)$$

$$m[g] = 0.1\left[\frac{mol}{L}\right] \times 0.5[L] \times 120\left[\frac{g}{mol}\right] = 6g$$

(b) Just like for Example A2, the only change in the calculation is the use of 156 instead of 120, thus:

$$m[g] = 0.1\left[\frac{mol}{L}\right] \times 0.5[L] \times 156\left[\frac{g}{mol}\right] = 7.8g$$

2. According to the definition of % wt/vol, as found in Table 2.1:

$$\%\frac{w}{v} = \frac{\text{weight solute }[g] \times 100}{\text{total volume }[mL]}$$

$$\text{weight of solute} = \%\frac{wt}{v} \times \frac{1}{100} \times \text{volume}$$

$$= 10\left[\frac{g}{mL}\right] \times \frac{1}{100} \times 150[mL] = 15g$$

3. Determine NaOH's molarity in a 40% wt/vol solution: Due to NaOH's equivalence of 1, molarity equals normality. Use the definition of wt % to obtain the mass of NaOH in 1 L and Eq. (2.9) to calculate normality:

$$\%\frac{w}{v} = \frac{\text{weight solute }[g] \times 100}{\text{total volume }[mL]}$$

$$\text{weight of solute }[g] = \%\frac{w}{v} \times \frac{1}{100} \times \text{volume}$$

$$= 40\left[\frac{g}{mL}\right] \times \frac{1}{100} \times 1000[mL]$$

$$= 400[g]$$

$$n[mol] = \frac{m[g]}{MW\left[\frac{g}{mol}\right]} = \frac{400}{40} = 10[mol](\text{in }1L)$$

Thus, the molarity and the normality of this solution are 10.

4. The number of equivalents for H_2SO_4 is 2, because it can donate 2 H^+, and so the normality is two times the molarity. The mL of NaOH can be found through inserting into Eq. (2.19):

$$\text{mL of NaOH} = \frac{\text{mL of sulfuric acid} \times N \text{ of sulfuric acid}}{N \text{ of NaOH}}$$

$$\text{mL of NaOH} = \frac{200 \times 4}{10} = 80\,mL$$

5. For HCl, normality and molarity are equal, because 1 H^+ is released per molecule HCl. The problem can be solved like Example A3 by using Eq. (2.13) to calculate M of concentrated HCl, followed by Eq. (2.15):

$$M\left[\frac{mol}{L}\right] = \frac{d \times 1000\left[\frac{g}{L}\right]}{MW\left[\frac{g}{mol}\right]} \times \%\frac{wt}{wt} \quad (2.13)$$

$$M\left[\frac{mol}{L}\right] = \frac{1.2 \times 1000\left[\frac{g}{L}\right]}{36.5\left[\frac{g}{mol}\right]} \times 0.37 = 12.16\left[\frac{mol}{L}\right]$$

$$M_1 \times v_1 = M_2 \times v_2 \quad (2.15)$$

v of concentrated $HCl[L] =$

$$\frac{\text{vol of diluted HCl }[L] \times M \text{ of diluted HCl}\left[\frac{mol}{L}\right]}{M \text{ of concentrated HCl}\left[\frac{mol}{L}\right]}$$

$$v \text{ of concentrated HCl}[L] = \frac{0.25[L] \times 2\left[\frac{mol}{L}\right]}{12.16\left[\frac{mol}{L}\right]}$$

$$= 0.041L \text{ or } 41mL$$

6. Just like for Example A3 and A5, determine the molarity of the concentrated acetic acid with Eq. (2.13) (ignore %wt/wt):

$$M \text{ acetic acid} = \frac{1.05 \times 1000}{60.06}\left[\frac{g \times mol}{g \times L}\right] = 17.5\left[\frac{mol}{L}\right]$$

The desired amount is 0.04 mol; thus, take the amount calculated below with Eq. (2.16) and dilute to 1 L:

$$v[L] = \frac{n}{M}\left[\frac{mol \times L}{mol}\right] = \frac{0.04}{17.5} = 0.0023\,L$$

7. The weight of acetic acid in 1 L of solution can be found analogously to Problem 2:

$$\%\frac{w}{v} = \frac{\text{weight solute }[g] \times 100}{\text{total volume }[mL]}$$

$$\text{weight of acetic acid} = \%\frac{wt}{v} \times \frac{1}{100} \times \text{volume}$$

$$= 1\left[\frac{g}{mL}\right] \times \frac{1}{100} \times 1000[mL] = 10g$$

The corresponding number of moles of 10 g can be found using Eq. (2.9):

$$n[\text{mol}] = \frac{m[\text{g}]}{MW\left[\dfrac{g}{mol}\right]} = \frac{10}{60.02} = 0.167\,[\text{mol}](\text{in}\,1\,\text{L})$$

This already answers the question: A 1% acetic acid solution contains 0.167 mol per L, not 0.1 mol per L.

8. This problem can be solved analogously to Problem 14:

$$\%\frac{w}{v} = \frac{\text{weight solute}\,[\text{g}] \times 100}{\text{total volume}\,[\text{mL}]}$$

$$\text{weight of sodium hydroxide} = \%\frac{wt}{v} \times \frac{1}{100} \times \text{volume}$$

$$= 10\left[\frac{g}{mL}\right] \times \frac{1}{100} \times 1000\,[\text{mL}]$$

$$= 100\,\text{g}$$

The corresponding number of moles of 100 g can be found using Eq. (2.9):

$$n[\text{mol}] = \frac{m[\text{g}]}{MW\left[\dfrac{g}{mol}\right]} = \frac{10}{40} = 0.25\,[\text{mol}](\text{in}\,1\,\text{L})$$

This already answers the question: A 10% sodium hydroxide solution contains 0.25 mol per L, not 1 mol per L.

9. The normality of $K_2Cr_2O_7$ is 6 times the molarity. Use (Eqs. 2.1 and 2.9) to calculate the molarity and then multiply with the number of equivalents to obtain normality:

$$\text{Molarity}\,(M)\left[\frac{mol}{L}\right] =$$

$$\frac{\text{number of moles}\,(n)[\text{mol}]}{\text{vol}\,(v)[\text{L}]} \qquad (2.1)$$

$$n[\text{mol}] = \frac{m}{MW}\left[\frac{g \times mol}{g}\right] \qquad (2.9)$$

$$M\left[\frac{mol}{L}\right] = \frac{\dfrac{0.2\,[\text{g}]}{294.187\left[\dfrac{g}{mol}\right]}}{0.1\,[\text{L}]} = 0.0068\left[\frac{mol}{L}\right]$$

$$N\left[\frac{\text{equivalents}}{L}\right] = M \times \text{number of equivalents}$$

$$= 0.0068 \times 6 = 0.04\left[\frac{\text{equivalents}}{L}\right]$$

10. The molecular weight of KHP is 204.22 g/mol, and as it only contains one unionized carboxyl group, its number of equivalents is 1 and its molarity equals its normality. According to Eq. (2.5), 100 mL would contain:

$$m[\text{g}] = M\left[\frac{mol}{L}\right] \times v[\text{L}] \times MW\left[\frac{g}{mol}\right] \qquad (2.5)$$

$$m[\text{g}] = 0.1 \times 0.1 \times 204.22 = 2.0422\,\text{g}$$

11. The first step is to find the desired amount of Ca in the 1000 mL. As listed in Table 2.1, ppm corresponds to:

$$\text{ppm} = \frac{\text{mg solute}}{\text{kg solution}}$$

$$\text{Thus, } 1000\,\text{ppm} = \frac{1000\,\text{mg Ca}}{\text{kg standard solution}}$$

For our example, the solution's density can be assumed to be 1, and thus there would be 1000 mg Ca/L or 1 g Ca/L. If 110.98 g of $CaCl_2$ contain 40.078 g Ca, then 1 g Ca is supplied by:

$$m[\text{g}] \text{ of } CaCl_2 = \frac{1 \times 110.98}{40.078} = 2.7691\,\text{g}$$

12. First, the correct ratio of sodium acetate/acetic acid needs to be determined; similar to Example problem C3, $[A^-]$ is expressed through $[AH]$ to have only one unknown quantity in the equation and then inserted into Eq. (2.2):

$$\left[A^-\right] = 0.1 - \left[AH\right]$$

$$5.5 = 4.76 + \log\frac{0.1 - \left[AH\right]}{\left[AH\right]}$$

$$10^{(5.5 - 4.76)} = \frac{0.1 - \left[AH\right]}{\left[AH\right]}$$

$$5.5 \times \left[AH\right] = 0.1 - \left[AH\right]$$

$$[AH] = \frac{0.1}{(5.5 + 1)} = 0.0154\ \left[A^-\right] = 0.0846$$

Preparing the buffer by approach 1 (Chap 2, Sect. 2.3) (1 L of 0.1 M solutions of sodium acetate and acetic acid):

$$m \text{ of sodium acetate}\,[\text{g}] = MW \times M = 82 \times 0.1 = 8.2\,[\text{g}]$$

$$v \text{ of acetic acid}\,[\text{mL}] = \frac{MW \times M}{d} = \frac{60.02 \times 0.1}{1.05} = 5.7\,[\text{mL}]$$

Mix them so that the resulting concentrations correspond to 0.0846 M of sodium acetate and 0.0154 M of acetic acid using Eq. (2.15):

$$M_1 \times v_1 = M_2 \times v_2 \qquad (2.15)$$

v of sodium acetate solution $[mL]$

$$= \frac{M \text{ of } \left[A^-\right] \text{ in buffer} \times v \text{ of buffer}}{M \text{ of stock}}$$

$$= \frac{0.0846 \times 0.25}{0.1} = 211.5 [mL]$$

v of acetic acid solution $[mL]$

$$= \frac{M \text{ of } [AH] \text{ in buffer} \times v \text{ of buffer}}{M \text{ of stock}}$$

$$= \frac{0.0154 \times 0.25}{0.1} = 38.5 [mL]$$

Preparing the buffer via approach 2 (Chap 2, Sect. 2.3) (directly dissolve appropriate amounts of sodium acetate and acetic acid in 250 mL):

m of sodium acetate $[g]$

$$= M \text{ of } \left[A^-\right] \text{ in buffer} \left[\frac{mol}{L}\right]$$

$$\times v \text{ of buffer} [L] \times MW \left[\frac{g}{mol}\right] \qquad (2.5)$$

m of sodium acetate $[g]$

$$= 0.25 [L] \times 0.0846 \left[\frac{mol}{L}\right] \times 82 \left[\frac{g}{mol}\right] = 1.73 [g]$$

vol of acetic acid $[mL] =$

$$\frac{\left\{ \begin{array}{c} \text{vol of buffer} [L] \times M \text{ of } [AH] \text{ in} \\ \text{buffer} \left[\frac{mol}{L}\right] \times MW \left[\frac{g}{mol}\right] \end{array} \right\}}{\left[\frac{g}{mL}\right]}$$

v of acetic acid $[mL] =$

$$\frac{0.25 [L] \times 0.0154 \left[\frac{\text{mol}}{L}\right] \times 60.02 \left[\frac{g}{\text{mol}}\right]}{1.05 \left[\frac{g}{mL}\right]} = 2.2 [mL]$$

These would be dissolved in, e.g., 200 mL, and then the pH adjusted and the volume brought up to 250 mL.

Preparing the buffer via approach 3 (Chap 2, Sect. 2.3) (dissolve the appropriate amount of acetic acid to yield 250 mL of a $0.1 M$ solution in < 250 mL, and then adjust the pH with NaOH of a high molarity, e.g., 6 or $10 M$ to pH 5.5):

v of acetic acid $[mL]$

$$= \frac{v \text{ of buffer} [L] \times M \text{ of buffer} \left[\frac{mol}{L}\right] \times MW \left[\frac{g}{mol}\right]}{\left[\frac{g}{mL}\right]}$$

v of acetic acid $[mL]$

$$= \frac{0.25 [L] \times 0.1 \left[\frac{mol}{L}\right] \times 60.02 \left[\frac{g}{mol}\right]}{1.05 \left[\frac{g}{mL}\right]} = 1.43 [mL]$$

13. The molarities of Na_2EDTA and $MgSO_4$ are found by rearranging Eq. (2.5):

$$m [g] = M \left[\frac{mol}{L}\right] \times v [L] \times MW \left[\frac{g}{mol}\right] \qquad (2.5)$$

$$M \left[\frac{mol}{L}\right] = \frac{m [g]}{v [L] \times MW \left[\frac{g}{mol}\right]}$$

$$M \left[\frac{mol}{L}\right] \text{ of } Na_2 EDTA = \frac{1.179 [g]}{0.25 [L] \times 372.24 \left[\frac{g}{mol}\right]}$$

$$= 0.0127 \left[\frac{mol}{L}\right]$$

$$M \left[\frac{mol}{L}\right] \text{ of } MgSO_4 = \frac{0.78 [g]}{0.25 [L] \times 246.47 \left[\frac{g}{mol}\right]}$$

$$= 0.0127 \left[\frac{mol}{L}\right]$$

To calculate the buffer pH, determine molarity of NH_4Cl using Eqs. (2.1 and 2.3) and NH_3 with Eq. (2.27):

M of NH_4Cl in buffer $\left[\frac{mol}{L}\right] =$

$$\frac{m [g]}{MW \left[\frac{g}{mol}\right] \times v [L]} = \frac{16.9}{53.49 \times 0.25} = 1.26 \left[\frac{mol}{L}\right]$$

$$M \text{ of conc.} NH_3 \left[\frac{mol}{L}\right] = \frac{\left[\frac{g}{mL}\right] \times 1000}{MW \left[\frac{g}{mol}\right]} \times \% wt$$

$$= \frac{0.88 \times 1000}{17} \times 0.28 = 14.5 \left[\frac{mol}{L}\right]$$

$$M \text{ in buffer} \left[\frac{\text{mol}}{\text{L}} \right]$$

$$= \frac{M \text{ of stock solutions} \left[\frac{\text{mol}}{\text{L}} \right] \times v \text{ of stock solutions} \left[\text{L} \right]}{v \text{ of buffer} \left[\text{L} \right]}$$

$$(2.27)$$

$$M \text{ of } NH_3 \text{ in buffer} \left[\frac{\text{mol}}{\text{L}} \right]$$

$$= \frac{M \text{ of conc.} NH_3 \left[\frac{\text{mol}}{\text{L}} \right] \times v \text{ of conc.} NH_3 \left[\text{L} \right]}{v \text{ of buffer} \left[\text{L} \right]}$$

$$M \text{ of } NH_3 \text{ in buffer} \left[\frac{\text{mol}}{\text{L}} \right]$$

$$= \frac{14.5 \left[\frac{\text{mol}}{\text{L}} \right] \times 0.143 \left[\text{L} \right]}{0.25 \left[\text{L} \right]} = 8.29 \left[\frac{\text{mol}}{\text{L}} \right]$$

The normal form of the Henderson-Hasselbalch equation may be used after calculating the pK_a of NH_4^+. NH_4Cl acts as the acid, NH_4OH is the base (NH_4OH is just another way of writing NH_3 in water). Note: The actual pK_a and pK_b may be slightly different because of the added salts affecting the ionic strength.

$$pK_a \text{ of } NH_4^+ = 14 - pK_b = 14 - 4.74 = 9.26$$

$$pH = 9.26 + \log \frac{8.29}{1.26} = 9.26 + 0.82 = 10.08$$

14. (a) Use Eq. (2.5) to obtain the masses:

$$m[\text{g}] = M \left[\frac{\text{mol}}{\text{L}} \right] \times v[\text{L}] \times MW \left[\frac{\text{g}}{\text{mol}} \right] \quad (2.5)$$

$$m[\text{g}] \text{ of } NaH_2PO_4 \cdot H_2O = 0.2 \times 0.5 \times 138 = 13.8[\text{g}]$$

$$m[\text{g}] \text{ of } Na_2HPO_4 \cdot 7H_2O = 0.2 \times 0.5 \times 268 = 26.8[\text{g}]$$

(b) Use Eq. (2.38) to express $[A^-]$ through $[AH]$ to substitute into Eq. (2.25); find the pK_a in Table 2.2:

$$0.1 = \left[A^- \right] + \left[AH \right]$$

$$\left[A^- \right] = 0.1 - \left[AH \right]$$

$$6.2 = 6.71 + \log \frac{0.1 - AH}{AH}$$

$$-0.51 = \log \frac{0.1 - [AH]}{[AH]}$$

$$0.309 = \frac{0.1 - [AH]}{[AH]}$$

$$[AH] \times (0.309 + 1) = 0.1$$

$$[AH] \left[\frac{\text{mol}}{\text{L}} \right] = \frac{0.1}{1.219} = 0.0764 \left[\frac{\text{mol}}{\text{L}} \right] \left[A^- \right] \left[\frac{\text{mol}}{\text{L}} \right]$$

$$= 1 - 0.0764 = 0.0236 \left[\frac{\text{mol}}{\text{L}} \right]$$

The molarities of the stock solutions are 0.2 $\left[\frac{\text{mol}}{\text{L}} \right]$. To find the volumes to mix, use Eq. (2.27):

$$M_1 \times v_1 = M_2 \times v_2 \quad (2.27)$$

$$v \text{ of } NaH_2PO_4 \text{ stock solution} \left[\text{L} \right] =$$

$$\frac{M \text{ of } NaH_2PO_4 \text{ in buffer} \times v \text{ of buffer}}{M \text{ of stock solution}}$$

$$v \text{ of } NaH_2PO_4 \text{ stock solution} \left[\text{L} \right] = \frac{0.0764 \times 0.2}{0.2}$$

$$= 0.0764 \left[\text{L} \right] \text{ or } 76 \text{ mL}$$

$$v \text{ of } Na_2HPO_4 \text{ stock solution} \left[\text{L} \right] =$$

$$\frac{M \text{ of } Na_2HPO_4 \text{ in buffer} \times v \text{ of buffer}}{M \text{ of stock solution}}$$

$$v \text{ of } NaH_2PO_4 \text{ stock solution} \left[\text{L} \right] = \frac{0.0236 \times 0.2}{0.2}$$

$$= 0.024 \left[\text{L} \right] \text{ or } 24 \text{ mL}$$

(c) This problem is similar to Example C3. 1 mL of $6M$ NaOH supplies Eq. (2.2):

$$n \text{ of NaOH} \left[\text{mol} \right] = M \times v = 6 \times 0.001 = 0.006 \left[\text{mol} \right]$$

The amounts of NaH_2PO_4 and Na_2HPO_4 are found through Eq. (2.2) (use either the buffer molarity or values from the stock solutions calculated for Problem12b):

$$n \text{ of } NaH_2PO_4 \left[\text{mol} \right] = M \times v = 0.2 \times 0.076$$

$$= 0.015 \left[\text{mol} \right]$$

$$n \text{ of } NaH_2PO_4 \left[\text{mol} \right] = M \times v = 0.2 \times 0.024$$

$$= 0.0048 \left[\text{mol} \right]$$

Addition of NaOH changes the ratio by increasing the amount of Na_2HPO_4 and decreasing NaH_2PO_4:

n of NaH_2PO_4 after HCl $[mol] = 0.015 - 0.006$
$$= 0.009 [mol]$$

n of Na_2HPO_4 after HCl $[mol] = 0.0048 + 0.006$
$$= 0.0108 [mol]$$

Substitute these values into Eq. (2.25) to find the new pH. (Note: You may insert molar ratios, and you do not need to convert to concentrations, because the ratio would stay the same):

$$pH = 6.71 + \log \frac{0.0108}{0.009} = 6.77$$

15. To find the pH at 25°C, the acid/base ratio needs to be substituted into Eq. (2.25):

$$pH = 8.06 + \log \frac{4}{1}$$

$$pH = 8.06 - 0.6 = 7.46$$

Calculate the pK_a at 60 °C with Eq. (2.61), and then insert into Eq. (2.25):

$$pK_a = 8.06 - (0.023 \times (60 - 25)) = 7.26$$

$$pH = 7.26 - 0.6 = 6.65$$

16. This problem is analogous to Example C3. Find the ratio of acid to base through Eqs. (2.25 and 2.38), and then use Eq. (2.52) to find the volume of formic acid and Eq. (2.5) to find the mass of ammonium formate:

$$0.01 = [A^-] + [AH]$$

$$[A^-] = 0.01 - [AH]$$

$$3.5 = 3.75 + \log \frac{0.01 - [AH]}{[AH]}$$

$$-0.25 = \log \frac{0.01 - [AH]}{[AH]}$$

$$0.562 = \frac{0.01 - [AH]}{[AH]}$$

$$[AH] = \frac{0.01}{1.562} = 0.0064 \ [A^-] = 0.0036$$

$$v[mL] = \frac{M \times v \times MW}{d} \qquad (2.52)$$

v of formic acid $[mL]$
$$= \frac{M \times v \text{ of buffer} \times MW}{d}$$
$$= \frac{0.0064 \left[\frac{\text{mol}}{\text{L}}\right] \times 46 \left[\frac{\text{g}}{\text{mol}}\right] \times 1 [\text{L}]}{1.22 \left[\frac{\text{g}}{\text{mL}}\right]} = 0.24 [mL]$$

m of ammonium formate $[g]$
$$= M \times v \times MW \text{ of buffer}$$
$$= 0.0036 \left[\frac{\text{mol}}{\text{L}}\right] \times 63.06 \left[\frac{\text{g}}{\text{mol}}\right] \times 1 [\text{L}] = 0.227 [g]$$

Answers to Practice Problems in Chap. 3, Dilutions and Concentrations

Andrew P. Neilson (✉) • *Sean F. O'Keefe*

Department of Food Science and Technology,
Virginia Polytechnic Institute and State University,
Blacksburg, VA, USA
e-mail: andrewn@vt.edu; okeefes@vt.edu

S.S. Nielsen, *Food Analysis Laboratory Manual*, Food Science Text Series,
DOI 10.1007/978-3-319-44127-6_30, © Springer International Publishing 2017

1. A diagram of this scheme is shown *here*.

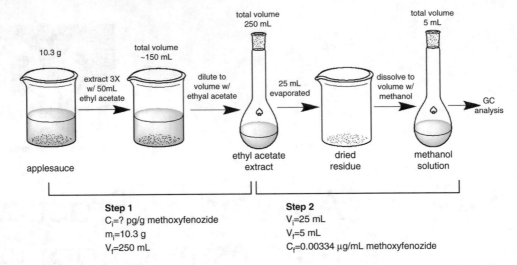

Step 1
C_i=? pg/g methoxyfenozide
m_i=10.3 g
V_f=250 mL

Step 2
V_i=25 mL
V_f=5 mL
C_f=0.00334 µg/mL methoxyfenozide

The calculations are performed as follows:

$$C_i = C_f \left(\frac{m \text{ or } V_2}{m \text{ or } V_1} \right) \left(\frac{m \text{ or } V_4}{m \text{ or } V_3} \right) \cdots \left(\frac{m \text{ or } V_k}{m \text{ or } V_{k-1}} \right)$$

$$C_i = \frac{0.00334 \text{ µg methoxy fenozide}}{\text{mL methanol solution}} \left(\frac{250 \text{ mL ethylacetate extract}}{10.3 \text{ g applesauce}} \right) \left(\frac{5 \text{ mL methanol solution}}{25 \text{ mL ethylactetate extract}} \right)$$

$$= \frac{0.0162 \text{ µg methoxy fenozide}}{\text{g applesauce}}$$

$$X = Cm = \left(\frac{0.0162 \text{ µg methoxy fenozide}}{\text{g applesauce}} \right) (113 \text{ g applesauce}) = 1.83 \text{ µg methoxy fenozide}$$

The concentration of methoxyfenozide in the applesauce is 0.0162 µg/g, and the total amount of methoxyfenozide in the entire applesauce cup is 1.83 µg.

2. A diagram of this scheme is shown *below* (note that the first dilution is "dilute to," while the last two dilutions are "dilute with").

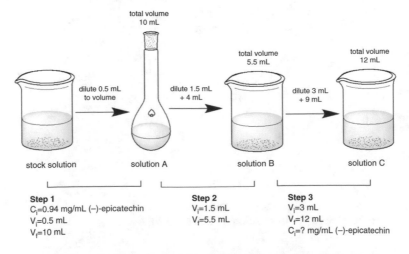

Step 1
C_i=0.94 mg/mL (−)-epicatechin
V_i=0.5 mL
V_f=10 mL

Step 2
V_i=1.5 mL
V_f=5.5 mL

Step 3
V_i=3 mL
V_f=12 mL
C_i=? mg/mL (−)-epicatechin

The calculations are performed as follows:

$$C_f = C_i \left(\frac{m \text{ or } V_1}{m \text{ or } V_2} \right) \left(\frac{m \text{ or } V_3}{m \text{ or } V_4} \right) \cdots \left(\frac{m \text{ or } V_{k-1}}{m \text{ or } V_k} \right)$$

$$C_f = \frac{0.94 \text{ mg EC}}{\text{mL stock solution}} \left(\frac{0.5 \text{ mL stock solution}}{10 \text{ mL solution } A} \right) \left(\frac{1.5 \text{ mL solution } A}{5.5 \text{ mL solution } B} \right) \left(\frac{3 \text{ mL solution } B}{12 \text{ mL solution } C} \right) = \frac{0.00320 \text{ mg EC}}{\text{mL solution } C}$$

$$C_f = \frac{0.00320 \text{ mg EC}}{\text{mL solution } C} \left(\frac{1 \text{ g EC}}{1000 \text{ mg EC}} \right) \left(\frac{1 \text{ mol EC}}{290.26 \text{ g EC}} \right) \left(\frac{1{,}000{,}000 \, \mu\text{mol EC}}{\text{mol EC}} \right) \left(\frac{1000 \text{ mL}}{1 \text{ L}} \right) = \frac{11.0 \, \mu\text{mol EC}}{\text{L}} = 11.0 \, \mu\text{M EC}$$

$$DF_\pounds = \frac{C_f}{C_i} = \frac{\dfrac{0.00320 \text{ mg EC}}{\text{mL}}}{\dfrac{0.94 \text{ mg EC}}{\text{mL}}} = 0.00341$$

$$\text{dilution "fold" or "} X \text{"} = \frac{1}{DF} = \frac{1}{0.00341} = 293$$

The concentration of solution C is 0.00320 mg (−)-epicatechin/mL [or 11.0 μM (−)-epicatechin]. The dilution factor is 0.00341, a 293-fold (293×) dilution of the stock solution.

3. A diagram of this scheme is shown *below* (note that all of these are "dilute with," the standards are prepared in parallel, and 100 μL = 0.1 mL).

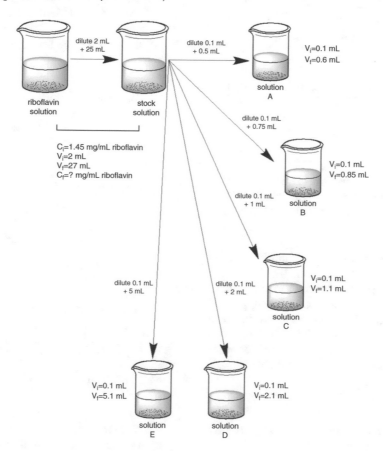

The calculations are performed as shown below. For the stock solution:

$$C_f = C_i\left(\frac{V_i}{V_f}\right) = \frac{1.45\,\text{mg riboflavin}}{\text{mL}}\left(\frac{2\,\text{mL}}{27\,\text{mL}}\right)$$

$$= \frac{0.107\,\text{mg riboflavin}}{\text{mL}}$$

For the standard solutions:

$$A \quad C_f = C_i\left(\frac{V_i}{V_f}\right) = \frac{0.107\,\text{mg riboflavin}}{\text{mL}}\left(\frac{0.1\,\text{mL}}{0.6\,\text{mL}}\right)$$

$$= \frac{0.0179\,\text{mg riboflavin}}{\text{mL}}$$

$$B \quad C_f = C_i\left(\frac{V_i}{V_f}\right) = \frac{0.107\,\text{mg riboflavin}}{\text{mL}}\left(\frac{0.1\,\text{mL}}{0.85\,\text{mL}}\right)$$

$$= \frac{0.0126\,\text{mg riboflavin}}{\text{mL}}$$

$$C \quad C_f = C_i\left(\frac{V_i}{V_f}\right) = \frac{0.107\,\text{mg riboflavin}}{\text{mL}}\left(\frac{0.1\,\text{mL}}{1.1\,\text{mL}}\right)$$

$$= \frac{0.00976\,\text{mg riboflavin}}{\text{mL}}$$

$$D \quad C_f = C_i\left(\frac{V_i}{V_f}\right) = \frac{0.107\,\text{mg riboflavin}}{\text{mL}}\left(\frac{0.1\,\text{mL}}{2.1\,\text{mL}}\right)$$

$$= \frac{0.00511\,\text{mg riboflavin}}{\text{mL}}$$

$$E \quad C_f = C_i\left(\frac{V_i}{V_f}\right) = \frac{0.107\,\text{mg riboflavin}}{\text{mL}}\left(\frac{0.1\,\text{mL}}{5.1\,\text{mL}}\right)$$

$$= \frac{0.00211\,\text{mg riboflavin}}{\text{mL}}$$

The riboflavin concentration in the stock solution and the five standards is 0.107, 0.0179, 0.0126, 0.00976, 0.00511, and 0.00211 mg/mL, respectively.

4. First, 160 g/L = 160 mg/mL. Therefore, large dilutions are needed to get from 160 mg/mL to 0–0.5 mg/mL. From this point, there are several approaches that could be used. One approach would be to dilute the stock solution down to the most concentrated standard (0.5 mg/mL) and then do further dilutions from that concentration. To do this, we calculate the DF:

$$DF = \frac{C_f}{C_i} = \frac{0.5\,\text{mg/mL}}{160\,\text{mg/mL}} = \frac{1}{320} = 0.003125$$

From this DF, we can calculate the ratio of volumes needed. Recall that:

$$DF = \frac{V_i}{V_f} = \frac{1}{320}$$

Therefore, we need to find a ratio of volumes equal to 1/320 to do this dilution. You do not have a 320 mL volumetric flask to do a simple 1 mL into 320 mL total dilution. However, you could use a 250 mL volumetric flask for the final volume and calculate the starting volume of the stock solution needed:

$$\frac{V_i}{V_f} = \frac{1}{320}$$

$$V_i = \frac{V_f}{320} = \frac{250\,\text{mL}}{320} = 0.781\,\text{mL}$$

Therefore, the 0.5 mL standard is made by diluting 0.781 mL (using a 1 mL adjustable pipettor) to 250 mL final volume. You could use a variety of different dilutions to get the same final concentrations, as long as you don't use less than 0.2 mL as your starting volume (e.g., you could also dilute 0.313 mL to 100 mL and get the same concentration). From the 0.5 mg/mL standard, the other standards (0–0.3 mg/mL) can easily be made in 2 mL volumes by combining variable volumes of the 0.5 mg/mL standard with water to achieve a total volume of 2 mL. These volumes are calculated as follows:

$$C_iV_i = C_fV_f \;\rightarrow\; V_i = \frac{C_fV_f}{C_i}$$

For 0 mg/mL $V_i = \dfrac{C_fV_f}{C_i}$

$$= \frac{(0\,\text{mg/mL})(2\,\text{mL})}{0.5\,\text{mg/mL}} = 0\,\text{mL}$$

For 0.1 mg/mL $V_i = \dfrac{C_fV_f}{C_i}$

$$= \frac{\left(0.1\dfrac{\text{mg}}{\text{mL}}\right)(2\,\text{mL})}{0.5\dfrac{\text{mg}}{\text{mL}}} = 0.4\,\text{mL}$$

and so forth for each solution. The volume of water is the amount needed to bring the starting volume up to 2 mL:

For 0 mg/mL: $\quad V_{\text{water}} = 2\,\text{mL} - 0\,\text{mL} = 2\,\text{mL}$

For 0.1 mg/mL: $\quad V_{\text{water}} = 2\,\text{mL} - 0.4\,\text{mL}$
$$= 1.6\,\text{mL}$$

and so forth. The dilutions from the diluted 0.5 mg/mL stock solution are shown in Table 30.1:

30.1 table	Dilution example for a standard curve			
Anthocyanins (mg/mL)	Diluted stock (mL)	Water (mL)	Total volume (mL)	
0.5	2	0	2.0	
0.3	1.2	0.8	2.0	
0.2	0.8	1.2	2.0	
0.1	0.4	1.6	2.0	
0	0	2	2.0	

A diagram of this scheme is shown *here*:

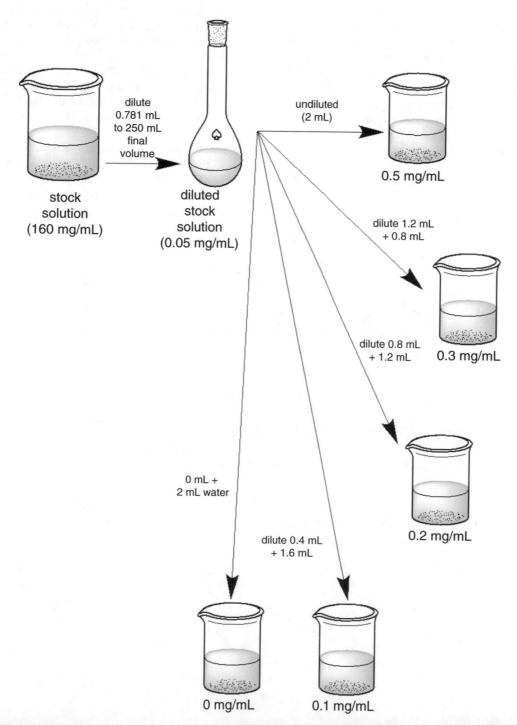

5. We know that the anthocyanin concentration of the juice is probably somewhere between 750 and 3000 µg/mL (0.75–3 mg/mL). However, it could literally be anywhere within this range. Samples with 0.75 and 3 mg/mL anthocyanins would require very different dilutions to get near the center of the standard curve (~0.25 mg/mL)—how do we handle this? The solution is to design a dilution scheme with various dilutions of the same sample, which will likely yield at least 1 dilution in the acceptable range that can be used for quantification. To do this, we assume that the anthocyanin concentration in the juice could be on the extremes (~0.75 and ~3 mg/mL) and right in the middle of those extremes (~1.125 mg/mL). Then, we can calculate 3 different dilutions (assuming the 3 different anthocyanin levels and diluting to the middle of the curve) so that we will have at least 1 useable sample regardless of the actual sample concentration. Let's assume that we want a final sample volume of 10 mL (we could pick any volume that corresponds to a volumetric flask we have available, but it wouldn't make sense to make 50–1000 mL of diluted sample if we only need 2 mL for the analysis):

Assume 0.75 mg / mL:

$$V_i = \frac{C_f V_f}{C_i} = \frac{(0.25\,\text{mg/mL})(10\,\text{mL})}{0.75\,\text{mg/mL}} = 3.33\,\text{mL}$$

Assume 1.125 mg / mL:

$$V_i = \frac{C_f V_f}{C_i} = \frac{(0.25\,\text{mg/mL})(10\,\text{mL})}{1.125\,\text{mg/mL}} = 2.22\,\text{mL}$$

Assume 3 mg / mL:

$$V_i = \frac{C_f V_f}{C_i} = \frac{(0.25\,\text{mg/mL})(10\,\text{mL})}{3\,\text{mg/mL}} = 0.833\,\text{mL}$$

Therefore, we would make three dilutions using 3.33 mL, 2.22 mL, and 0.833 mL of the juice and dilute each to a final volume of 10 mL. We would then analyze the standard solutions and the 3 dilutions and use the diluted sample (and corresponding dilution factor) with an analytical response within the range of the standard curve to calculate the anthocyanin concentration in the juice.

6. This problem is set up as shown below. Note that all of the minerals from 2.8 mL milk are diluted to a total volume of 50 mL, so these steps can be simplified somewhat in the calculation as shown in the figure.

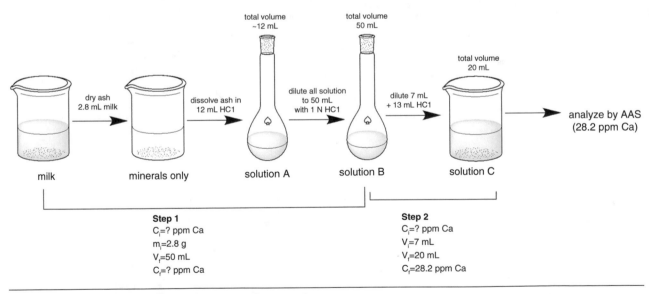

The problem can then be set up and solved as a multistep dilution as follows:

$$C_i = C_f \left(\frac{m\,\text{or}\,V_2}{m\,\text{or}\,V_1} \right) \left(\frac{m\,\text{or}\,V_4}{m\,\text{or}\,V_3} \right) \cdots \left(\frac{m\,\text{or}\,V_k}{m\,\text{or}\,V_{k-1}} \right)$$

$$C_i = 28.2\,\text{ppm Ca} \left(\frac{50\,\text{mL}}{2.8\,\text{mL}} \right) \left(\frac{20\,\text{mL}}{7\,\text{mL}} \right) = 1440\,\text{ppm Ca}$$

7. First, you need to get the sample and standards into the same units. Convert the sample caffeine concentration to mM:

$$\frac{170\,mg\,caffeine}{400\,mL}\left(\frac{1000\,mL}{1\,L}\right)\left(\frac{1\,g}{1000\,mg}\right)\left(\frac{1\,mol\,caffeine}{194.2\,caffeine}\right)\left(\frac{1\times10^6\,\mu mol\,caffeine}{1\,mol\,caffeine}\right)=\frac{2190\,\mu mol\,caffeine}{L}=2190\,\mu M\,caffeine$$

Then, determine the starting volume needed to dilute to 250 mL total volume and obtain a concentration at the center of the standard curve (50 µM caffeine):

$$C_iV_i = C_fV_f \rightarrow$$
$$V_i = \frac{C_fV_f}{C_i} = \frac{(50\,m\text{M caffeine})(250\,\text{mL})}{(2190\,m\text{M caffeine})}$$
$$= 5.71\,\text{mL}$$

Therefore, if 5.17 mL of the energy drink is diluted to 250 mL final volume, the caffeine concentration will be ~50 µM, which is right in the middle of your standard curve.

31
chapter

Answers to Practice Problems in Chap. 4, Use of Statistics in Food Analysis

Andrew P. Neilson (✉) • *Sean F. O'Keefe*
Department of Food Science and Technology,
Virginia Polytechnic Institute and State University,
Blacksburg, VA, USA
e-mail: andrewn@vt.edu; okeefes@vt.edu

S.S. Nielsen, *Food Analysis Laboratory Manual*, Food Science Text Series,
DOI 10.1007/978-3-319-44127-6_31, © Springer International Publishing 2017

1. What is the mean of the observations (mg/cup)?

$$\bar{x} = \frac{\sum x_i}{n} = \frac{x_1 + x_2 + \cdots + x_{n-1} + x_n}{n} = \frac{1967.1\,\text{mg}/\text{cup}}{6}$$
$$= 327.85\,\text{mg}/\text{cup, round to}\,328\,\text{mg}/\text{cup}$$

What is the standard deviation of the observations (mg/cup)? Since we have < 30 observations, the correct formula for sample standard deviation is:

$$\text{SD}_{n-1} = \sqrt{\frac{\sum(x_i - \bar{x})^2}{n-1}} = \sqrt{\frac{\sum(x_i - 327.85\,\text{mg}/\text{cup})^2}{6-1}}$$
$$= \sqrt{\frac{460.215\,\text{mg}^2/\text{cup}^2}{5}} = 9.593904\,\text{mg}/\text{cup}$$

Calculate the 96% confidence interval for the true population mean (use t-score, not Z score, because n is small). We know that the formula for a CI is:

$$\text{CI}: \bar{x} \pm t_{\frac{\alpha}{2},\,\text{df}=n-1} \times \frac{\text{SD}}{\sqrt{n}}$$

We know the mean, SD, and n already, so we just need the t-score. First, calculate df:

$$\text{df} = n - 1 = 6 - 1 = 5$$

Next calculate C and $\alpha/2$:

$$\text{for 96\% CI},\ C = 0.96 \rightarrow \frac{\infty}{2} = \frac{1-C}{2} = \frac{1-0.96}{2} = 0.02$$

Then, go to the t-table and find the t-score that corresponds to df = 5, $\alpha/2 = 0.02$:

$$t_{\frac{\alpha}{2},\,\text{df}=n-1} = t_{0.02,5} = 2.757$$

Putting it all together:

$$\text{CI}: \bar{x} \pm t_{\frac{\alpha}{2},\,\text{df}=n-1} \times \frac{\text{SD}}{\sqrt{n}}$$

$$327.85\,\text{mg}/\text{cup} \pm 2.757 \times \frac{9.593904\,\text{mg}/\text{cup}}{\sqrt{6}}$$

$$\rightarrow 327.85\,\text{mg}/\text{cup} \pm 10.7983\,\text{mg}/\text{cup}$$

What are the upper and lower limits of the 96% confidence interval for the population mean?

The upper limit of the CI is:

$$\bar{x} + t_{\frac{\alpha}{2},\,\text{df}=n-1} \times \frac{\text{SD}}{\sqrt{n}} \rightarrow 327.85\frac{\text{mg}}{\text{cup}} + 10.7983\frac{\text{mg}}{\text{cup}}$$
$$= 338.648\,\text{mg}/\text{cup}$$

The lower limit of the CI is:

$$\bar{x} - t_{\frac{\alpha}{2},\,\text{df}=n-1} \times \frac{\text{SD}}{\sqrt{n}} \rightarrow 327.85\frac{\text{mg}}{\text{cup}} - 10.7983\frac{\text{mg}}{\text{cup}}$$
$$= 317.052\,\text{mg}/\text{cup}$$

Use a t-test to determine if the sample mean provides strong enough evidence that the population is "out of spec" (i.e., the actual population mean $\neq$ 343 mg cup). Since we have mean a "target" ($\mu = 343$ mg/cup) that we want to compare a sample against, we use a one sample t-test:

$$t_{\text{obs}} = \frac{|\bar{x} - \mu|}{\frac{\text{SD}}{\sqrt{n}}}$$

Plugging in the mean, SD, n, and μ values:

$$t_{\text{obs}} = \frac{|\bar{x} - \mu|}{\frac{\text{SD}}{\sqrt{n}}} = \frac{|327.85\,\text{mg}/\text{cup} - 343\,\text{mg}/\text{cup}|}{\frac{9.593904\,\text{mg}/\text{cup}}{\sqrt{6}}}$$
$$= \frac{|-15.15\,\text{mg}/\text{cup}|}{3.91669\,\text{mg}/\text{cup}} = 3.868$$

Based on t_{obs}, is there sufficient evidence that the population is "out of spec" (99% confidence)?

The decision rule is that we need to compare $t_{\text{df},\,\alpha/2}$ vs. t_{obs}. Find $t_{\text{df},\,\alpha/2}$:

for 99% confidence,

$$C = 0.99 \rightarrow \frac{\infty}{2} = \frac{1-C}{2} = \frac{1-0.99}{2} = 0.005$$
$$\text{df} = n - 1 = 6 - 1 = 5$$

From the t-score table shown, $t_{5,\,0.005} = 4.032$.
Next, compare $t_{\text{df},\,\alpha/2}$ vs. t_{obs}:

if $t_{\text{obs}} > t_{\text{critical}} \rightarrow$ there IS strong evidence that the true pop mean $\neq \mu$

if $t_{\text{obs}} < t_{\text{critical}} \rightarrow$ there IS NOT strong evidence that the true pop mean $\neq \mu$

In this case, since t_{obs} (3.868) < t_{critical} (4.032), there IS NOT sufficient evidence that the population is "out of spec" with 99% confidence.

2. You should use a two-sample t-test to determine if the means are statistically different. We need the mean and the $\text{SD}n_{-1}$ for each population. Using your calculator or Excel, we get:

Line 1: $n = 5$, $\bar{x} = 86.98$, and SD$= 0.81670068$
Line B: $n = 5$, $\bar{x} = 89.02$, and SD$= 0.195$ 0.192353841

Next, calculate the pooled variance ($s_p{}^2$) from the two sample standard deviations:

$$\text{pooled variance} = s_p^2 = \frac{(n_1 - 1)\text{SD}_1^2 + (n_2 - 1)\text{SD}_2^2}{n_1 + n_2 - 2}$$

$$= \frac{(5-1)(0.81670068)^2 + (5-1)(0.192353841)^2}{5+5-2} = 0.352$$

Now, calculate t_{obs}:

$$t_{\text{obs}} = \frac{|\bar{x}_1 - \bar{x}_2|}{\sqrt{s_p^2 \left(\dfrac{1}{n_1} + \dfrac{1}{n_2}\right)}} = \frac{|86.98 - 89.02|}{\sqrt{0.352 \times \left(\dfrac{1}{5} + \dfrac{1}{5}\right)}}$$

$$= \frac{|-2.04|}{\sqrt{0.1408}} = 5.43662$$

Based on the *t*-test, is there strong evidence that the sample means are statistically different? For a two-sample *t*-test, the decision rule is:

$$t_{obs} > t_{\frac{\alpha}{2}, \, \text{df} = n_1 + n_2 - 2} \rightarrow \text{strong evidence that means are significantly different}$$

$$t_{obs} < t_{\frac{\alpha}{2}, \, \text{df} = n_1 + n_2 - 2} \rightarrow \text{insufficient evidence that means are significantly different}$$

$$95\% \text{ confidence} \rightarrow C = 0.95, \quad \frac{\alpha}{2} = \frac{1-C}{2} = \frac{1-0.95}{2}$$

$$= \frac{0.05}{2} = 0.025$$

$$\text{df} = n_1 + n_2 - 2 = 5 = 5 - 2 = 8$$

Then, go to the *t*-table and find:

$$t_{\text{critical}}\left(t_{\frac{\alpha}{2}, \, \text{df} = n_1 + n_2 - 2}\right): \ t_{0.025, 8} = 2.306$$

Since t_{obs} (5.44) $> t_{\text{critical}}$ (2.31), there is strong evidence that the means differ (95 % confidence).

Correction to: Food Analysis Laboratory Manual

S. Suzanne Nielsen

CORRECTION TO:

S.S. Nielsen, *Food Analysis Laboratory Manual*, **Food Science Text Series,**
https://doi.org/10.1007/978-3-319-44127-6

An error in the production process unfortunately led to publication of the book before incorporating the below corrections. This has now been corrected and approved by the Editor.

p. 35	*Col 1, 3.4.2, line 6*	*M = mass" changed to "m = mass"*
p. 67	Col 2, 5.2.1, step 5, line 4	8" changed to "2/3"
p. 67	Col 2, 5.2.1, step 5, line 5	oz" changed to "cup"
p. 67	Col 2, 5.2.1, step 5, line 7	227" changed to "170 g"
p. 69	Col 1, step 7, 2nd equation	Change denominator from "227" changed to "170"
p. 69	Col 2, step 7, 3rd equation	286" changed to "382"
p. 69	Col 2, step 8, line 4	286" changed to "382"
p. 69	Col 2, step 8, line 6	286" changed to "382"
p. 69	Col 2, step 8, line 16	data" changed to "% wt/wt for Dairy Calcium"
p. 139	Col 2, step 7, line 2	Part 5" changed to "Part 6"
p. 140	Col 2, 3 lines above step 2	99.14 ug/mL" changed to "99140 µg/mL"
p. 144	Col 1, line 4	Deleted entire sentence "The US Food and Drug Administration requires the vitamin C content to be listed on the nutrition label of foods."
p. 146	Col 2, top line of Titer eq	5 mL" changed to "50 mL"
p. 168	Col 1, line 3	300-600 ppm" changed to "30-600 ppm"
p. 168	Col 1, line 4	Deleted "30-600 mg/L"

The updated online version of this book is available at
https://doi.org/10.1007/978-3-319-44127-6

S. S. Nielsen, *Food Analysis Laboratory Manual*, Food Science Text Series,
https://doi.org/10.1007/978-3-319-44127-6_32, © Springer International Publishing 2019

p. 188	Col 2, question 2, line 2	"name" changed to "nature"
p. 190	Col 1, line 7	W in denominator from "W" changed to "W x 1000"
p. 191	Col 1, sect. 22.4.9, %FFA eq.	W in denominator from "W" changed to "W x 1000"
p. 191	Col 2, sect. 22.5.4, last 4 lines	Change 0.2 N to 0.1 N
		Change 50 g to 25, add "pentahydrate" after "sodium thiosulfate", and change "l" to "L", to read "Dissolve ca. 25 g sodium thiosulfate pentahydrate in 1 L dd water

Printed in the United States
By Bookmasters